Advances in Hard-to-Cut Materials

Advances in Hard-to-Cut Materials

Manufacturing, Properties, Process Mechanics and Evaluation of Surface Integrity

Special Issue Editors

Grzegorz M. Królczyk
Radosław W. Maruda
Szymon Wojciechowski

MDPI • Basel • Beijing • Wuhan • Barcelona • Belgrade • Manchester • Tokyo • Cluj • Tianjin

Special Issue Editors

Grzegorz M. Królczyk
Opole University of Technology
Poland

Radosław W. Maruda
University of Zielona Gora
Poland

Szymon Wojciechowski
Poznan University of Technology
Poland

Editorial Office
MDPI
St. Alban-Anlage 66
4052 Basel, Switzerland

This is a reprint of articles from the Special Issue published online in the open access journal *Materials* (ISSN 1996-1944) (available at: https://www.mdpi.com/journal/materials/special_issues/hcm).

For citation purposes, cite each article independently as indicated on the article page online and as indicated below:

LastName, A.A.; LastName, B.B.; LastName, C.C. Article Title. *Journal Name* **Year**, *Article Number*, Page Range.

ISBN 978-3-03928-354-5 (Pbk)
ISBN 978-3-03928-355-2 (PDF)

© 2020 by the authors. Articles in this book are Open Access and distributed under the Creative Commons Attribution (CC BY) license, which allows users to download, copy and build upon published articles, as long as the author and publisher are properly credited, which ensures maximum dissemination and a wider impact of our publications.

The book as a whole is distributed by MDPI under the terms and conditions of the Creative Commons license CC BY-NC-ND.

Contents

About the Special Issue Editors . . . . **vii**

Szymon Wojciechowski, Grzegorz M. Królczyk and Radosław W. Maruda
Advances in Hard–to–Cut Materials: Manufacturing, Properties, Process Mechanics and Evaluation of Surface Integrity
Reprinted from: *Materials* **2020**, *13*, 612, doi:10.3390/ma13030612 . . . . **1**

Janusz Kluczyński, Lucjan Śnieżek, Krzysztof Grzelak and Janusz Mierzyński
The Influence of Exposure Energy Density on Porosity and Microhardness of the SLM Additive Manufactured Elements
Reprinted from: *Materials* **2018**, *11*, 2304, doi:10.3390/ma11112304 . . . . **7**

Chander Prakash, Sunpreet Singh, Munish Kumar Gupta, Mozammel Mia, Grzegorz Królczyk and Navneet Khanna
Synthesis, Characterization, Corrosion Resistance and In-Vitro Bioactivity Behavior of Biodegradable Mg–Zn–Mn–(Si–HA) Composite for Orthopaedic Applications
Reprinted from: *Materials* **2018**, *11*, 1602, doi:10.3390/ma11091602 . . . . **17**

Lei Guo, Xinrong Zhang, Shibin Chen and Jizhuang Hui
An Experimental Study on the Precision Abrasive Machining Process of Hard and Brittle Materials with Ultraviolet-Resin Bond Diamond Abrasive Tools
Reprinted from: *Materials* **2019**, *12*, 125, doi:10.3390/ma12010125 . . . . **37**

Chander Prakash, Sunpreet Singh, Catalin Iulian Pruncu, Vinod Mishra, Grzegorz Królczyk, Danil Yurievich Pimenov and Alokesh Pramanik
Surface Modification of Ti-6Al-4V Alloy by Electrical Discharge Coating Process Using Partially Sintered Ti-Nb Electrode
Reprinted from: *Materials* **2019**, *12*, 1006, doi:10.3390/ma12071006 . . . . **49**

Navneet Khanna, Jay Airao, Munish Kumar Gupta, Qinghua Song, Zhanqiang Liu, Mozammel Mia, Radoslaw Maruda and Grzegorz Krolczyk
Optimization of Power Consumption Associated with Surface Roughness in Ultrasonic Assisted Turning of Nimonic-90 Using Hybrid Particle Swarm-Simplex Method
Reprinted from: *Materials* **2019**, *12*, 3418, doi:10.3390/ma12203418 . . . . **65**

Rongkai Tan, Xuesen Zhao, Tao Sun, Xicong Zou and Zhenjiang Hu
Experimental Investigation on Micro-Groove Manufacturing of Ti-6Al-4V Alloy by Using Ultrasonic Elliptical Vibration Assisted Cutting
Reprinted from: *Materials* **2019**, *12*, 3086, doi:10.3390/ma12193086 . . . . **85**

Munish Kumar Gupta, Muhammad Jamil, Xiaojuan Wang, Qinghua Song, Zhanqiang Liu, Mozammel Mia, Hussein Hegab, Aqib Mashood Khan, Alberto Garcia Collado, Catalin Iulian Pruncu and G.M. Shah Imran
Performance Evaluation of Vegetable Oil-Based Nano-Cutting Fluids in Environmentally Friendly Machining of Inconel-800 Alloy
Reprinted from: *Materials* **2019**, *12*, 2792, doi:10.3390/ma12172792 . . . . **101**

Sunpreet Singh, Chander Prakash, Parvesh Antil, Rupinder Singh, Grzegorz Królczyk and Catalin I. Pruncu
Dimensionless Analysis for Investigating the Quality Characteristics of Aluminium Matrix Composites Prepared through Fused Deposition Modelling Assisted Investment Casting
Reprinted from: *Materials* **2019**, *12*, 1907, doi:10.3390/ma12121907 . . . . **121**

Irene Buj-Corral, Jose-Antonio Ortiz-Marzo, Lluís Costa-Herrero, Joan Vivancos-Calvet and Carmelo Luis-Pérez
Optimal Machining Strategy Selection in Ball-End Milling of Hardened Steels for Injection Molds
Reprinted from: *Materials* **2019**, *12*, 860, doi:10.3390/ma12060860 137

Mozammel Mia, Grzegorz Królczyk, Radosław Maruda and Szymon Wojciechowski
Intelligent Optimization of Hard-Turning Parameters Using Evolutionary Algorithms for Smart Manufacturing
Reprinted from: *Materials* **2019**, *12*, 879, doi:10.3390/ma12060879 151

Paweł Twardowski and Martyna Wiciak-Pikuła
Prediction of Tool Wear Using Artificial Neural Networks during Turning of Hardened Steel
Reprinted from: *Materials* **2019**, *12*, 3091, doi:10.3390/ma12193091 163

Mohammad Uddin, Animesh Basak, Alokesh Pramanik, Sunpreet Singh, Grzegorz M. Krolczyk and Chander Prakash
Evaluating Hole Quality in Drilling of Al 6061 Alloys
Reprinted from: *Materials* **2018**, *11*, 2443, doi:10.3390/ma11122443 179

Tomasz Bartkowiak and Christopher A. Brown
Multiscale 3D Curvature Analysis of Processed Surface Textures of Aluminum Alloy 6061 T6
Reprinted from: *Materials* **2019**, *12*, 257, doi:10.3390/ma12020257 193

About the Special Issue Editors

Grzegorz M. Królczyk is Professor and Vice-Rector for Research and Development at Opole University of Technology and author or co-author of 180 scientific publications (100 JCR papers) as well as nearly 30 studies and industrial applications. His main directions of scientific activity are in the analysis and improvement of manufacturing processes, surface metrology, and surface engineering. His research focuses on sustainable manufacturing as a tool for the practical implementation of the concept of social responsibility in the area of machining. A member of several scientific organizations, including an expert of the Section of Technology of the Committee on Machine Building of the Polish Academy of Sciences. In addition, he is a member of several editorial committees of scientific journals. He participated in advisory and opinion forming bodies, including the advisory team of the Minister of Science and Higher Education. Krolczyk is co-author of two patent applications, and his scientific activities have been rewarded numerous times both in Poland and around the world.

Radosław W. Maruda PhD Eng is Head of the Department of Materials, Technology and Maintenance of Machines at the Institute of Mechanical Engineering, Faculty of Mechanical Engineering, University of Zielona Gora. Professor Radoslaw Maruda main interest is in machining. He has gained experience in the industry, where he has been working as a technical consultant in various projects. His academic research aims to determine the effectiveness of the use of organic methods of cooling (minimum quantity lubrication and cooling minimum quantity lubrication) in planning. The impact of EP and AW additives, introduced into the liquid cooling lubricant, is also examined in his research. In particular, the research aims to minimize machining defects and strives to increase the precision of machine parts within production processes. The main research topics concern surface metrology, including surface analysis in terms of their functionality and tool life. The analyzed surfaces are generated in the process of turning, milling, water jet cutting, or welding. In addition, sustainable production and clean production are also considered. Professor Radoslaw Maruda is the author and co-author of approx. 79 scientific papers.

Szymon Wojciechowski PhD is Associate Professor and Head of the Laboratory for Precise Machining at the Faculty of Mechanical Engineering, Poznan University of Technology, Poland. His scientific interests mainly concern the modeling and research of dynamic phenomena and technological effects of precise/microcutting of difficult-to-cut materials. This scientific work has resulted in authorship or co-authorship of over 70 publications, including 40 indexed in Web of Science and Scopus databases (e.g., publications in such journals as *International Journal of Machine Tools and Manufacture, Measurement, Journal of Cleaner Production, Composites Part A, Applied Surface Science*), participation in numerous scientific projects, patent applications, and industrial implementations, as well as many Polish and international awards for his scientific activities. Moreover, Professor Szymon Wojciechowski is an appointed expert in several Polish and European scientific organizations, including the Polish Ministry of Science and Higher Education and Hungarian National Research Development and Innovation Office.

Editorial

Advances in Hard–to–Cut Materials: Manufacturing, Properties, Process Mechanics and Evaluation of Surface Integrity

Szymon Wojciechowski [1,*], Grzegorz M. Królczyk [2] and Radosław W. Maruda [3]

[1] Faculty of Mechanical Engineering and Management, Poznan University of Technology, 3 Piotrowo St., 60-965 Poznan, Poland
[2] Department of Manufacturing Engineering and Production Automation, Faculty of Mechanical Engineering, Opole University of Technology, 5 Mikolajczyka Street, 45-271 Opole, Poland; g.krolczyk@po.opole.pl
[3] Faculty of Mechanical Engineering, University of Zielona Gora, Prof. Z. Szafrana Street 4, 65-516 Zielona Gora, Poland; r.maruda@ibem.uz.zgora.pl
* Correspondence: sjwojciechowski@o2.pl

Received: 29 December 2019; Accepted: 28 January 2020; Published: 30 January 2020

Abstract: The rapid growth of a modern industry results in a growing demand for construction materials with excellent operational properties. However, the improved features of these materials can significantly hinder their manufacturing, therefore they can be defined as hard–to–cut. The main difficulties during the manufacturing/processing of hard–to–cut materials are attributed to their high hardness and abrasion resistance, high strength at room or elevated temperatures, increased thermal conductivity, as well as their resistance to oxidation and corrosion. Nowadays the group of hard–to–cut materials includes the metallic materials, composites, as well as ceramics. This special issue, "Advances in Hard–to–Cut Materials: Manufacturing, Properties, Process Mechanics and Evaluation of Surface Integrity" provides a collection of research papers regarding the various problems correlated with hard–to–cut materials. The analysis of these studies reveals primary directions regarding the developments in manufacturing methods, and the characterization and optimization of hard–to–cut materials.

Keywords: hard–to–cut materials; machining; additive manufacturing; mechanics; surface integrity

---

Nowadays, in many industrial branches, the growing demand for construction materials with excellent operational and mechanical properties is observed. Especially in the aerospace, biomedical, electronic and automotive industries, construction materials with high hardness, abrasion resistance, a high strength in a range of various temperatures, increased thermal conductivity, as well as resistance to oxidation and corrosion, are very often employed. Unfortunately, these unique features significantly deteriorate the machinability of these materials, and thus they are defined as hard–to–cut.

The major problems occurring during the machining of hard–to–cut materials include the high values of cutting forces, high levels of vibrations in machining systems, the concentration of heat, the growth of cutting temperature, rapid tool wear and the risk of catastrophic tool failure, as well as frequent stability loss and a significant deterioration in surface finish.

The group of hard–to–cut materials is extensive and still expanding, attributed to the development of novel manufacturing techniques (e.g., additive technologies). Currently, the group of hard–to–cut materials includes hardened and stainless steels, titanium, cobalt and nickel alloys, composites and ceramics, as well as the hard clads fabricated by additive techniques.

This special issue, "Advances in Hard–to–Cut Materials: Manufacturing, Properties, Process Mechanics and Evaluation of Surface Integrity" provides the collection of thirteen research articles presenting recent activity and developments in this field. Studying these works reveals the current problems and research directions concerning hard–to–cut materials. Among these, the novel production and machining techniques and the production/machining optimization methods, as well as the novel measurement/characterization techniques, can be identified (Figure 1).

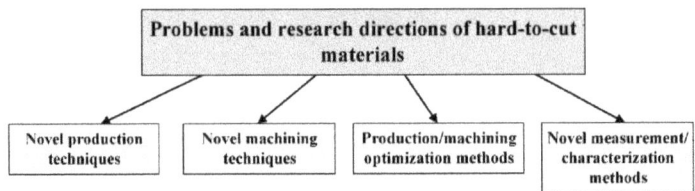

Figure 1. Current major problems and directions concerning the hard–to–cut materials.

The problems regarding the application of novel manufacturing techniques for hard–to–cut materials are presented in four papers. Kluczyński et al. [1], investigated the porosity and the microhardness of 316L austenitic steel, manufactured with the application of selective laser melting (SLM) additive technology. The authors have revealed that microstructure porosity is affected by the hatching distance and exposure velocity. As the hatching distance increases, the microstructure porosity of this element increases, and the decrease in exposure velocity causes a decline in porosity level. Moreover, an increase in microhardness with an increase in the exposure energy density was observed. This observation can be connected with the combined effect of grain refinement strengthening (Hall–Petch relation) and grain boundary strengthening. Prakash et al. [2] developed a method for the production of porous Mg–based biodegradable structures, based on the hybridization of elemental alloying and spark plasma sintering technology. The authors employed suitable proportions of silicon (Si) and hydroxyapatite (HA) to enhance the mechanical, chemical, and geometrical features. They found that the addition of HA and Si elements affects the improvement of structural porosity, with a low elastic modulus and hardness of the Mg–Zn–Mn matrix, respectively. Moreover, the addition of both HA and Si elements refined the grain structure and improved the hardness of the as–fabricated structures. Authors have also detected the formation of various biocompatible phases, whose appearance enhances the corrosion performance and biomechanical integrity of manufactured structures. Guo et al. [3] proposed ultraviolet–curable resin bonding for a precision abrasive machining tool, aiming to deliver a rapid, flexible, economical, and environment–friendly additive manufacturing process to replace the hot press and sintering process. Authors have employed a customized ultraviolet curing system based on the Machine UV–100, and the Dymax 5000 flood ultraviolet curing system used for the initial material properties test of the cured ultraviolet–curable resin composites. The manufactured precision abrasive machining tool consisted of an ultraviolet–curable epoxy resin 425 as a bond and monocrystalline diamond grains as abrasives. Authors have proved that the application of an abrasive machining tool equipped with the ultraviolet–curable resin bonding during lapping process enabled an approximately 10% lower surface roughness parameter Ra and 25% less weight loss of the workpiece than those obtained in the iron plate lapping process. Prakash et al. [4], in their study, employed two methods (electric discharge coating (EDC) and electric discharge machining processes (EDM)) to coat a composite layer TiO_2–TiC–NbO–NbC on the Ti–64 alloy. The conducted research revealed that the application of the EDC process with a high peak current and high Nb–powder concentration enabled the formation of a crack–free thick layer (215 µm) on the workpiece surface. Moreover, further inspections have shown that the obtained coating has a high hardness and adhesion strength, which enables it to enhance the wear resistance of the Ti–64 alloy.

This collection of papers also presents that—apart from the novel manufacturing technologies—the current research direction of hard–to–cut materials involves novel machining techniques. This scientific problem matter is covered in three papers. Khanna et al. [5] employed the ultrasonic–assisted turning (UAT) process of the Nimonic–90 superalloy in order to replace the conventional cutting and obtain improved technological effects. The results showed that the ultrasonic–assisted turning process affects the reduction in surface roughness and power consumption values as compared with the conventional turning process. This is correlated with the micro–chipping effect induced by UAT process kinematics. Besides, the chips formed during the ultrasonic–assisted turning were regular and fragmented when compared to those obtained from the conventional turning process. The ultrasonic–assisted machining has been also applied by Tan et al. [6] to the micro–groove manufacturing in the Ti–6Al–4V alloy. The application of this kind of machining process aims to minimize the level of material swelling and springback and improve the machining quality. The experimental results proved that the material swelling and springback were significantly reduced and the surface integrity was substantially improved during the ultrasonic elliptical vibration–assisted cutting process in comparison to the conventional cutting process. Apart from vibration–assisted cutting, the novel methods of machining related to hard–to–cut materials involve also the application of nano–cutting fluids. Gupta et al. [7] employed different nano–cutting fluids (aluminum oxide (Al_2O_3), molybdenum disulfide (MoS_2), and graphite) during the turning of the Inconel 800 alloy under the minimum quantity lubrication (MQL) conditions. The obtained results reveal that the MoS_2– and graphite–based nanofluids can affect the improvement in cutting effects, especially at the high cutting speed values. Moreover, the overall performance of graphite–based nanofluids is better in terms of good lubrication and cooling properties. The presence of small quantities of graphite in vegetable oil significantly improves the machining characteristics of Inconel–800 alloy as compared with the two other nanofluids.

The next important research direction regarding the hard–to–cut materials includes the production/machining optimization methods. These problems are covered in four research papers. Singh et al. [8] applied the Vashy–Buckingham π–theorem for the selection of input parameters of the fused deposition modeling assisted by the investment casting process, enabling the obtainment of optimal hardness, dimensional accuracy, and surface roughness of manufactured aluminum matrix composite (AMC). The validation of the proposed models, conducted on the basis of the ANOVA method, proves their applicability to the optimization of aluminum matrix composite manufacturing during the fused deposition modeling assisted by the investment casting. Buj–Corral et al. [9] employed a central composite design to model the behavior of surface roughness during ball end milling of hot work–hardened tool steel W–Nr, consisting of a two level factorial design with four factors ($2^4 = 16$ experiments), and four central points. The conducted studies have shown that the radial depth of the cut was the most relevant factor on Ra and Rt for both climb and conventional milling. However, the axial depth of cut, cutting speed and feed per tooth have a slight influence on surface roughness within the investigated range. Mia et al. [10] proposed the application of evolutionary–based algorithms (teaching–learning–based optimization and bacterial foraging optimization) for the optimization of the hardened high–carbon steel AISI 1060 turning process. It was found that teaching–learning–based optimization (TLBO) was found to be superior to the bacteria foraging optimization (BFO) in terms of better convergence and a shorter time of computation—hence, the TLBO is recommended during the optimization of hard turning processes. The hardened steel optimization problems were also investigated by Twardowski and Wiciak–Pikuła [11]. They predicted the tool wear during turning of hardened 100Cr6 steel with the application of multilayer perceptron (MLP)–based artificial neural networks. The obtained results show that selection of the number of neurons in the hidden layer and activation function in the hidden and initial layers significantly affect the reliability of tool wear prediction. Alterations in the model structure at the beginning of its formulation help to achieve the assessments at a satisfactory level. Therefore, the artificial neural network with a multilayer perceptron is an effective method for predicting tool condition during the machining of hard–to–cut materials.

Ultimately, the developments in production, machining and optimization techniques regarding the hard–to–cut materials also entail advancements in metrological description and characterization. As part of this subject, the two research papers were published. Uddin et al. [12] applied a multi–dimensional evaluation of hole quality in an Al6061 alloy after drilling. Authors have employed the novel octagonal–ellipse load cell set–up for measurements of feed force and torque. Moreover, the tests also involved SEM analyses of a drill–bits after cutting and measurements of hole diameter errors with the application of a machine tool probe. Bartkowiak and Brown [13] proposed the novel multiscale method for calculating curvature tensors on measured surface topographies of a 6061 T6 alloy. The curvature tensors were calculated as functions of scale, i.e., size, and position from a regular, orthogonal array of measured heights. Moreover, in the derivations, vectors normal to the measured surface were calculated first, then the eigenvalue problem was solved for the curvature tensor. The validity of these methods has been proven by the high consistency of the results with expectations of manufactured surfaces. These expectations included the nature of the curvature and their orientations relative to manufactured features on the surfaces.

The knowledge contained in papers covered in this special issue can be helpful for the efficient selection of manufacturing and characterization methods, as well as the conditions, strategies and types of tools used during the machining of hard–to–cut materials, allowing the improvement of manufacturing performance and economics.

Funding: This research received no external funding.

Conflicts of Interest: The authors declare no conflict of interest.

References

1. Kluczyński, J.; Śnieżek, L.; Grzelak, K.; Mierzyński, J. The Influence of Exposure Energy Density on Porosity and Microhardness of the SLM Additive Manufactured Elements. *Materials* **2019**, *11*, 2304. [CrossRef] [PubMed]
2. Prakash, C.; Singh, S.; Gupta, M.K.; Mia, M.; Królczyk, G.M.; Khanna, N. Synthesis, Characterization, Corrosion Resistance and In–Vitro Bioactivity Behavior of Biodegradable Mg–Zn–Mn–(Si–HA) Composite for Orthopaedic Applications. *Materials* **2018**, *11*, 1602. [CrossRef] [PubMed]
3. Guo, L.; Zhang, X.; Chen, S.; Hui, J. An Experimental Study on the Precision Abrasive Machining Process of Hard and Brittle Materials with Ultraviolet–Resin Bond Diamond Abrasive Tools. *Materials* **2019**, *12*, 125. [CrossRef] [PubMed]
4. Prakash, C.; Singh, S.; Pruncu, C.I.; Mishra, V.; Królczyk, G.M.; Pimenov, D.Y.; Pramanik, A. Surface Modification of Ti–6Al–4V Alloy by Electrical Discharge Coating Process Using Partially Sintered Ti–Nb Electrode. *Materials* **2019**, *12*, 1006. [CrossRef] [PubMed]
5. Khanna, N.; Airao, J.; Gupta, M.K.; Song, Q.; Liu, Z.; Mia, M.; Maruda, R.W.; Krolczyk, G.M. Optimization of Power Consumption Associated with Surface Roughness in Ultrasonic Assisted Turning of Nimonic–90 Using Hybrid Particle Swarm–Simplex Method. *Materials* **2019**, *12*, 3418. [CrossRef] [PubMed]
6. Tan, R.; Zhao, X.; Sun, T.; Zou, X.; Hu, Z. Experimental Investigation on Micro–Groove Manufacturing of Ti–6Al–4V Alloy by Using Ultrasonic Elliptical Vibration Assisted Cutting. *Materials* **2019**, *12*, 3086. [CrossRef] [PubMed]
7. Gupta, M.K.; Jamil, M.; Wang, X.; Song, Q.; Liu, Z.; Mia, M.; Hegab, H.; Khan, A.M.; Collado, A.G.; Pruncu, C.I.; et al. Performance Evaluation of Vegetable Oil–Based Nano–Cutting Fluids in Environmentally Friendly Machining of Inconel–800 Alloy. *Materials* **2019**, *12*, 2792. [CrossRef] [PubMed]
8. Singh, S.; Prakash, C.; Antil, P.; Singh, R.; Królczyk, G.M.; Pruncu, C.I. Dimensionless Analysis for Investigating the Quality Characteristics of Aluminium Matrix Composites Prepared through Fused Deposition Modelling Assisted Investment Casting. *Materials* **2019**, *12*, 1907. [CrossRef] [PubMed]
9. Buj-Corral, I.; Ortiz-Marzo, J.-A.; Costa-Herrero, L.; Vivancos-Calvet, J.; Luis-Pérez, C. Optimal Machining Strategy Selection in Ball-End Milling of Hardened Steels for Injection Molds. *Materials* **2019**, *12*, 860. [CrossRef] [PubMed]
10. Mia, M.; Królczyk, G.M.; Maruda, R.W.; Wojciechowski, S. Intelligent Optimization of Hard–Turning Parameters Using Evolutionary Algorithms for Smart Manufacturing. *Materials* **2019**, *12*, 879. [CrossRef] [PubMed]
11. Twardowski, P.; Wiciak-Pikuła, M. Prediction of Tool Wear Using Artificial Neural Networks during Turning of Hardened Steel. *Materials* **2019**, *12*, 3091. [CrossRef] [PubMed]

12. Uddin, M.; Basak, A.; Pramanik, A.; Singh, S.; Krolczyk, G.M.; Prakash, C. Evaluating Hole Quality in Drilling of Al 6061 Alloys. *Materials* **2018**, *11*, 2443. [CrossRef] [PubMed]
13. Bartkowiak, T.; Brown, C.A. Multiscale 3D Curvature Analysis of Processed Surface Textures of Aluminum Alloy 6061 T6. *Materials* **2019**, *12*, 257. [CrossRef] [PubMed]

© 2020 by the authors. Licensee MDPI, Basel, Switzerland. This article is an open access article distributed under the terms and conditions of the Creative Commons Attribution (CC BY) license (http://creativecommons.org/licenses/by/4.0/).

Article

The Influence of Exposure Energy Density on Porosity and Microhardness of the SLM Additive Manufactured Elements

Janusz Kluczyński *, Lucjan Śnieżek, Krzysztof Grzelak and Janusz Mierzyński

Institute of Machine Building, Faculty of Mechanical Engineering, Military University of Technology, 00-908 Warsaw 49, Poland; lucjan.sniezek@wat.edu.pl (L.S.); krzysztof.grzelak@wat.edu.pl (K.G.); janusz.mierzynski@wat.edu.pl (J.M.)
* Correspondence: janusz.kluczynski@wat.edu.pl; Tel.: +48-725-456-619

Received: 19 October 2018; Accepted: 14 November 2018; Published: 16 November 2018

Abstract: Selective laser melting (SLM) is an additive manufacturing technique. It allows elements with very complex geometry to be produced using metallic powders. A geometry of manufacturing elements is based only on 3D computer-aided design (CAD) data. The metal powder is melted selectively layer by layer using an ytterbium laser. This paper contains the results of porosity and microhardness analysis made on specimens manufactured during a specially prepared process. Final analysis helped to discover connections between changing hatching distance, exposure speed and porosity. There were no significant differences in microhardness and porosity measurement results in the planes perpendicular and parallel to the machine building platform surface.

Keywords: additive manufacturing; SLM technology; porosity research; microhardness research

1. Introduction

In recent years an intensive development of additive manufacturing technology (AM) has been observed. This innovative technology is often called "3D printing". It became one of the leading automated production technologies and it seems to be as important as subtractive manufacturing, plastic forming or casting [1]. Selective laser melting (SLM) is one of the most popular additive manufacturing techniques. It is based on selective fusion of metallic powders using an ytterbium laser, where the manufacturing process is based on a "powder bed". During the last 10 years it has become one of the most developed AM technologies [2–9]. Regarding other additive manufacturing techniques, selective laser melting is characterized by:

- High-dimensional accuracy of the manufactured elements;
- Relatively low anisotropy of mechanical properties;
- A significant number of available materials;
- Low porosity of the manufactured elements.

The SLM process is based on low granulation powder (15–45 μm). The building job can be modified by changing different parameters which indirectly and/or directly affect the quality of the melted area. The possibilities to modify the manufacturing process in the SLM technique has created the possibility to conduct scientific research at many scientific and industry facilities [10–19]. One of the most common topics is the analysis of the process parameters which influence on the mechanical properties of manufactured elements [20–30]. In this paper, the influence of manufacturing process parameters on the porosity and microhardness of the additive manufactured elements was determined.

The modified parameters were:

- Laser power;
- Exposure velocity;
- Hatching distance.

Based on the research conducted, final conclusions were formulated and further research directions were defined.

2. Material

In this study, grade 316L austenitic steel (1.4404) was used. The material has been sourced by the SLM Solutions Group AG, Estlandring 4, 23560 Lübeck, Germany. Its density was 7.92 g/cm^3. The chemical composition of the analyzed steel is shown in Table 1.

Table 1. Chemical composition of 316L steel.

C	Mn	Si	P	S	N	Cr	Mo	Ni
				wt.%				
max. 0.03	max. 2.00	max. 0.75	max. 0.04	max. 0.03	max. 0.10	16.00–18.00	2.00–3.00	10.00–14.00

The material was manufactured using an argon atomization process. The powder particles (shown on Figure 1) have spherical or nearly spherical shapes with a particle size range between 15 µm to 45 µm. Also, satellite particles could be observed.

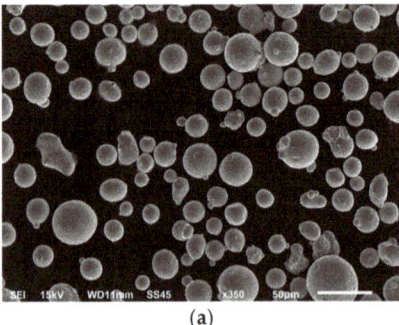
(a)

(b)

Figure 1. 316L powder scanning electron microscope (SEM) micrographs with (**a**) 50 µm scale and (**b**) 10 µm scale.

3. Experiments

Porosity and microhardness tests were carried out on specimens with the same geometry. Specimens had the form of cubes with a side length of 10 mm. These test parts were designed in such a way as to assure analysis of the distribution of mechanical properties in two different planes. The first was a plane parallel to the building platform surface, and the second one was a plane perpendicular to the platform surface.

The aforementioned planes are showed in Figure 2. As "xy" was named the plane parallel to the building platform surface, which is also normal to the direction of element growth (Z axis). The plane perpendicular to the building platform surface, which is also tangent to the direction of element growth (Z axis), is marked with "yz".

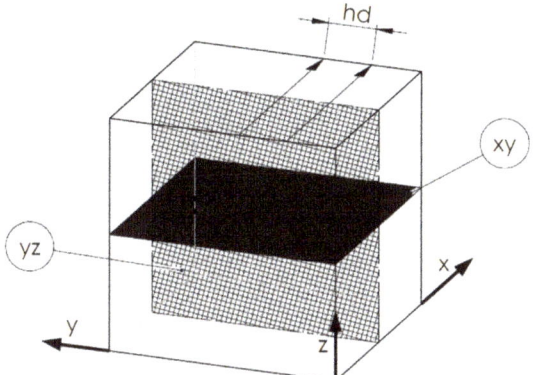

Figure 2. 3D model of a cubic sample, where: x—plane parallel to the building platform surface, y—plane perpendicular to the building platform surface, hd (hatching distance)—distance between the exposure vectors, Z—direction of growth (element building).

For each sample, different sets of process parameters were used, which are summarized in Table 2. Modified parameters were components of Equation (1) which affects the additive manufacturing energy density.

$$\rho_E \left[\frac{J}{mm^3}\right] = \frac{L_P[W]}{e_v\left[\frac{mm}{s}\right] \cdot h_d[mm] \cdot l_t[mm]} \quad (1)$$

where:

- L_P—laser power [W];
- e_v—exposure velocity [mm/s];
- h_d—hatching distance [mm];
- l_t—layer thickness [mm].

The modified parameters were the laser power, the exposure velocity, and the hatching distance. These specific components had been determined by the optical system and the energy source. It was caused by the possibility of analyzing the impact of modified parameters in a small range of its changes. One of the modified parameters was exposure velocity, also known as scanning speed. This determines the time of the laser exposure on each scanning line. Analysis of the influence of layer thickness on porosity and microhardness would be difficult to verify in this case for many reasons:

- Proper calibration of the powder reservoir (recouter);
- Inert gas flow speed;
- Clearance in the worm gear in the building platform leveling mechanism.

The manufacturing process parameters were changed within ±10% of the recommended value (item 1 in Table 2). The selected range of parameters modification was reached after consultation with specialists from the SLM Solutions company. In addition, parameters 28–30 (Table 2) differ significantly from the SLM System manufacturer's data. The reason for testing these parameters was the good mechanical property of specimens tested and described in [31]. The specimens (Figure 3) were created during a single process. The manufacturing file for the machine was prepared using the SLM Metal Build Processor module in the Magics software (version 19.0). All specimens were manufactured using 316L austenitic steel powder.

Table 2. Sets of analyzed production parameters.

Parameters Set	L_P [W]	e_v [mm/s]	h_d [mm]	ρ_E [J/mm^3]
1	190	900	0.12	58.64
2	190	990	0.12	53.31
3	190	810	0.12	65.16
4	200	900	0.12	61.73
5	200	990	0.12	56.12
6	200	810	0.12	68.59
7	180	900	0.12	55.56
8	180	990	0.12	50.51
9	180	810	0.12	61.73
10	190	900	0.13	54.13
11	190	990	0.13	49.21
12	190	810	0.13	60.15
13	200	900	0.13	56.98
14	200	990	0.13	51.80
15	200	810	0.13	63.31
16	180	900	0.13	51.28
17	180	990	0.13	46.62
18	180	810	0.13	56.98
19	190	900	0.11	63.97
20	190	990	0.11	58.16
21	190	810	0.11	71.08
22	200	900	0.11	67.34
23	200	990	0.11	61.22
24	200	810	0.11	74.82
25	180	900	0.11	60.61
26	180	990	0.11	55.10
27	180	810	0.11	67.34
28	150	400	0.08	156.25
29	150	700	0.06	119.05
30	120	300	0.08	166.67

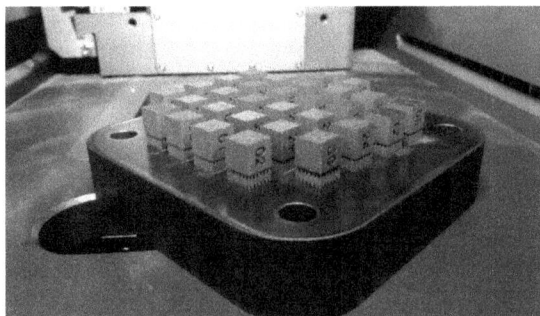

Figure 3. SLM 125HLs' building platform with manufactured specimens.

4. Porosity Analysis Results and Discussion

For each specimen the porosity was analyzed in the central part of the metallographic section. All visible pores were marked in both analyzed planes. The porosity was determined by images analyzed using a scanning electron microscope (SEM) (Figure 4).

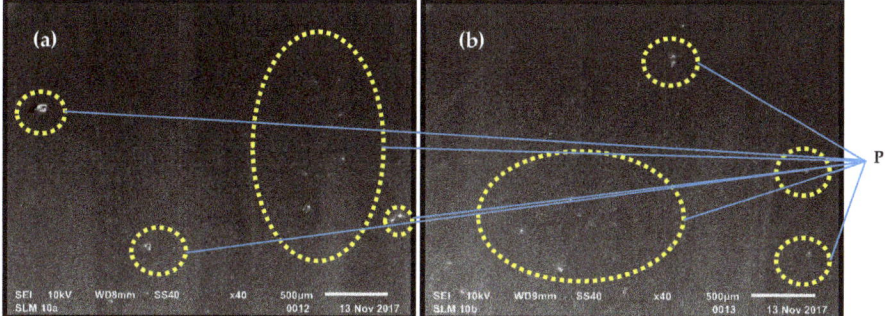

Figure 4. Image of visible pores in the plane parallel (**a**) and perpendicular (**b**) to building platform surface (areas of pores marked with the letter "P").

Porosity quantitative analysis were based on the microstructure images. It was carried out using a histogram check in GIMP software (version 2.0). The determination of porosity was based on the calculation of the Equation (2):

$$\rho[\%] = \frac{L_p}{L_c} \cdot 100\% \qquad (2)$$

where:

L_p—number of pixels in the contoured pores;
L_c—number of pixels of the image entire area.

The porosity analysis allowed to determine the influence of used laser power, hatching distance and exposure velocity (Figure 5). The analysis includes the groups of parameters in which only one was different from the parameters tested.

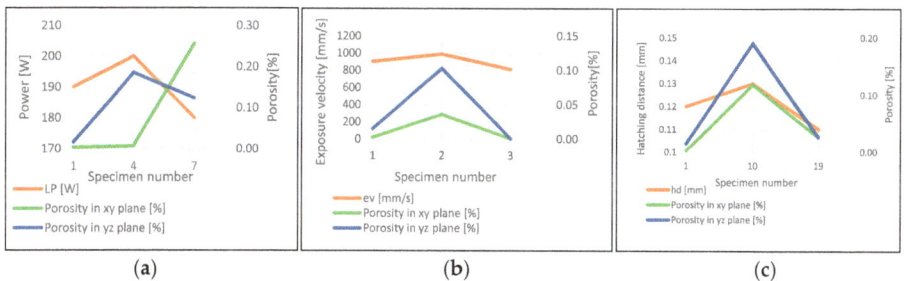

Figure 5. The influence of power (**a**), exposure velocity (**b**) and hatching distance (**c**) on porosity in the parallel (xy) and perpendicular plane (yz) to the building platform surface.

Based on the conducted analysis of the laser power influence graphs, the exposure velocity and the hatching distance (Figure 5), it can be noted that the power modification has no direct effect on the porosity changes. However, the influence of the other two parameters is noticeable. During changes to the exposure velocity in the range of ±10%, the porosity changes slightly—0.02%. A significant impact on the porosity can be seen when the hatching distance changes.

To emphasize the representation of the porosity changes, depending on the exposure velocity and hatching distance, proper diagrams were plotted (Figures 6 and 7).

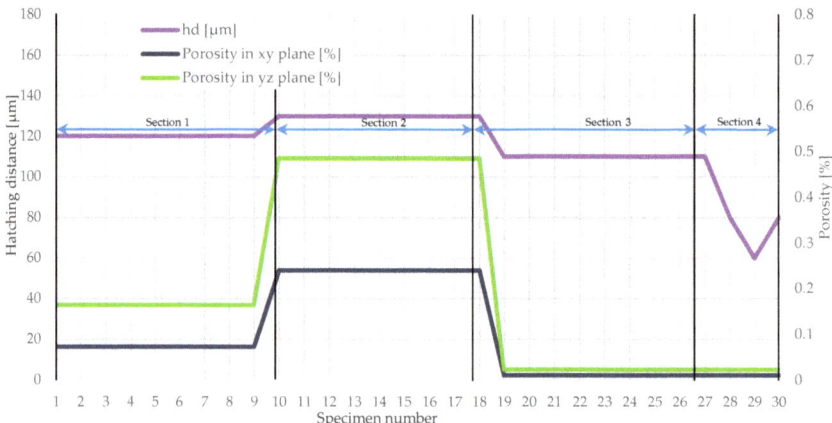

Figure 6. Variations of the average porosity in four ranges referenced to modified hatching distance in particular parameters group.

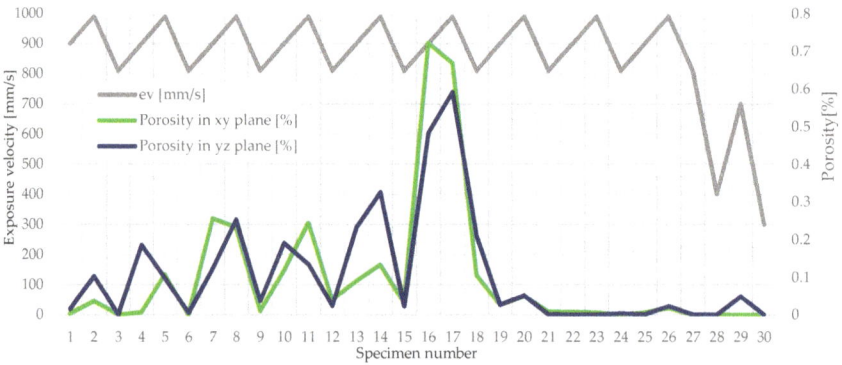

Figure 7. Porosity changes related to modified exposure velocity in particular parameters group.

This allows changes in the porosity in all specimens to be noticed and also the highest porosity peaks in the specimens to be noted. It was also observed that the exposure velocity and the hatching distance increase enhance the porosity growth (specimens 16 and 17 in Figure 7).

The hatching distance directly affects the porosity of the elements produced using SLM. The parameter increasing by 10% results in a nearly twice an increase in the proportion of pores. This phenomenon is caused by an increase in the distance between subsequent melt paths. This determines a reduction of the number of molten metallic powder particles. It can be observed in the parallel (xy) and perpendicular (yz) planes to the building platform surface.

A similar case can be observed when analyzing the influence of the exposure velocity, where porosity increasing with this parameter growth. It was noticed in both analyzed planes. This parameter does not affect the porosity as much as the change of the hatching distance, but with 10% change in the exposure velocity it is noticeable. Figures 6 and 7 show changes in porosity related to the modification of the hatching distance and the exposure velocity for all specimens. Noteworthy are the porosity peaks for specimens 16 and 17, where both the exposure velocity and the hatching distance have been increased.

Specimens manufactured using the unusual parameters proposed in [31] did not reveal better porosity and microhardness. These settings (marked in Section 4 in Figure 7) gave very similar properties to the rest of the manufactured elements. The main feature of the settings from the study [31] was the significantly smaller hatching. Lowering this value negatively affects the process's efficiency.

5. Microhardness Analysis Results and Discussion

The microhardness analysis was carried out on the same specimens used in porosity research. For each of the planes, a different configuration of the distribution of measurement points was adopted.

In the plane parallel to the building platform ("xy" in Figure 2), the influence of the linear exposure method on microhardness changes in five parallel rows was checked (Figure 8).

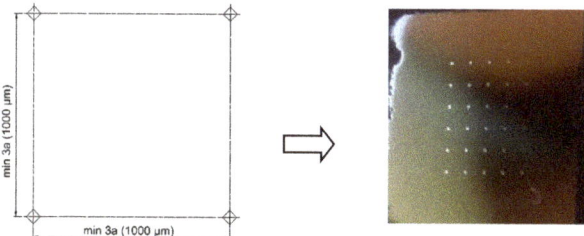

Figure 8. Distribution of the measurement points on a plane parallel to the building platform surface.

In the plane perpendicular to the building platform surface ("yz" in Figure 2), the effect of layers solidification (along Z axis) on microhardness changes in three parallel rows was checked (Figure 9). Similar to the porosity research, the influence of the laser power, the exposure velocity and the hatching distance (Figure 10) on the microhardness distribution in the specimens was determined. Also, in the case of microhardness analysis, the groups of parameters in which only one of the tested parameters changed were the laser power, the exposure velocity, and the hatching distance.

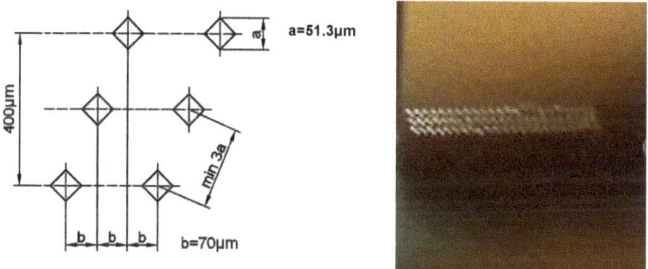

Figure 9. Distribution of the measurement points on a plane perpendicular to the building platform surface.

Proper preparation of the microhardness research allowed the lack of direct impact of parameter changes to be observed. The only observed dependence is the effect of the exposure velocity on microhardness, where microhardness decreases with the increase of the exposure velocity. However, it is insignificant and fits within the limits of measurement error. The lack of direct dependence between microhardness and one of the modified parameters was the reason for further analysis. The diagram of exposure energy density affect on microhardness was prepared. In sets where the parameters are changing in the range of $\pm 10\%$ from the recommended value, there is a noticeable relationship between the exposure energy density and the microhardness change (Figure 11).

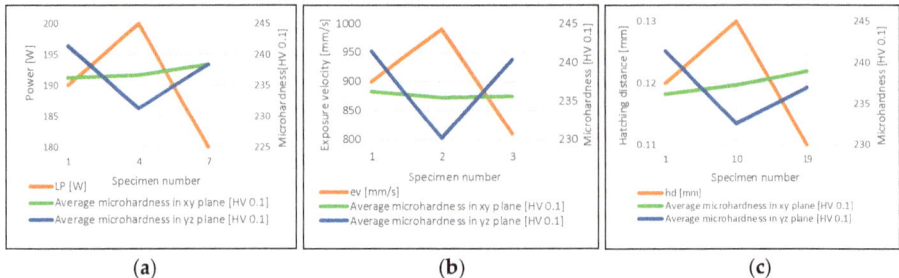

Figure 10. The influence of power (**a**), exposure velocity (**b**) and hatching distance (**c**) on microhardness in the parallel (xy) and perpendicular plane (yz) to the building platform surface.

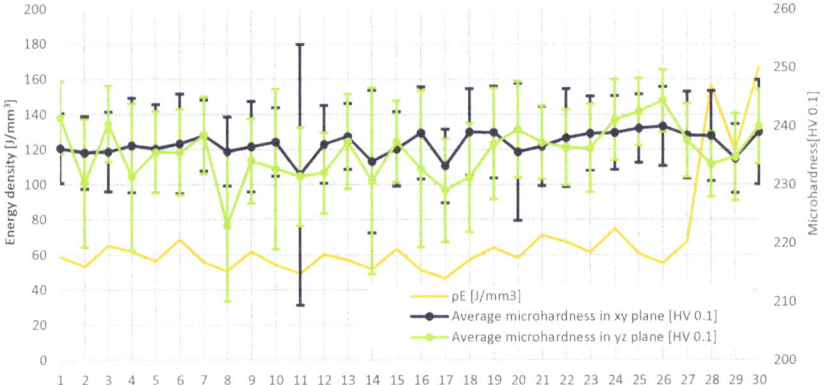

Figure 11. The influence of exposure energy density on microhardness in particular groups of parameters.

Microhardness distribution on both of the measured planes helped reveal that in a range of modification of parameters by ±10% of the nominal value, microhardness slightly increases with the growth of the exposure energy density. Those changes could be connected only with exposure velocity. As recorded in Figure 11, microhardness changes are related only to the change in the exposure speed—which affects the exposure energy density. This statement is valid for the range of parameter changes within ±10% of the nominal value of the parameters only. It can be concluded that the influence of modifying manufacturing parameters on microhardness is not as important as in the case of porosity. The main reason that there are no significant changes of microhardness when parameters change is too low a range of changed laser power. In [16] a dependence between laser power, hardness ($HV_{0.5}$) and exposure time could be noticed.

6. Final Conclusions

Analysis of changes in the laser power, exposure velocity and hatching distance allowed identification of the influence of these parameters on porosity and microhardness of specimens additive manufactured using the SLM technique. The research allowed the following conclusions to be drawn:

- There are no significant differences in microhardness and porosity measurement results in the planes perpendicular and parallel to the machine building platform surface. The main reason for the lack of visible changes of microhardness is to the low range of the changed parameters: laser power and exposure velocity;

- The hatching distance has a significant influence on the porosity of the manufactured elements. As the hatching distance increases, the microstructure porosity of this element increases;
- Exposure velocity changes affect the additive manufactured element's porosity. Lowering the exposure velocity cause the porosity to decrease;
- The relationship between exposure energy density changes and microhardness was identified. In the range of ±10% of the nominal value of the parameters, an increase of microhardness with an increase of the exposure energy density was observed. The microhardness increase is connected with the combined effect of grain refinement strengthening (Hall–Petch relation) and grain boundary strengthening [22];
- Conducted analyses of porosity and microhardness allowed for the selection of 5 groups of parameters which will be used to produce specimens for further research.

Author Contributions: Conceptualization was performed by all authors; Methodology, J.K.; K.G.; J.M.; Software, J.K.; Formal Analysis, L.S.; K.G.; Investigation, J.K.; J.M.; Writing-Original Draft Preparation, J.K.; J.M.; Writing-Review & Editing, J.K.; Visualization, J.K.; Supervision, L.S.; K.G.; Project Administration, L.S.

Funding: This research received no external funding.

Conflicts of Interest: The authors declare no conflict of interest.

References

1. Siemiński, P.; Budzik, G. *Additive Manufacturing. 3D Printing. 3D Printers*; Warsaw University of Technology Publishing House: Warsaw, Poland, 2015. (In Polish)
2. Badiru, A.B.; Valencia, V.V.; Liu, D. *Additive Manufacturing Handbook: Product Development for the Defense Industry*, 1st ed.; CRC Press: Baca Raton, FL, USA, 24 May 2017.
3. Zadpoor, A.A. Frontiers of additively manufactured metallic materials. *Materials* **2018**, *11*, 1566. [CrossRef] [PubMed]
4. Sayeed Ahmed, G.; Algarni, S. Development and FE thermal analysis of a radially grooved brake disc developed through direct metal laser sintering. *Materials* **2018**, *11*, 1211. [CrossRef] [PubMed]
5. Demir, A.G.; Previtali, B. Additive manufacturing of cardiovascular CoCr stents by selective laser melting. *Mater. Des.* **2017**, *119*, 338–350. [CrossRef]
6. Fousova, M.; Dvorsky, D.; Vronka, M.; Vojtech, D.; Lejcek, P. The use of selective laser melting to increase the performance of AlSi9Cu3Fe alloy. *Materials* **2018**, *11*, 1918. [CrossRef] [PubMed]
7. AlMangour, B.; Grzesiak, D.; Yang, J.M. In-situ formation of novel TiC-particle-reinforced 316L stainless steel bulk-form composites by selective laser melting. *J. Alloy Compd.* **2017**, *706*, 409–418. [CrossRef]
8. Han, X.; Sawada, T.; Schille, C.; Schweizer, E.; Scheideler, L.; Geis-Gerstorfer, J.; Rupp, F.; Spintzyk, S. Comparative analysis of mechanical properties and metal-ceramic bond strength of Co-Cr dental alloy fabricated by different manufacturing processes. *Materials* **2018**, *11*, 1801. [CrossRef] [PubMed]
9. Wang, X.; Kustov, S.; Van Humbeeck, J. A short review on the microstructure, transformation behavior and functional properties of NiTi shape memory alloys fabricated by selective laser melting. *Materials* **2018**, *11*, 1683. [CrossRef] [PubMed]
10. Zhong, Y.; Liu, L.; Wikman, S.; Cui, D.; Shen, Z. Intragranular cellular segregation network structure strengthening 316L stainless steel prepared by selective laser melting. *J. Nucl. Mater.* **2016**, *470*, 170–178. [CrossRef]
11. Lu, Y.; Gan, Y.; Lin, J.; Guo, S.; Wu, S.; Lin, J. Effect of laser speeds on the mechanical property and corrosion resistance of CoCrW alloy fabricated by SLM. *Rapid Prototyp. J.* **2017**, *23*, 28–33. [CrossRef]
12. Riemer, A.; Leuders, S.; Thöne, M.; Richard, H.A.; Tröster, T.; Niendorf, T. On the fatigue crack growth behavior in 316L stainless steel manufactured by selective laser melting. *Eng. Fract. Mech.* **2014**, *120*, 15–25. [CrossRef]
13. Fousova, M.; Vojtech, D.; Doubrava, K.; Daniel, M.; Lin, C.F. Influence of inherent surface and internal defects on mechanical properties of additively manufactured Ti6Al4V alloy: Comparison between selective laser melting and electron beam melting. *Materials* **2018**, *11*, 537. [CrossRef] [PubMed]

14. Bassoli, E.; Denti, L. Assay of secondary anisotropy in additively manufactured alloys for dental applications. *Materials* **2018**, *11*, 1831. [CrossRef] [PubMed]
15. AlMangour, B.; Grzesiak, D.; Cheng, J.; Ertas, Y. Thermal behavior of the molten pool, microstructural evolution, and tribological performance during selective laser melting of TiC/316L stainless steel nanocomposites: Experimental and simulation methods. *J. Mater. Process. Technol.* **2018**, *257*, 288–301. [CrossRef]
16. Metelkova, J.; Kinds, Y.; Kempen, K.; de Formanoir, C.; Witvrouw, A.; Van Hooreweder, B. On the influence of laser defocusing in Selective Laser Melting of 316L. *Addit. Manuf.* **2018**, *23*, 161–169. [CrossRef]
17. Yasa, E.; Kruth, J.P. Microstructural investigation of selective laser melting 316L stainless steel parts exposed to laser re-melting. *Procedia Eng.* **2011**, *19*, 389–395. [CrossRef]
18. Sun, J.; Yang, Z.; Yang, Y.; Wang, D. Research on the microstructure and properties of the overhanging structure formed by selective laser melting. *Rapid Prototyp. J.* **2017**, *23*, 904–910. [CrossRef]
19. Morgan, R.; Sutcliffe, C.J.; O'Neill, W. Density analysis of direct metal laser re-melted 316L stainless steel cubic primitives. *J. Mater. Sci.* **2004**, *39*, 1195–1205. [CrossRef]
20. Mahmoudi, M.; Elwany, A.; Yadollahi, A.; Thompson, S.M.; Bian, L.; Shamsaei, N. Mechanical properties and microstructural characterization of selective laser melted 17-4 PH stainless steel. *Rapid Prototyp. J.* **2017**, *23*, 280–294. [CrossRef]
21. Delgado, J.; Ciurana, J.; Rodríguez, C.A. Influence of process parameters on part quality and mechanical properties for DMLS and SLM with iron-based materials. *Int. J. Adv. Manuf. Technol.* **2012**, *60*, 601–610. [CrossRef]
22. Kluczyński, J. Strength Properties Analysis of Additive Manufactured Elements SLM Technology. In Proceedings of the Knowledge and Innovations materals from the 3rd Young Scientists Conference, Warsaw, Poland, November 2015.
23. Lu, Y.; Zhang, Y.; Cong, M.; Li, X.; Xu, W.; Song, L. Microstructures, mechanical and corrosion properties of the extruded AZ31-xCaO alloys. *Materials* **2018**, *11*, 1467. [CrossRef] [PubMed]
24. Yadollahi, A.; Sjamsaei, N.; Thompson, S.M.; Seely, D.W. Effects of process time interval and heat treatment on the mechanical and microstructural properties of direct laser deposited 316L stainless steel. *Mater. Sci. Eng. A* **2015**, *644*, 171–183. [CrossRef]
25. Zhu, Y.; Peng, T.; Jia, G.; Zhang, H.; Xu, S.; Yang, H. Electrical energy consumption and mechanical properties of selective-laser-melting-produced 316L stainless steel samples using various processing parameters. *J. Clean. Prod.* **2019**, *208*, 77–85. [CrossRef]
26. Liverani, E.; Toschi, S.; Ceschini, L.; Fortunato, A. Effect of selective laser melting (SLM) process parameters on microstructure and mechanical properties of 316L austenitic stainless steel. *J. Mater. Process. Technol.* **2017**, *249*, 255–263. [CrossRef]
27. AlMangour, B.; Grzesiak, D.; Borkar, T.; Yang, J.M. Densification behavior, microstructural evolution, and mechanical properties of TiC/316L stainless steel nanocomposites fabricated by selective laser melting. *Mater. Des.* **2018**, *138*, 119–128. [CrossRef]
28. Alsalla, H.H.; Smith, C.; Hao, L. Effect of build orientation on the surface quality, microstructure and mechanical properties of selective laser melting 316L stainless steel. *Rapid Prototyp. J.* **2018**, *24*, 9–17. [CrossRef]
29. Huang, C.; Lin, X.; Yang, H.; Liu, F.; Huang, W. Microstructure and tribological properties of laser forming repaired 34CrNiMo6 steel. *Materials* **2018**, *11*, 1722. [CrossRef] [PubMed]
30. Bültmann, J.; Merkt, S.; Hammer, C.; Hinke, C.; Prahl, U. Scalability of the mechanical properties of selective laser melting produced micro-struts. *J. Laser Appl.* **2015**, *27*, S29206. [CrossRef]
31. Wang, D.; Liu, Y.; Yang, Y.; Xiao, D. Theoretical and experimental study on Surface roughness of 316L stainless steel metal parts obtained through selective laser melting. *Rapid Prototyp. J.* **2016**, *22*, 706–716. [CrossRef]

© 2018 by the authors. Licensee MDPI, Basel, Switzerland. This article is an open access article distributed under the terms and conditions of the Creative Commons Attribution (CC BY) license (http://creativecommons.org/licenses/by/4.0/).

Article

Synthesis, Characterization, Corrosion Resistance and In-Vitro Bioactivity Behavior of Biodegradable Mg–Zn–Mn–(Si–HA) Composite for Orthopaedic Applications

Chander Prakash [1], Sunpreet Singh [1], Munish Kumar Gupta [2], Mozammel Mia [3,*], Grzegorz Królczyk [4] and Navneet Khanna [5]

[1] School of Mechanical Engineering, Lovely Professional University, Phagwara, Punjab 144411, India; chander.mechengg@gmail.com (C.P.); snprt.singh@gmail.com (S.S.)
[2] Mechanical Engineering Department, National Institute of Technology, Hamirpur 177005, India; munishguptanit@gmail.com
[3] Mechanical and Production Engineering, Ahsanullah University of Science and Technology, Dhaka 1208, Bangladesh
[4] Department of Manufacturing Engineering and Automation, Opole University of Technology, 76 Proszkowska St., 45-758 Opole, Poland; g.krolczyk@po.opole.pl
[5] Mechanical Engineering, Institute of Infrastructure, Technology, Research and Management (IITRAM), Gujarat 380026, India; navneetkhanna@iitram.ac.in
* Correspondence: mozammelmiaipe@gmail.com

Received: 17 August 2018; Accepted: 31 August 2018; Published: 3 September 2018

Abstract: Recently, magnesium (Mg) has gained attention as a potential material for orthopedics devices, owing to the combination of its biodegradability and similar mechanical characteristics to those of bones. However, the rapid decay rate of Mg alloy is one of the critical barriers amongst its widespread applications that have provided numerous research scopes to the scientists. In this present, porous Mg-based biodegradable structures have been fabricated through the hybridization of elemental alloying and spark plasma sintering technology. As key alloying elements, the suitable proportions of silicon (Si) and hydroxyapatite (HA) are used to enhance the mechanical, chemical, and geometrical features. It has been found that the addition of HA and Si element results in higher degree of structural porosity with low elastic modulus and hardness of the Mg–Zn–Mn matrix, respectively. Further, addition of both HA and Si elements has refined the grain structure and improved the hardness of the as-fabricated structures. Moreover, the characterization results validate the formation of various biocompatible phases, which enhances the corrosion performance and biomechanical integrity. Moreover, the fabricated composites show an excellent bioactivity and offer a channel/interface to MG-63 cells for attachment, proliferation and differentiation. The overall results of the present study advocate the usefulness of developed structures for orthopedics applications.

Keywords: magnesium; alloying; spark plasma sintering; elastic modulus; corrosion resistance; bioactivity

1. Introduction

The demand of artificial organs and other biomedical devices has increased drastically during the recent decades. Commonly used and successful implant materials are stainless steel (SS), cobalt–chromium (Co–Cr), titanium (Ti) and their alloys/composites [1]. However, they may have exceeded their full potential because of their drawbacks. Firstly, the Young's modulus of the aforementioned materials (110–200 GPa) is higher than that of the bone (7–25 GPa), which causes stress shielding [2]. As a result, the bone resorption occurs, which causes implant loosening and failure. Secondly, the implants made up of these biomaterials are unbiodegradable and after bone healing, the implants should be taken out from the body by performing a second surgery [3]. Due to adverse repercussions of non-degradable materials, feasibility of developing biodegradable materials has attracted the greatest interest. Recently, magnesium (Mg) and its alloys have gained a great deal of attention as a promising and potential biodegradable material for the fabrication of bone fixation accessories, because of their high biological properties [4]. However, the poor corrosion resistance of Mg is one of the most critical barriers, owing to which it degrades very rapidly after implantation [5,6]. Over the years, several methodologies were used to control the degradation rate of Mg and its alloys [7–9]. Elemental alloying has been reported as the most effective technique to improve corrosion resistance and mechanical properties of Mg alloys [10].

Elemental alloying is a powder metallurgical process, in which metallic powder particles are mechanically alloyed and subsequently sintered by appropriate techniques. In the past, alloying of safe elements, such as zinc (Zn), aluminium (Al), silver (Ag), yttrium (Y), zirconium (Zr), neodymium (Nd), silicon (Si), manganese (Mn), titanium dioxide (TiO_2), and calcium (Ca), were selected in response to increase the corrosion resistance and biological function of Mg [11–17]. Although the element Al in Mg composites improves mechanical properties, released Al^+ ions cause Alzheimer's disease and muscle fiber damage [18–20]. It was reported that Zr causes very serious diseases, such as liver, lung, and breast cancer [21]. Zhang et al. [22] reported that alloying Nd and Y in Mg alloys disrupt the growth of tissues around the implant. Li et al. [23] observed that the alloying of Ca reduces the degradation rate and improves biomechanical integrity in a corrosive medium. Moreover, Ca is a base element of human bone, which stimulates the new tissue growth and accelerates the bone healing process. The alloying of Zn and Mn in the Mg matrix enhances both elasticity and corrosion resistance [10]. Recently, Ben-Hamu et al. [24] reported that Si has proved to be an essential element being alloyed to develop tissues and immune systems. The developed Mg–Si composites possess required mechanical properties, low ductility, and high strength. Moreover, polygonal-shape Mg_2Si intermetallics inhibit the corrosion more effectively compared to the Chinese script.

Recently, the application of spark plasma sintering (SPS) technique for the synthesis of Mg-based alloys and composites with improved mechano-biological, antibacterial and corrosion performance has been reported. Sunil et al. [25] developed biodegradable Mg–hydroxyapatite (HA) composites by the SPS technique and studied the consequence of HA weight % on corrosion resistance of the developed composites. The Mg–10%HA composite exhibits best corrosion resistance and high hardness. Zheng et al. [26] synthesized a Mg–Al–Zn alloy by SPS, which possesses a maximum microhardness of 140 HV, a compressive yield strength of 442.3 MPa, and an ultimate strength of 546 MPa, which are comparatively higher than those values of conventional Mg alloys. Zhang et al. [27] studied the effect of Ca and Zn on a Mg–Si composite, and it was found that the addition of Ca and Zn to the Mg–Si alloy improved the bio-corrosion resistance and shows very good biocompatibility. In vitro analysis revealed that excellent adhesion and growth of osteoblastic cell has been observed and in vivo results suggested that the alloy has good biocompatibility. The Mg–Zn–Mn–Ca alloy developed by the elemental alloying and SPS technique exhibits high yield strength (58–69 MPa), strong tensile strength (177–205 MPa), and strong hardness (49–53 Hv) [28]. The effect of HA along with Zn and Mn on the microstructure, corrosion performance and mechanical properties of Mg alloy was reported. The alloying of HA (5 wt %) improves the corrosion resistance of Mg [29,30]. Further, Prakash et al. [31] investigated the effect of mechanical alloying-assisted SPS process (MA-SPS) parameters on structural porosity, elastic modulus, and hardness of the composite. Multi-objective particle swarm optimization (MO-PSO) has been utilized to determine the optimal setting of MA-SPS to sinter mechanically tuned biocompatible composites with improved corrosion properties.

It is clear that many studies, in the past, reported on design, development and synthesis of Mg alloy alloyed with Mn, and Zn using various fabrication techniques, with the aim of controlling the degradation rate. However, to the best of authors' knowledge, limited work is available on hybrid alloying of Si and HA and their effects on mechanical, corrosion properties, degradation and bioactivity analysis of Mg–Zn–Mn alloys. This paper is aimed at studying the synthesis, characterization, corrosion and cell response of Mg–Zn–Mn–(Si–HA) composites fabricated via the MA-SPS technique. The key expectation is that the fabricated porous composite will exhibit an improved biomechanical integrity while offering increased corrosion resistance to delay the degradation and improved bioactivity for orthopedic applications.

2. Materials and Methods

2.1. Mechanical Alloying and Consolidation of Spark Plasma Sintering

In this work, high-purity (~99.9%) elemental powders of Mg, Mn, Zn, Si, and HA were used to synthesize Mg–Zn–Mn–Si–HA composites. The chemical composition of the proposed bio-composites in wt % is listed in Table 1. The required powders were weighed and MA has been carried out using planetary ball mill (Fritsch Pulverisette 7, M/s. Fritsch, Germany.) with SS vial and SS balls with a diameter of 5 mm. The powder mixture was mechanically alloyed for about 12 h at 300 rpm with a ball/powder ratio of 10/1. Stearic acid (0.1 gm) was used to prevent agglomeration and excessive cold welding of powders. The blended powders were preheated at 100 °C for 1 h, in the argon atmosphere, in order to remove the moisture. Then, the blended powder was consolidated by the SPS process (SPS-5000 machine; model: Dr. Sinter SPS-625, Fuji Electronic Industrial Co. Ltd., Tsurugashima, Japan). The SPS was carried out at a heating rate of 50 °C/min (for a holding time of 5 min), under vacuum, and at different sintering temperatures and pressure conditions as illustrated in Table 2, as per the procedure reported elsewhere [29–31]. Figure 1 presents the fabrication route for the synthesis of Mg–Zn–Mn–(HA–Si) alloy. A graphite die was used for the sintering and the solid compacts of 20 mm in diameter and 4 mm in thickness were synthesized. The objective of changing the temperature and pressure level is to investigate their effect on the porosity, relative density, elastic modulus, and micro-hardness.

Table 1. Composition of alloying elements in wt % as-proposed for bio-composites.

Composite	Composition	Alloying Element Composition, wt %				
		Zn	Mn	Si	HA	Mg
Type-I	Mg–Zn–Mn–HA	1	5		10	Bal.
Type-II	Mg–Zn–Mn–Si	1	5	10		Bal.
Type-III	Mg–Zn–Mn–HA–Si	1	5	10	10	Bal.

Table 2. Process parameters of the mechanical alloying assisted SPS and their levels.

Process Parameters	Symbol	Units	Levels
Type of alloying element	A_e		HA, Si, Si–HA
Milling time, h	T_m	h	4, 8, 12
Sintering pressure	P_s	MPa	30, 40, 50
Sintering temperature	T_s	°C	350, 400, 450
Heating rate		°C/min	50
Holding time		Min	5
Atmosphere			Argon

2.2. Metallurgical and Mechanical Characteristics

The grain size and the lattice-strain of mechanically alloyed powder were determined by the Williamson–Hall method, as expressed in Equation (1):

$$B\cos\theta = \frac{K\lambda}{D} 2\epsilon \sin\theta \qquad (1)$$

where D is the crystal size, K is the shape factor (assume to be 0.9), λ is the wavelength of X-ray, B is the full width at half maximum, ε is the lattice strain, and θ is the Bragg angle [32,33].

The samples for microstructure examination were cut from the sintered compacts by low-speed diamond cutter, and then samples were well polished using emery paper, diamond paste, and napped cloth. The microstructure and morphology of composite were investigated by FE-SEM (Field-Emission Scanning Electron Microscope). The elemental composition was determined with an EDS (Energy Dispersive Spectrometer) detector coupled with the FE-SEM. The phases present in the synthesized composites were studied by X-ray diffraction (XRD) with CuKα radiation at an incident angle range of 20–80°. The elastic modulus and hardness of the as-developed composites were determined via a nano-indentation technique (model: Hyistron TI-950 indentation system, Bruker's, Minneapolis, MN, USA) via the Oliver–Pharr approach by using the Berkovich tip at 1000 µN [34].

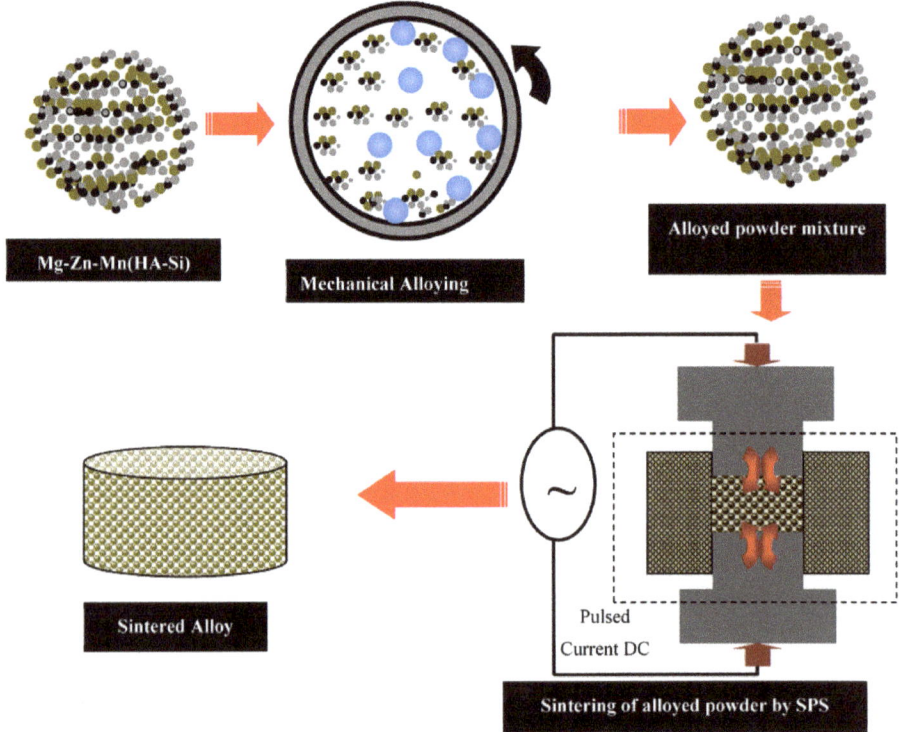

Figure 1. Fabrication route for synthesis of the Mg–Zn–Mn–(HA-Si) alloy.

2.3. Potentiodynamic Corrosion and Degradation Test

The corrosion characteristics of the as-synthesized composites were analysed by the potentiodynamic polarization test through an electrochemical system (Gamry 1000E, Potentiostat/Galvanostat, Gamry Instruments, Warminster, PA, USA) in simulated body fluids (9 g/L NaCl, 0.24 g/L $CaCl_2$, 0.43 g/L KCl, and 0.2 g/L $NaHCO_3$ at pH 7.2). The as-synthesized specimens, graphite rode, and saturated-carmol-electrode (SCE) were treated as the test electrode, the counter, and the reference electrode, respectively. The tests were performed at 37 °C to simulate the physiological environment. The corrosion characteristics were determined according to the approach reported in previous studies [35,36]. The simulated body fluid (SBF) test was conducted to find out the degradation rate of the specimens after 3, 7, and 14 days. The samples were well polished and dipped into the SBF solution in sterilized vials as per the ASTM-G31-72 standard, as reported in [31,37]. After a predetermined time period of immersion, the samples were retrieved from the glass vial, cleaned by water, and dehydrated into the desiccators for 24 h. The degradation rate was determined by the weight loss due to Mg^{2+} ion release in the SBF solution. The degraded surface was investigated by the FE-SEM and ESD techniques. The degradation behavior of as-synthesized composites was also measured and determined by the released concentration of Mg^{2+} molecules/ions in physiological environment during the immersion test, as per the procedure adopted elsewhere [38].

2.4. In Vitro Bioactivity Test

The cell culture, MTT, and differentiation assays were performed to examine the bioactivity and biocompatibility of the as-sintered porous composites using human MG-63 osteoblasts cell lines. The samples were sliced into 5 mm in diameter and 3 mm in thickness according to the geometry of a 96-well culture plat. The cells were cultured in a flask containing Dulbecco's Modified Eagle Medium supplemented with 10% bovine serum Sigma-Aldrich, (SIGMA, St. Louis, MO, USA) and 1 vol % penicillin (Invitrogen, Thermo Fisher Scientific corporation, Waltham, MA, USA) in an incubator at 37 °C and 5% CO_2 until confluent. The confluent cells were seeded on the Mg composites at a cell density of 1×10^5 cells/cm^2. The cell proliferation was evaluated using MTT assay (3-(4,5-dimethylthiazol-2-yl)-2,5-diphenyltetrazolium bromide) based on the conversion of MTT substrate to formazan by viable cells. At given time points, the culture medium was removed, and the MTT reagent (50 mL per well, thiazolyl blue tetrazolium bromide (M2128, Sigma-Aldrich, SIGMA, St. Louis, MO, USA) was added to the culture plate and incubated at 37 °C for 4 h. Then, the MTT reagent was removed and dimethyl sulfoxide (50 mL) was added to each well to dissolve the formazan crystals. The results of the MTT assay were expressed as a measure of optical density that was determined at a wavelength of 570 nm. Cell proliferation was also evaluated by determining the DNA content [39]. For staining the live cells, acetoxymethyl (AM) ester (Calcein, Molecular Probes, Crailsheim, Germany) was used, which is a fluorescent indicator. The cell distribution growth on the sample surface was analyzed using a florescent microscope (FM, Scope. A1, Carl Zeiss, Thornwood, NY, USA). After the cultivation period of 48 h, the adherent cells were fixed with 3.7 vol % paraformaldehyde for 10 min and permeabilized with 0.1 vol % Triton X-100 (in PBS) for 10 min at room temperature [40]. At incubation periods of 1, 3 and 7 days, the cultured-specimens were withdrawn from the physiological environment and subjected to fixation using the glutaraldehyde solution and then dehydrated using a series of ethanol. Cell differentiation was evaluated using cellular alkaline phosphatase-specific activity [orthophosphoric monoester phosphohydrolase, alkaline; E.C. 3.1.3.1] as an early differentiation marker and osteocalcin content in the conditioned media as a late differentiation marker. Alkaline phosphatase activity was assayed from the release of p-nitrophenol from p-nitrophenylphosphate at pH 10.2, as previously described. Activity values were normalized to the protein content, which was detected as colorimetric cuprous cations in biuret reaction (BCA Protein Assay Kit, Pierce Biotechnology Inc., Rockford, IL, USA) at 570 nm (Microplate reader, BioRad Laboratories Inc., Hercules, CA, USA) [41]. All experiments were repeated three times to ensure validity of the observations. Analysis of variance (ANOVA) and the significant difference between groups was determined using the Student's *t* test at a 95% confidence interval. A *p* value of less than 0.05 was considered as statistically significant.

3. Results and Discussion

3.1. Powder Morphology

Figure 2 presents the SEM micrograph and associated EDS spectrum of powder particles before MA. The HA powder particles used were of 0.5 µm (irregular), whereas others exhibited spherical morphology with an average size of 25 µm. It has been found that there was no powder loss incurring during the alloying process as the sample size before and after the alloying was recorded to be

10 gm. However, the size of the grains and lattice strain of alloyed powder were determined by the Williamson–Hall method [33]. The size and morphology of powder particles were changed with milling time. Figure 3 presents the variation in the grain size of powder particles during MA and morphology of powder particles after MA of 12 h. Figure 3a shows how the grain size and lattice strain varied with the milling time. As the milling time increased, the powder particle size was reduced. The particle size decreased to 250 nm and the lattice strain was about 0.22%, after milling for 4 h. On the other hand, after milling of 12 h, the powder size was reduced notably to 75 nm and the lattice strain was 0.14%. The lattice strain increased, as the milling time increased. The increase in lattice strain is attributed to the increase in the lattice imperfections, such as grain boundaries and dislocation density. The morphology of the mixture observed by FE-SEM (JEOL 7600F, Tokyo, Japan) showed that no diffusion occurred at higher localized temperatures over the processing span. Figure 3b–d show the morphology of alloyed powder after 12 h of milling and it can be clearly seen that the powder size was significantly reduced to <75 nm.

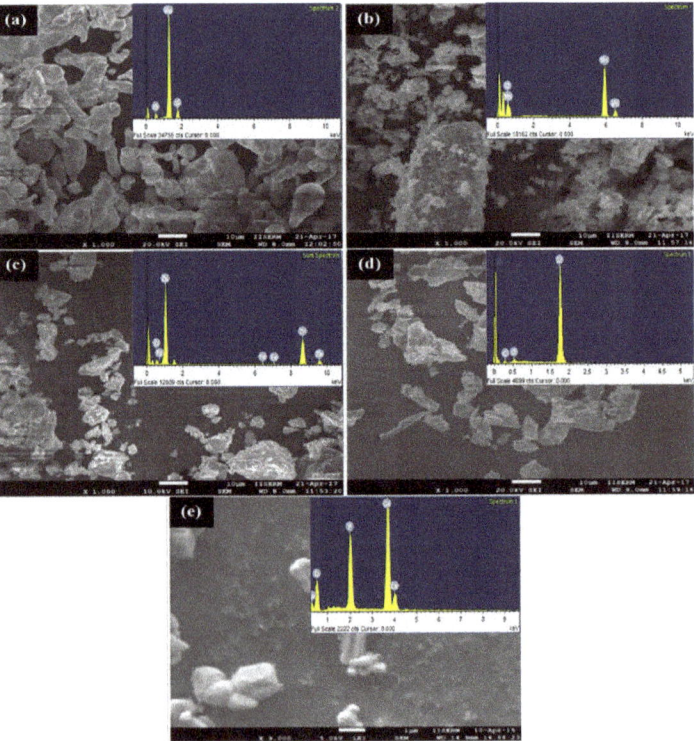

Figure 2. SEM micrographs and EDS spectra of raw powders: (**a**) magnesium; (**b**) manganese; (**c**) zinc; (**d**) silicon and (**e**) hydroxyapatite.

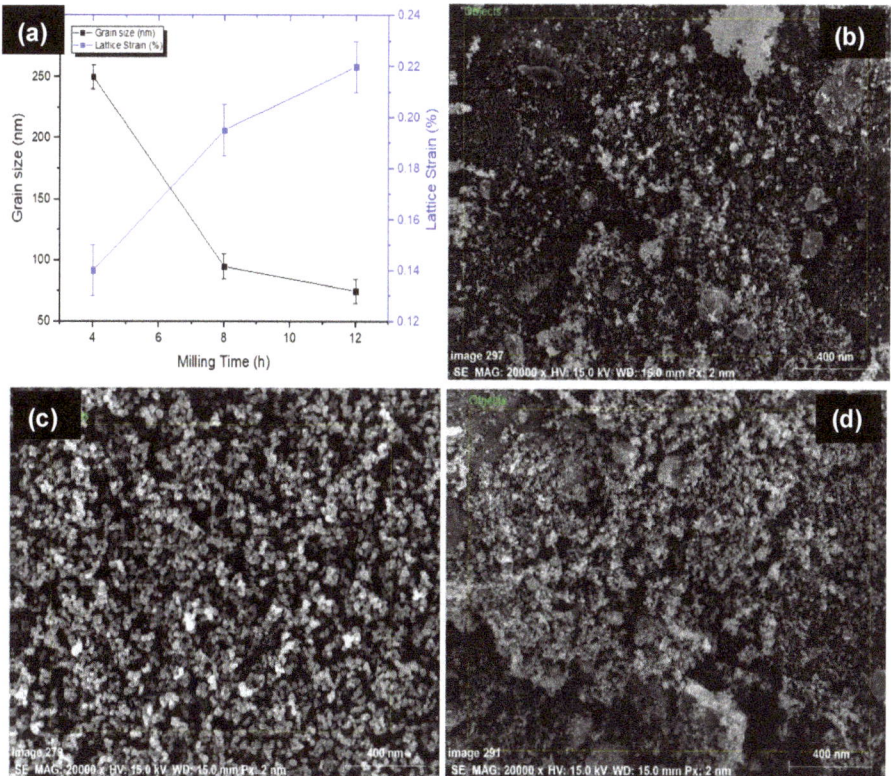

Figure 3. (**a**) Evaluation of grain size and lattice strain as a function of time; powder morphology and size after ball milling for 12 h: (**b**) Mg–Zn–Mn–Si; (**c**) Mg–Zn–Mn–HA; (**d**) Mg–Zn–Mn–Si–HA.

3.2. Microstructure

The structural morphology of the synthesized composites was directly dependent on the sintering temperature and applied pressure. As the sintering temperature and pressure increased, the densification of sintered green compact was increased. The high value of sintering pressure induces the high driving force, which helps in densifying or compacting the powder particles. Reportedly elevated sintering temperature assists the coalescence of the powder and reduces the porosity [30,31]. Figure 4a presents the sintering of powder particles during the SPS process. During the SPS process, thermal energy was generated due to electrical sparks between the powder particles and the contact area caused partial melting of the grain boundary of powder while uniaxially applied pressure densified the powder mixture (Figure 4b). The process of densification and solidification formed the final sintered compact. Figure 4c shows the mass transformation during the SPS process and the phenomena of partial diffusion and welding of powder particles as presented by Zheng et al. [32]. Three types of Mg-based composites, Mg–Zn–Mn–HA (Type-I), Mg–Zn–Mn–Si (Type-II), and Mg–Zn–Mn–Si–HA(Type-III) were synthesized. Figure 5 presents the microstructures and EDS spectra of all type of composites at a sintering pressure and a temperature of 40 MPa and 400 °C, respectively. Evidently, all three composites were completely densified and exhibited a low degree of structural porosity. With the change in element alloying composition (Si and HA), a distinct morphology can be observed in the composites. A thin and sharp needle-like laminar structure was observed as-distributed along the grain boundaries in Type-I composite (Figure 5a).

Sunil et al. reported similar observations on HA addition in Mg composite, which enhances corrosion resistance [25]. This is attributed to the fact that the individual Mg flakes bonded together with HA and formed the layer-by-layer laminar structure in the form of needle (MgCaO).

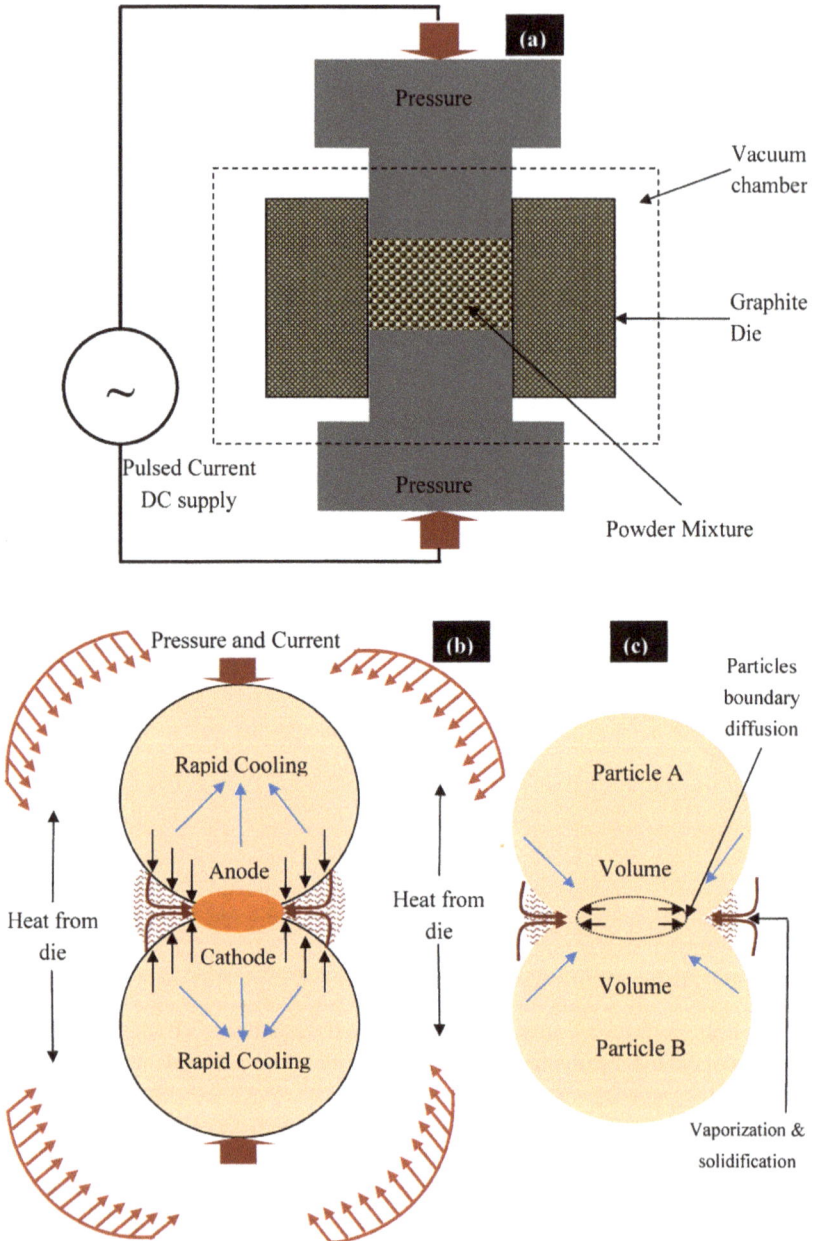

Figure 4. (**a**) Schematic representation of SPS technique and (**b**,**c**) mechanism of sintering of powder particles during the SPS process.

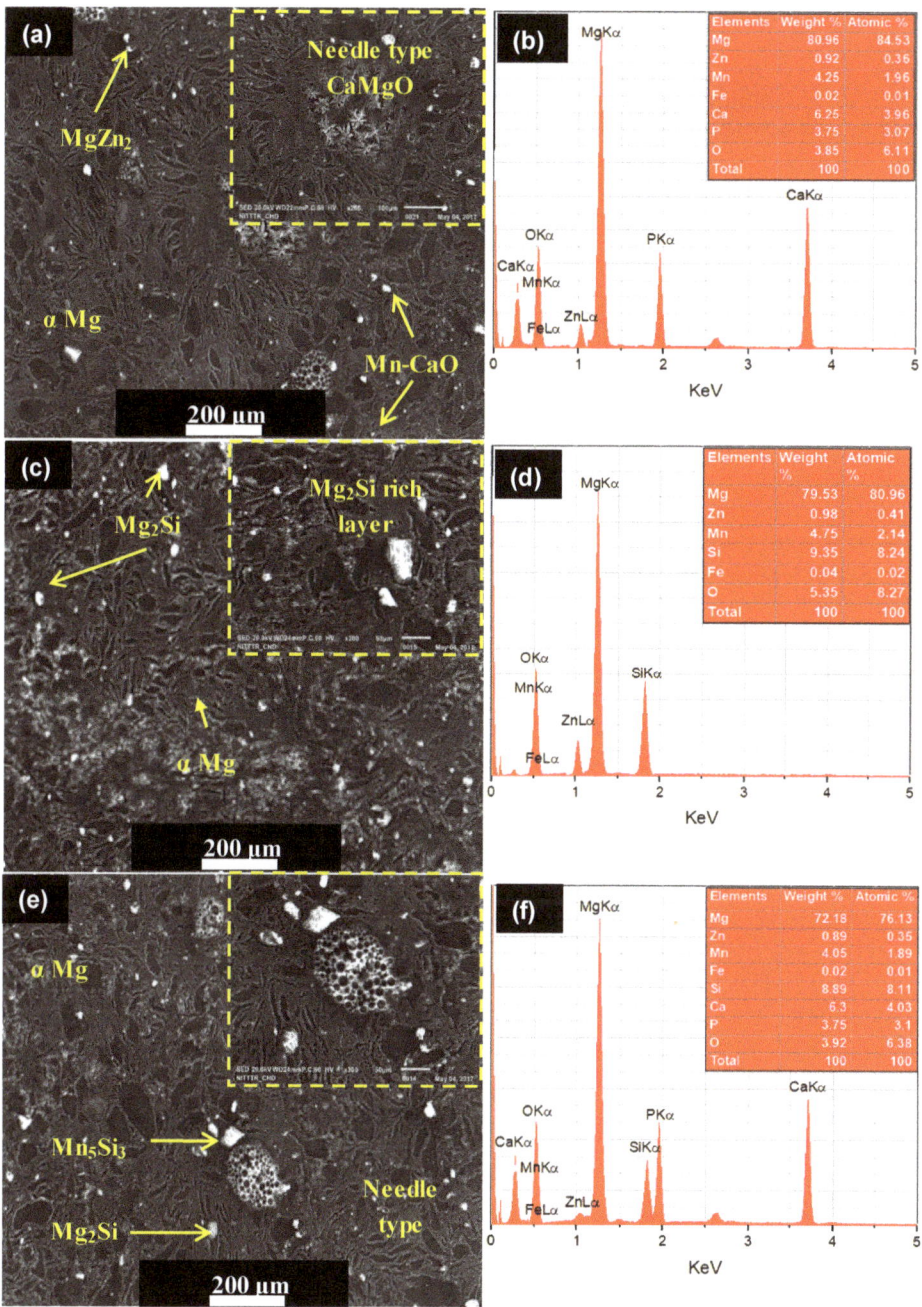

Figure 5. SEM micrographs and associated EDS spectra of as-fabricated composites sintered at a pressure of 50 MPa and a sintering temperature of 400 °C: (**a**,**b**) Mg–Zn–Mn–Si; (**c**,**d**) Mg–Zn–Mn–HA; and (**e**,**f**) Mg–Zn–Mn–Si–HA.

EDS analysis indicated the element composition of Type-I composite and Fe, Ca, P, and O elements appeared with other elements (Mg, Zn, and Mn), as can be observed in Figure 5b. The peak intensities of Ca, P, and O elements confirmed the uniform distribution of HA in the composite. The SPS does not allow the oxidation during the sintering process. This is because the finer powder particle reacts quickly at room temperature. Therefore, the possible reaction behind the appearance of O element in the as-fabricated alloys resulted from handling of powder sample after milling and before sintering. The uniformly distributed HA in the composite leads to increase in the corrosion resistance [29–31]. On the other hand, when Si was used as an alloying element instead of HA, the typical change in the structure has been witnessed. When Si was used instead of HA, a mixture of discontinuous laminar and eutectic structure was observed. The microstructure of Type-II composite mainly comprised α–Mg, $MgZn_2$, MnSi, and Mg_2Si stages, as shown in Figure 5c. The Zn and $MgZn_2$ existed as a hexagonally packed structure and a secondary phase, respectively. The $MgZn_2$ phases were observed as an agglomeration of the compact fleck. The intermetallics Mn_5Si_3 and Mg_2Si phases appeared in polygonal shape and can be clearly identified at the high magnification (×300). Ben-Hamu et al. observed the similar microstructure [24]. The associated EDS spectra confirmed the appearance of Si with other elements (Mg, Zn, Mn, Fe, and O), as illustrated in Figure 5d. When Si and HA were added in the Mg–Zn–Mn composite, the microstructure showed different morphologies (refer to Figure 5e). When HA and Si were used jointly as alloying elements, the appearance of needle-like structure can be clearly seen. Dark, gray and bright phases were identified as Mg matrix, CaMgSi, and Mg_2Si phase. The typical eutectic structure disappeared and needle-like MgCaO phases formed. EDS analysis indicates the element composition of Type-III composite and Fe, Si, Ca, P, and O elements appeared with other elements (Mg, Zn, and Mn), as can be observed in Figure 5f.

The XRD patterns of all types of sintered composites are presented in Figure 6. It can be observed that all sintered composites had the same XRD pattern; however, their respective peak intensities changed with the weight percentages of Si and HA. Biocompatible and biomimetic phases were identified in the sintered composites. MgCaO, Mn–CaO, and CaMgZn phases were observed in the Type-I composite. The Type-II composite comprised Mg_2Si, $Mg0.97Zn0.03$, and Mn_5Si_3 phases. Mg_2Si was expected to enhance the corrosion resistance. The Type-III composite showed CaMgSi, Mg_2Si, Mn_5Si_3, Mn–CaO, and CaMgZn phases, which are beneficial to form the apatite growth and improve the bioactivity.

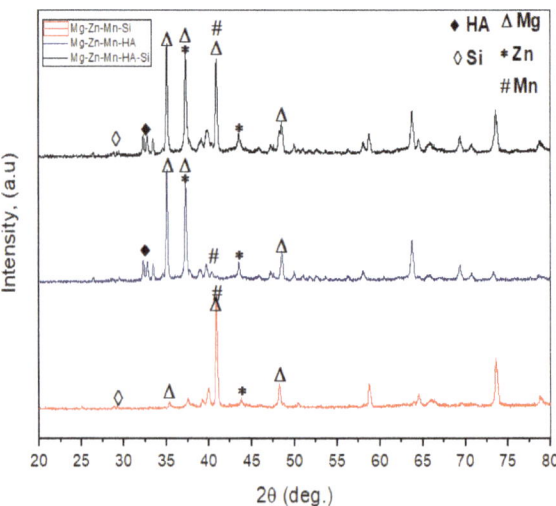

Figure 6. X-ray diffraction patterns of all types of as-fabricated Mg composites.

3.3. Mechanical Properties

Figure 7 shows the distinctive loading/unloading plots for all types of sintered composites. Table 3 presents mechanical properties of all three composite composites. The Type-I composite (Mg–Zn–Mn–HA) showed low elastic modulus and hardness, which were estimated to be 32 GPa and 0.54 GPA, respectively. The Type-I composite exhibited low hardness due to high degree of structural porosity. The high degree of porosity in structure causes the reduction in mechanical properties of the bulk material. When Si was used as an alloying element instead of HA, the densification of bulk increased and the mechanical properties of compact were improved in terms of hardness and elastic modulus. The Type-II composite (Mg–Zn–Mn–Si) offered high values of elastic modulus and hardness, which were estimated to be about 45 GPa and 1.97 GPa, respectively. When Si and HA were used jointly as alloying elements, the degree level of porosity increased again, which led to the reduction in hardness and elastic modulus again. The hardness of as-synthesized alloys was higher than the pure Mg. The increase in the hardness of bulk material is due to cold hardening of Mg as well as due to the presence of HA and MgO at inter-laminar sites. Notably, the elastic modulus and hardness for the Type-III composite were 39 GPa and 1.18 GPa, which were smaller than the Type-II composite but higher than the Type-I composite, as seen in Table 3.

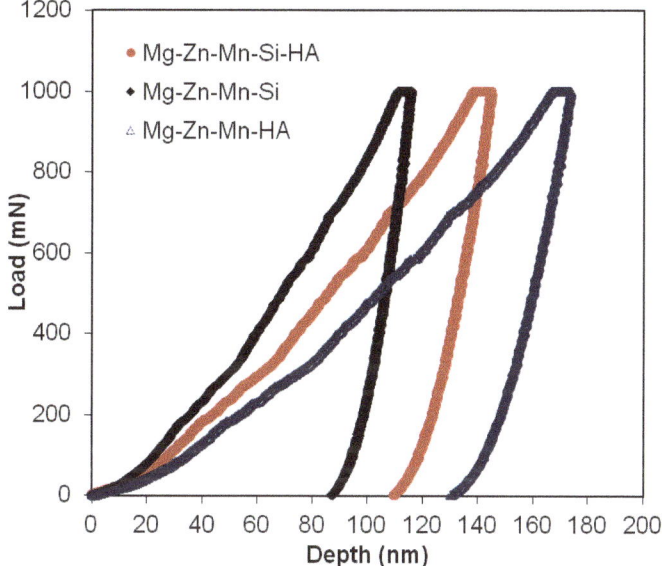

Figure 7. Load–depth curves of as-fabricated all types of Mg composites.

Table 3. Elastic modulus and hardness of the sintered biocomposites.

Mg Alloys	Mechanical Properties			
	Elastic Modulus, E (GPa)		Hardness, H (GPa)	
	Mean of Sample Group	Standard Deviation	Mean of Sample Group	Standard Deviation
Mg–Zn–Mn–HA	32	1.58	0.54	0.02
Mg–Zn–Mn–Si	45	2.64	1.97	0.03
Mg–Zn–Mn–Si–HA	39	1.98	1.18	0.02

3.4. In Vitro Corrosion and Degradation Analysis

In vitro corrosion characteristics and degradation behavior of the as-fabricated composites were assessed by a Tafel extrapolation method and an immersion test. Figure 8 presents the corrosion characteristics and degradation behavior of as-fabricated composites. Figure 8a illustrates a comparison of corrosion Tafel polarization curves of all synthesized composites. Table 4 presents the determined corrosion characteristics, such as the corrosion potential (E_{corr}), corrosion current density (I_{corr}), polarization resistance (R_p), and corrosion rate (C_R) for all types of materials. From the investigation, it can be seen that cathodic and anodic reactions obtained were same for all types of specimens, which are the typical characteristics of passive behavior. The corrosion parameters for Mg–Zn–Mn, such as I_{corr} and E_{corr}, were measured to be 19.5 µA/cm^2 and −1.2 mV, respectively. The corrosion current density was very low as compared to all specimens; the samples had least corrosion resistance and the increased degradation was 1.98 mm/year. When the Si was alloyed in Mg–Zn–Mn, the corresponding current density and corrosion potential were measured around 7.7 µA/cm^2 and −1.27 mV, respectively. The Type-II composite possessed higher corrosion resistance as compared to Mg–Zn–Mn and the corrosion rate was measured to be around 1.45 mm/year. However, the Type-II alloy still had low corrosion resistance and the alloying of Si element was less preventive from corrosion. When HA was used as an alloying element, the hyperbolic curve was shifted slightly towards the lower current density, and the corresponding current density and corrosion potential were measured to be around 3.5 µA/cm^2 and −1.13 mV, respectively. The Type-I composite possessed better corrosion resistance as compared to Mg–Zn–Mn and the Type-II composite. The corresponding corrosion rate was measured to be around 0.97 mm/year. The alloying of HA element in Mg–Zn–Mn increased the corrosion resistance. This is attributed to the formation of corrosion barrier phases (CaMg and Mg0.97Zn0.03) in the composite that promoted the apatite layer growth on the composite surface, which resisted the degradation/corrosion of composite in the SBF medium. The corrosion morphology of Mg–Zn–Mn–HA composite samples was found less corroded as compared to the Mg–Zn–Mn–Si composite (Figure 8b). On the other hand, when both HA and Si were used as alloying elements, excellent corrosion resistance was offered by the specimen, and the corresponding current density and corrosion potential were measured to be around 0.98 µA/cm^2 and −1.17 mV, respectively. The Mg–Zn–Mn–Si–HA composite possessed better corrosion resistance as compared to all other types of as-sintered composites, and the corrosion rate was measured to be around 0.15 mm/year. The above observed finding suggested that the Type-III composite can hold up the degradation rate at a pace that matches the period of bone healing, which is the prime objective of the current study.

Table 4. Corrosion parameters determined by the Tafel extrapolation method.

Parameters	Mg Alloys			
	Mg–Zn–Mn	Mg–Zn–Mn–Si	Mg–Zn–Mn–HA	Mg–Zn–Mn–Si–HA
I_{corr} (µA/cm^2)	22.7	7.7	3.3	0.98
E_{corr} (mV)	−1.27	−1.27	−1.13	−1.17
C_R (mm/year)	1.98	1.45	0.97	0.15

Figure 8b represents the degradation behavior of the as-fabricated Mg composite specimens in SBF. It has been found that during the initial period, the degradation rate of all-sintered alloys was high, but no further effects were seen after 28 days. Comparatively, the degradation rate of Type-II alloy was high as compared to the Type-I and Type-III composites. When 10% HA and Si were used as a reinforcement, the rate of mass deposition of apatite layer was high as compared to the Type-I and Type-II composite samples. Figure 8c illustrates the Mg^{2+} concentration in the SBF solution. In the early phase of immersion test (up to 7 days), the release of Mg^{2+} was higher, but after 7 days, the release rate of Mg^{2+} began reducing as a result of deposition of a thick apatite layer on the surface of specimens. The Mg^{2+} dissolution was found larger for Mg–Zn–Mn–Si specimens among all types of composites, presenting high degradation, which showed a similar trend as found in the degradation rate (Figure 8b). Furthermore, when Si and HA elements were used as alloying elements, a very significant and drastic reduction in the dissolution of Mg^{2+} ion was found, as can be seen from Figure 8c.

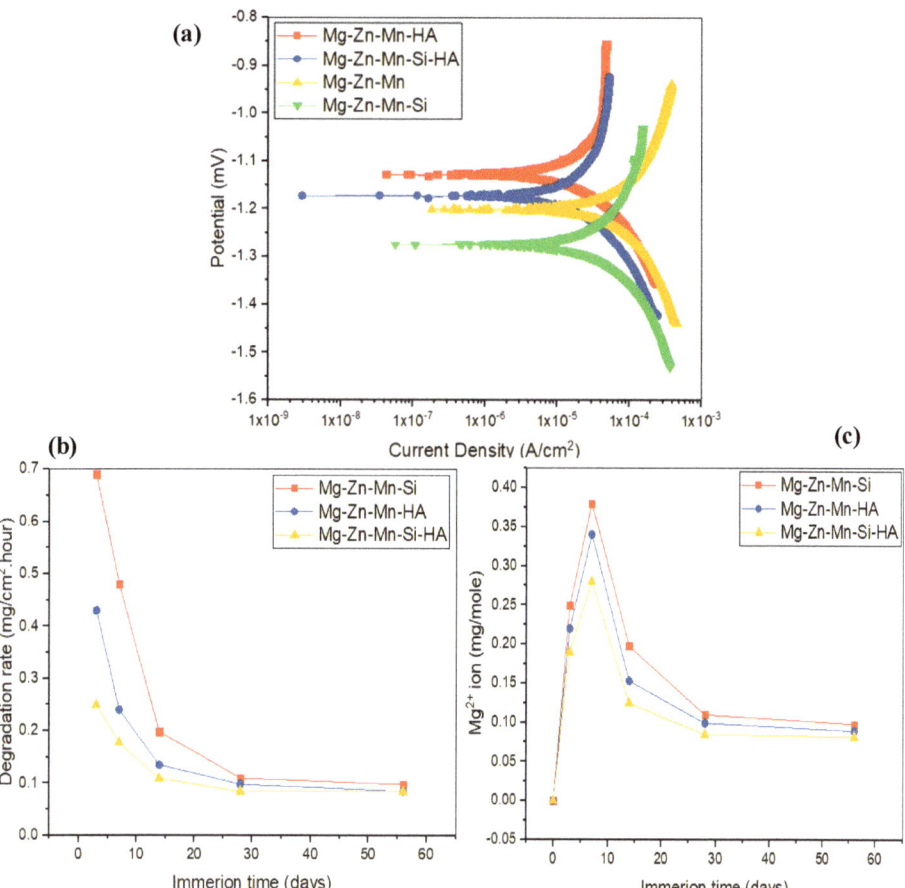

Figure 8. (a) Potential dynamic polarization curves of Mg–Zn–Mn, Mg–Zn–Mn–Si, Mg–Zn–Mn–HA, and Mg–Zn–Mn–Si–HA alloys at (37 ± 1) °C; (b) degradation rate of composites as a function of time, and (c) concentrations of Mg^{2+} of composites in the simulated body fluid (SBF) medium.

Figure 9 shows the corroded morphologies and EDS spectra of the all types of samples after 28 days of immersion in the SBF solution. The Type-II alloy surface was found to be highly corroded. This is because the developed apatite layer on the composite surface was weaker and therefore degraded rapidly in the SBF medium. The apatite layer was shredded due to its highly porous nature and degradation took place in the form of pulverized fine particles, as can be seen in Figure 9a. Open holes, cracks, and shredded layers were clearly seen on the corroded surface due to release of H_2 gas and Mg^{2+} ions. The shredding of apatite and traces of pulverized Ca and P particles can be easily identified as holes/cracks. The growth of apatite layer formation was confirmed by EDS-analysis, as can be seen in Figure 9b. When HA was used as an alloying element, the composite sample was less corroded as compared to the Mg–Zn–Mn–Si composite. The apatite layer growth on the composite (Type-I) surface was high as compared to the Type-II composite, which resisted the degradation/corrosion of composite in the SBF medium. This is attributed to the formation of corrosion barrier phases (CaMg and Mg0.97 Zn0.03) in the composite. The corrosion morphology of Mg–Zn–Mn–HA composite sample was seen in Figure 9c. Still, open holes, shredding of apatite layer and traces of pulverised materials were found and high peaks of Ca and P in the associated EDS spectrum confirmed the formation of the thick layer of apatite growth, as can be seen in Figure 9d. Samples with hybrid and proportionate filling of HA and Si elements showed better corrosion resistance. The Mg–Zn–Mn–Si–HA composite had the least corrosion rate as compared to Mg–Zn–Mn–Si and Mg–Zn–Mn–HA composites, as can be seen in Figure 9e. This is because a very thick apatite layer was developed on the composites' surface, which resisted it from degradation and the presence of CaMgSi, Mg_2Si, Mn_5Si_3, Mn–CaO, CaMgZn, and MnSi phases fortified the mechano-corrosion and biological properties. Figure 9f presents the corroded surface morphology of the Mg–Zn–Mn–Si–HA composite and a pulverized surface with comparatively less holes was observed.

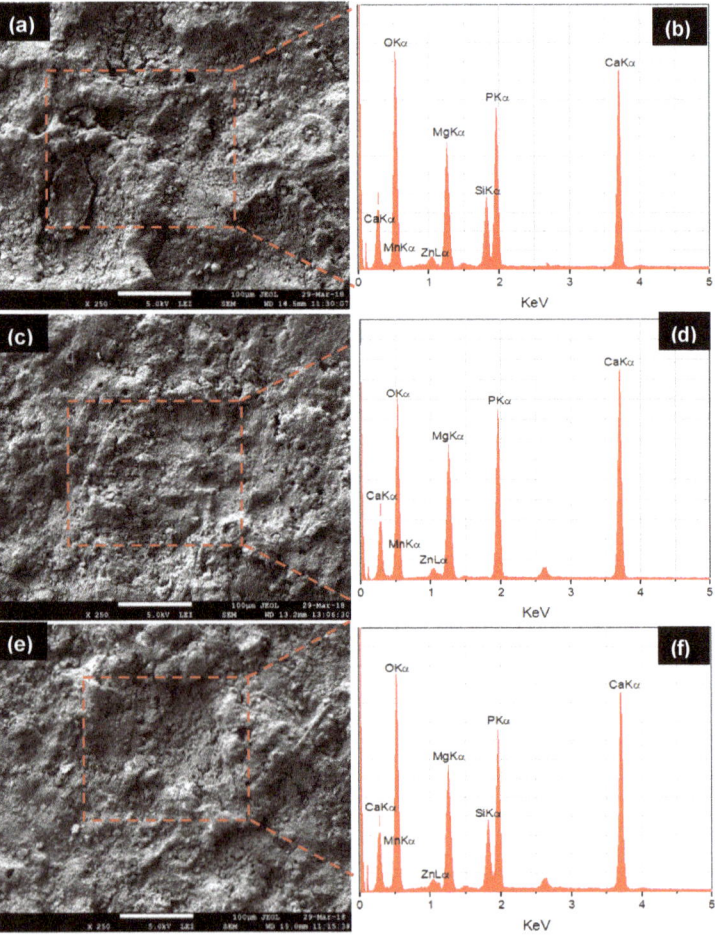

Figure 9. SEM micrographs and EDS spectra of the degraded morphology of composites after 28 days in the SBF immersion.

3.5. In Vitro Biocompatibility Assessment

The structural morphology and elemental composition of implant played a very important role in establishing the bio-mechanical bonding between the implant surface and surrounding tissues. A number of studies reported that HA has significant influence on the adhesion and growth of cells [29,35,38]. Recently, the alloying of Si and HA elements is found favourable for the enhancement of bioactivity of Mg alloys and composites [29–31]. Figure 10 presents the fluorescent fluorescence staining, cell attachment, proliferation activities and differentiation activities of osetoblatic cell (MG-63) on the as-synthesized Mg–Zn–Mn–(HA-Si) composites. With the increase in incubation time, the adhesion and proliferation of MG-63 cells increased significantly.

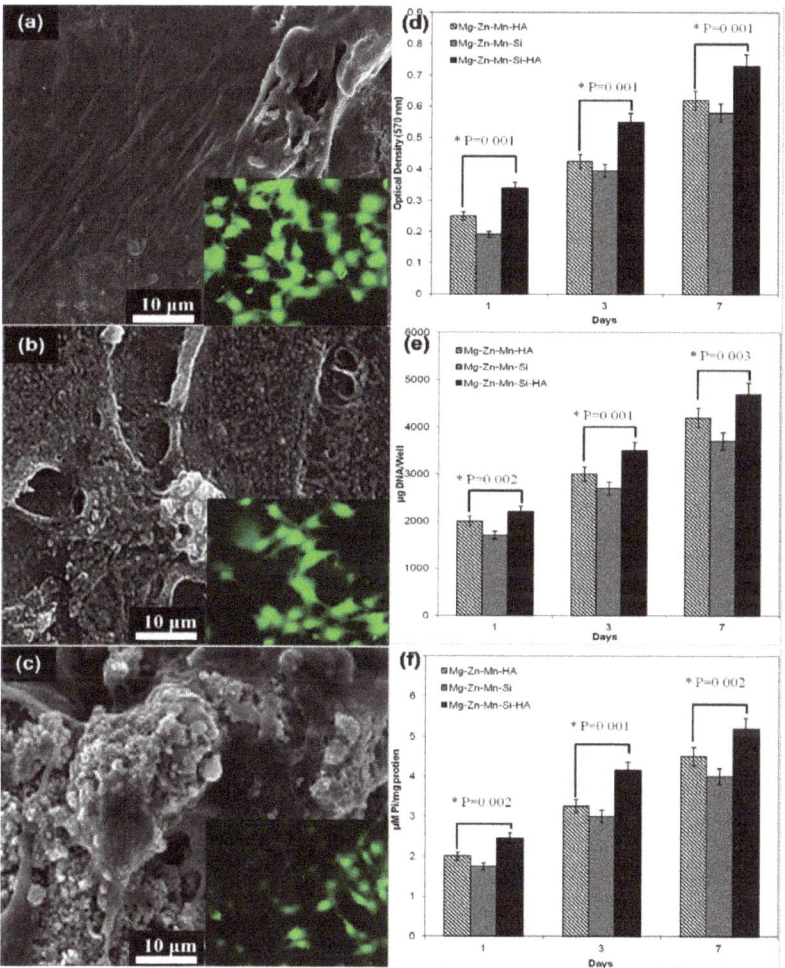

Figure 10. MG-63 cell adhesion after 24 h: (**a**) Mg–Zn–Si; (**b**) Mg–Zn–Mn–HA and (**c**) Mg–Zn–Mn–Si–HA surface and cell proliferation and differentiation: (**d**) MTT assay; (**e**) DNA content and (**f**) alkaline phosphatase-specific (ALP) activity of MG-63 cells determined on Days 1, 3, and 7 (individual group was statistically highly significant ($p < 0.001$)).

Figure 10a–c present the attached cell morphology and fluorescent staining on the Type-I, Type-II, and type-III composites, respectively. Fluorescent staining is generally used to indicate intracellular esterase activity present in viable cells. Dense and evenly dispersed multi-layered cells with large nuclei were observed for all samples; however, in the case of Mg–Zn–Mn–(HA–Si) samples, there were larger numbers of living cells. Compositionally, the reinforcement of HA had very significant impact on the apatite-inducing ability and bioactivity of implant. Moreover, spontaneous formation of bio-compatible phases of composition CaMgSi, Mg_2Si, Mn_5Si_3, Mn–CaO, CaMgZn, and MnSi, provided a biomimetic inert layer on the alloy surface, which accelerated the bone adhesion, proliferation, growth and differentiation of MG-63 cell line. Moreover, porous structure leads to the formation of hydrophilic surface and provides a vehicle and mechanical anchoring sites to interact with cells. In the current study, the as-sintered porous Mg–Zn–Mn–(Si–HA) alloys possessed micro-scale pore structures ranging from 20–50 μm mimicking human bone, which met the requirement of osseiointergation. After coming in the contact with the composite surface, the MG-63 cells started adhering on the surface, and after 24 h, cells started spreading. The shape of cells mainly elongated and polygonal which indicated that cells were well adhered, spread and proliferated. Polygonal-shape cells represented the excellent adhesion and growth on the as-synthesized composites surface. A number of activities, such as filopodias, lamellipodia, and peripheral ruffles, were seen. Figure 9d–f present the cell proliferation, DNA content, and alkaline phosphatase-specific (ALP) differentiation activities. All observed data was statically analyzed at a 95% confidence level using ANOVA, and individual group was statistically highly significant ($p < 0.001$) for each treatment (different alloy compositions) at different time intervals (days). Higher numbers of cells were grown on the Type-III composite surface. The optical density showed the proliferation of MG-63 cells on the composite test specimens, as presented in Figure 9d. The Type-III composite surface possessed a higher cell proliferation rate. This is attributed to the presence of Si and HA elements, which enhanced the bone formation process. Moreover, the structural porosities escalated the surface energy, which promoted protein absorption and cell growth. The DNA content on the specimen's surface increased with the increase in the proliferation rate, as can be seen that the Type-III composite specimens had a higher proportion of DNA content (Figure 9e). The ALP-type differentiation activities of MG-63 cells were presented in Figure 9f. The serum level of ALP activity was found significantly higher in the Type-III composite specimens, compared with the Type-II and Type-I composite specimens.

4. Conclusions

Biomimetic, biodegradable, low elastic and mechanically tuned Mg–Zn–Mn–(Si–HA) composites were fabricated by the element alloying and SPS technique. The investigation revealed that pore characteristics of size ranging from 25–50 μm, and 20–30% porosity has been achieved by adding HA and Si from 5 wt %. The Mg–Zn–Mn–(Si–HA) alloys possessed not only porous structure, but also possessed low elastic modulus ranging from 15 to 30 GPa that helped in reducing the stress shielding effect. Further, the developed alloys attained reasonable hardness ranging from 86–200 HV. The alloying of HA and Si elements led to the formation of biomimetic and biocompatible phases, such as CaMg, $MgSi_2$, Mg–Zn, Mn–Si, Mn–CaO, Mn–P, Ca–Mn–O, and CaMgSi in the porous layers, which enhanced the corrosion characteristics of the alloys. Moreover, the appearance of Ca, P and O elements in the EDS spectrum conferred the bioactivity of the as-synthesized alloys. The in vitro bioactivity results indicated that the Mg–Zn–Mn–HA–Si alloy had excellent biocompatibility and promoted cell adhesion, growth, proliferation, and differentiation.

Author Contributions: Conceptualization, C.P. and S.S.; methodology, C.P. and S.S.; software, C.P. and S.S.; validation, C.P., S.S., M.K.G., M.M. and G.K.; formal analysis, C.P.; investigation, C.P., S.S., M.K.G., M.M., G.K.; N.K.; resources, C.P. and S.S.; data curation, C.P. and S.S.; writing of the original draft preparation, C.P., S.S., M.K.G., M.M., G.K.; N.K.; visualization, M.K.G., M.M. and G.K.; supervision, G.K; writing of review and editing, M.M., G.K.; N.K.

Funding: This research received no external funding.

Conflicts of Interest: The authors declare no conflicts of interest.

References

1. Geetha, M.; Singh, A.K.; Asokamani, R.; Gogia, A.K. Ti based biomaterials, the ultimate choice for orthopaedic implants—A review. *Prog. Mater. Sci.* **2009**, *54*, 397–425. [CrossRef]
2. Prakash, C.; Kansal, H.K.; Pabla, B.S.; Puri, S.; Aggarwal, A. Electric discharge machining—A potential choice for surface modification of metallic implants for orthopedic applications: A review. *Proc. Inst. Mech. Eng. B J. Eng. Manuf.* **2016**, *230*, 331–353. [CrossRef]
3. Spoerke, E.D.; Murray, N.G.; Li, H.; Brinson, L.C.; Dunand, D.C.; Stupp, S.I. A bioactive titanium foam scaffold for bone repair. *Acta Biomater.* **2005**, *1*, 523–533. [CrossRef] [PubMed]
4. Staiger, M.P.; Pietak, A.M.; Huadmai, J.; Dias, G. Magnesium and its alloys as orthopedic biomaterials: A review. *Biomaterials* **2006**, *27*, 1728–1734. [CrossRef] [PubMed]
5. Song, G.L.; Atrens, A. Corrosion mechanisms of magnesium alloys. *Adv. Eng. Mater.* **1999**, *1*, 11–33. [CrossRef]
6. Song, G.; Atrens, A. Understanding magnesium corrosion—A framework for improved alloy performance. *Adv. Eng. Mater.* **2003**, *5*, 837–858. [CrossRef]
7. Uddin, M.S.; Hall, C.; Murphy, P. Surface treatments for controlling corrosion rate of biodegradable Mg and Mg-based alloy implants. *Sci. Technol. Adv. Mater.* **2015**, *16*, 053501. [CrossRef] [PubMed]
8. Abidin, N.I.Z.; Rolfe, B.; Owen, H.; Malisano, J.; Martin, D.; Hofstetter, J.; Uggowitzer, P.J.; Atrens, A. The in vivo and in vitro corrosion of high-purity magnesium and magnesium alloys WZ21 and AZ91. *Corros. Sci.* **2013**, *75*, 354–366. [CrossRef]
9. Uddin, M.S.; Rosman, H.; Hall, C.; Murphy, P. Enhancing the corrosion resistance of biodegradable Mg-based alloy by machining-induced surface integrity: Influence of machining parameters on surface roughness and hardness. *Int. J. Adv. Manuf. Technol.* **2017**, *90*, 2095–2108. [CrossRef]
10. Zhang, E.; Yin, D.; Xu, L.; Yang, L.; Yang, K. Microstructure, mechanical and corrosion properties and biocompatibility of Mg–Zn–Mn alloys for biomedical application. *Mater. Sci. Eng. C* **2009**, *29*, 987–993. [CrossRef]
11. Atrens, A.; Song, G.L.; Liu, M.; Shi, Z.M.; Cao, F.Y.; Dargusch, M.S. Review of recent developments in the field of magnesium corrosion. *Adv. Eng. Mater.* **2015**, *17*, 400–453. [CrossRef]
12. Atrens, A.; Song, G.L.; Cao, F.Y.; Shi, Z.M.; Bowen, P.K. Advances in Mg corrosion and research suggestions. *J. Magnes. Alloys* **2013**, *1*, 177–200. [CrossRef]
13. Radha, R.; Sreekanth, D. Insight of magnesium alloys and composites for orthopedic implant alications—A review. *J. Magnes. Alloys* **2017**, *5*, 286–312. [CrossRef]
14. Huang, X.; Han, G.; Huang, W. T6 Treatment and Its Effects on Corrosion Properties of an Mg–4Sn–4Zn–2Al Alloy. *Materials* **2018**, *11*, 628. [CrossRef] [PubMed]
15. Wang, J.; Jiang, W.; Guo, S.; Li, Y.; Ma, Y. The Effect of Rod-Shaped Long-Period Stacking Ordered Phases Evolution on Corrosion Behavior of Mg95. 33Zn2Y2. 67 Alloy. *Materials* **2018**, *11*, 815. [CrossRef] [PubMed]
16. Gavras, S.; Buzolin, R.H.; Subroto, T.; Stark, A.; Tolnai, D. The Effect of Zn Content on the Mechanical Properties of Mg–4Nd–xZn Alloys (x = 0, 3, 5 and 8 wt %). *Materials* **2018**, *11*, 1103. [CrossRef] [PubMed]
17. Cheng, W.; Zhang, Y.; Ma, S.; Arthanari, S.; Cui, Z.; Wang, H.X.; Wang, L. Tensile Properties and Corrosion Behavior of Extruded Low-Alloyed Mg–1Sn–1Al–1Zn Alloy: The Influence of Microstructural Characteristics. *Materials* **2018**, *11*, 1157. [CrossRef] [PubMed]
18. Ferreira, P.C.; Piai, K.D.A.; Takayanagui, A.M.M.; Segura-Muñoz, S.I. Aluminum as a risk factor for Alzheimer's disease. *Rev. Latinoam. Enfermagem.* **2008**, *16*, 51–157. [CrossRef]
19. Shingde, M.; Hughes, J.; Boadle, R.; Wills, E.J.; Pamphlett, R. Macrophagic myofasciitis associated with vaccine-derived aluminium. *Med. J. Aust.* **2005**, *183*, 145–146. [PubMed]
20. Luo, L.; Liu, Y.; Duan, M. Phase Formation of Mg–Zn–Gd Alloys on the Mg-rich Corner. *Materials* **2018**, *11*, 1351. [CrossRef] [PubMed]
21. Zhang, S.X.; Li, J.; Song, Y.; Zhao, C.L.; Zhang, X.N.; Xie, C.Y.; Zhang, Y.; Tao, H.R.; He, Y.H.; Jiang, Y.; et al. In vitro degradation, hemolysis and MC3T3-E1 cell adhesion of biodegradable Mg–Zn alloy. *Mater. Sci. Eng. C* **2009**, *29*, 1907–1912. [CrossRef]

22. Zhang, X.; Wang, Z.; Yuan, G.; Xue, Y. Improvement of mechanical properties and corrosion resistance of biodegradable Mg–Nd–Zn–Zr alloys by double extrusion. *Mater. Sci. Eng. B* **2012**, *177*, 1113–1119. [CrossRef]
23. Li, Z.; Gu, X.; Lou, S.; Zheng, Y. The development of binary Mg-Ca alloys for use as biodegradable materials within bone. *Biomaterials* **2009**, *29*, 1329–1344. [CrossRef] [PubMed]
24. Ben-Hamu, G.; Eliezer, D.; Shin, K.S. The role of Si and Ca on new wrought Mg–Zn–Mn based alloy. *Mater. Sci. Eng. A* **2007**, *447*, 35–43. [CrossRef]
25. Sunil, B.R.; Ganapathy, C.; Kumar, T.S.; Chakkingal, U. Processing and mechanical behavior of lamellar structured degradable magnesium–hydroxyapatite implants. *J. Mech. Behav. Biomed. Mater.* **2014**, *40*, 178–189. [CrossRef] [PubMed]
26. Zheng, B.L.; Ertorer, O.; Li, Y.; Zhou, Y.Z.; Mathaudhu, S.N.; Tsao, C.Y.A.; Lavernia, E.J. High strength, nano-structured Mg–Al–Zn alloy. *Mater. Sci. Eng. A* **2017**, *528*, 2180–2191. [CrossRef]
27. Zhang, E.; Yang, L.; Xu, J.; Chen, H. Microstructure, mechanical properties and bio-corrosion properties of Mg–Si (–Ca, Zn) alloy for biomedical application. *Acta Biomater.* **2010**, *6*, 1756–1762. [CrossRef] [PubMed]
28. Fu, J.; Liu, K.; Du, W.; Wang, Z.; Li, S.; Du, X. Microstructure and mechanical properties of the as-cast Mg–Zn–Mn–Ca alloys. *IOP Conf. Ser. Mater. Sci. Eng.* **2016**, *182*, 012053. [CrossRef]
29. Singh, B.P.; Singh, R.; Mehta, J.S.; Prakash, C. Fabrication of Biodegradable Low Elastic Porous Mg–Zn–Mn–HA Alloy by Spark Plasma Sintering for Orthopaedic Applications. *IOP Conf. Ser. Mater. Sci. Eng.* **2017**, *225*, 012050. [CrossRef]
30. Prakash, C.; Singh, S.; Verma, K.; Sidhu, S.S.; Singh, S. Synthesis and characterization of Mg–Zn–Mn–HA composite by spark plasma sintering process for orthopedic applications. *Vacuum* **2018**, *155*, 578–584. [CrossRef]
31. Prakash, C.; Singh, S.; Sidhu, S.S.; Pabla, B.S.; Uddin, M.S. Bio-inspired Low Elastic Biodegradable Mg–Zn–Mn–Si–HA Alloy Fabricated by Spark Plasma Sintering. *Mater. Manuf.* **2018**. [CrossRef]
32. Zheng, R.X.; Ma, F.M.; Xiao, W.L.; Ameyama, K.; Ma, C.L. Achieving enhanced strength in ultrafine lamellar structured Al2024 alloy via mechanical milling and spark plasma sintering. *Mater. Sci. Eng. A* **2017**, *687*, 155–163. [CrossRef]
33. Cullity, B.D. Elements of X-Ray Diffraction. *Am. J. Phys.* **1957**, *25*, 394. [CrossRef]
34. Oliver, W.C.; Pharr, G.M. An improved technique for determining hardness and elastic modulus using load and displacement sensing indentation experiments. *J. Mater. Res.* **1992**, *7*, 1564–1583. [CrossRef]
35. Prakash, C.; Uddin, M.S. Surface modification of β-phase Ti implant by hydroaxyapatite mixed electric discharge machining to enhance the corrosion resistance and in-vitro bioactivity. *Surf. Coat. Technol.* **2017**, *326*, 134–145. [CrossRef]
36. Prakash, C.; Singh, S.; Pabla, B.S.; Uddin, M.S. Synthesis, characterization, corrosion and bioactivity investigation of nano-HA coating deposited on biodegradable Mg–Zn–Mn alloy. *Surf. Coat. Technol.* **2018**, *346*, 9–18. [CrossRef]
37. Shi, Z.; Liu, M.; Atrens, A. Measurement of the corrosion rate of magnesium alloys using Tafel extrapolation. *Corros. Sci.* **2010**, *52*, 579–588. [CrossRef]
38. Jaiswal, S.; Kumar, R.M.; Gupta, P.; Kumaraswamy, M.; Roy, P.; Lahiri, D. Mechanical, corrosion and biocompatibility behaviour of Mg-3Zn-HA biodegradable composites for orthopaedic fixture accessories. *J. Mech. Behav. Biomed. Mater.* **2018**, *78*, 442–454. [CrossRef] [PubMed]
39. Prakash, C.; Kansal, H.K.; Pabla, B.S.; Puri, S. Processing and characterization of novel biomimetic nanoporous bioceramic surface on β-Ti implant by powder mixed electric discharge machining. *J. Mater. Eng. Perform.* **2015**, *24*, 3622–3633. [CrossRef]

40. Furko, M.; Havasi, V.; Kónya, Z.; Grünewald, A.; Detsch, R.; Boccaccini, A.R.; Balázsi, C. Development and characterization of multi-element doped hydroxyapatite bioceramic coatings on metallic implants for orthopedic applications. *Bol. Soc. Esp. Ceram. Vidr.* **2018**, *57*, 55–65. [CrossRef]
41. Gittens, R.A.; McLachlan, T.; Olivares-Navarrete, R.; Cai, Y.; Berner, S.; Tannenbaum, R.; Schwartz, Z.; Sandhage, K.H.; Boyan, B.D. The effects of combined micron-/submicron-scale surface roughness and nanoscale features on cell proliferation and differentiation. *Biomaterials* **2011**, *32*, 3395–3403. [CrossRef] [PubMed]

© 2018 by the authors. Licensee MDPI, Basel, Switzerland. This article is an open access article distributed under the terms and conditions of the Creative Commons Attribution (CC BY) license (http://creativecommons.org/licenses/by/4.0/).

Article

An Experimental Study on the Precision Abrasive Machining Process of Hard and Brittle Materials with Ultraviolet-Resin Bond Diamond Abrasive Tools

Lei Guo [1,2,*], Xinrong Zhang [1], Shibin Chen [1] and Jizhuang Hui [1]

[1] Key Laboratory of Road Construction Technology and Equipment, Chang'an University, South 2nd Ring, Xi'an 710064, Shannxi, China; zxrong@chd.edu.cn (X.Z.); sbchen@chd.edu.cn (S.C.); huijz@chd.edu.cn (J.H.)
[2] Shaanxi Fast Auto Drive Engineering Technology Research Center, West Avenue, Xi'an 710119, Shannxi, China
* Correspondence: lguo@chd.edu.cn; Tel.: +86-029-8233-4483

Received: 10 December 2018; Accepted: 26 December 2018; Published: 2 January 2019

Abstract: Ultraviolet-curable resin was introduced as a bonding agent into the fabrication process of precision abrasive machining tools in this study, aiming to deliver a rapid, flexible, economical, and environment-friendly additive manufacturing process to replace the hot press and sintering process with thermal-curable resin. A laboratory manufacturing process was established to develop an ultraviolet-curable resin bond diamond lapping plate, the machining performance of which on the ceramic workpiece was examined through a series of comparative experiments with slurry-based iron plate lapping. The machined surface roughness and weight loss of the workpieces were periodically recorded to evaluate the surface finish quality and the material removal rate. The promising results in terms of a 12% improvement in surface roughness and 25% reduction in material removal rate were obtained from the ultraviolet-curable resin plate-involved lapping process. A summarized hypothesis was drawn to describe the dynamically-balanced state of the hybrid precision abrasive machining process integrated both the two-body and three-body abrasion mode.

Keywords: abrasive machining; sapphire substrate; resin bond

1. Introduction

Hard and brittle materials in the forms of silicon, sapphire, glass, and different types of ceramics have gradually become one of the most broadly used materials in the modern industry. Thanks to their superior material properties in chemical, physical, optical, and electronic characteristics, hard and brittle materials can be utilized in various fields from the screen of cell phones to the optics cavity of laser gyros. However, the machining of this material is still challenging due to their extreme hardness, brittleness, and chemical stability. According to the specific application, the aim of hard and brittle material machining is not just to remove the material efficiently, but also to ensure a desirable surface quality in terms of surface flatness, surface roughness, and surface integrity. These characteristics are mainly determined by the material removal mechanism of the precision machining processes, like lapping and polishing, thereby the role of the machining tools that directly contact the materials in these machining processes significantly affect the output quality of the process.

Lapping plates fabricated with metals, such as cast iron (Fe) and copper (Cu), have been broadly used for semiconductor material abrasive machining processes. These metal plates are capable of providing a relatively higher material removal rate, and they are also easily manufactured economically. After lapping with these hard plates, a softer tool, like a tin (Sn)/lead (Pb) plate or metal-resin composite plate, is utilized in the polishing process to perform atomic-level material removal, and a high-quality surface finish can be obtained meanwhile as a result [1]. However, in addition to the machining tool as introduced, lapping and polishing processes are also primarily affected by the involved abrasive slurry, which is a mixture of the abrasive grains and liquid carrier that can be either oil-based or water-based. During the machining process, the abrasive grains from the slurry are rolling between the workpiece and base plate and the material is mainly removed in three-body abrasion mode. On the other hand, in some case of the lapping process, the abrasive grains are fixed on the plate through the fabrication process, so the material is mostly removed from the workpiece in two-body abrasion. According to Kim et al. [2], a relatively higher material removal rate can be obtained through two-body abrasion lapping, while a dense abrasive slurry is needed for three-body abrasion to achieve the same rate. As a result, both the process waste and cost increased significantly in the slurry-based lapping process, and environmental pollution could be another potential issue.

Recently, researchers focused their attention toward the two-body abrasion mode in the fixed abrasive lapping plate [3]. However, considering the surface finish quality, the two-body abrasion produces a rougher surface than three-body abrasion. For this reason, researchers around the world started to study the possibility of combining the two-body and three-body abrasion modes to integrate their advantages in machining efficiency and surface finish. Luo et al. [4] tried to develop a so-called semi-fixed abrasive tool with sol-gel technology to form a softer bond between the abrasive grains and bonding agent. During the machining process, the fall-off semi-fixed abrasive grains could work in three-body abrasion mode with the fixed grains working in two-body abrasion mode. The fabrication process was based on the cross-linking reaction of the sodium alginate (AGS) at certain conditions. Therefore, the reaction completion and process parameter control could potentially limit the application of this technology. Based on the conventional thermal press and sintering process, Pyun et al. [5] fabricated a high-performance copper-resin plate for sapphire machining to combine the two-body and three-body abrasion mode, and examined the influence of different amounts of curing agent as a function of resin weight. The interface between the Cu and resin and the hardness of the lapping plate were found to be the primary factor affecting the material removal, and thereby caused the temporary two-body abrasive transformed to three-body abrasive. All the investigations introduced above have shown promising results in terms of the material removal rate and nanoscale surface roughness on the machining of hard and brittle materials.

In this paper, we proposed a new abrasive machining tool fabricated with ultraviolet-curable resin and diamond abrasive grains. Resin bond is one of the most widely used bonding agents in the manufacturing of abrasive tools including grinding wheel, lapping plate and polishing pad. For the past few decades, the thermal-curable resin has been primarily selected as the bonding agent in the industry. However, the high-energy consumption, byproduct, and environmental issues stimulated the research in developing a more efficient manufacturing process. The advantages of prototyping technology attracted the attempts from researchers to testify to the feasibility of utilizing light-curable resin in the fabrication of abrasive tools [6,7]. The creative idea of the present research was based on the application of ultraviolet-curable resin prototyping technology, which helps us easily developed an abrasive tool with the capability of generating a hybrid material removal mode of two-body and three-body abrasion during the machining process. Compared with the conventional sintering process with thermal-curable resin, this novel technique significantly reduces the curing time and energy cost. Moreover, In order to verify the practical machining performance of this unique ultraviolet-curable resin bond abrasive tool, we conducted a group of comparative experiments on the technical ceramic workpiece, between the conventional iron plate lapping process and the ultraviolet-curable resin bond lapping plate. It is hoped that this study could be undertaken to help guide a new direction of the precision machining technology of hard and brittle materials.

2. Materials and Methods

2.1. Ultraviolet-Curable Resin

The ultraviolet-curable epoxy resin prepared for this study was supplied by Dymax Corporation (Torrington, CT, USA), labeled as light weld 425 optically clear structural adhesive. From our previous research on the feasibility of using ultraviolet-curable resin as a bonding agent in the manufacturing of abrasive tools [8], we compared the material properties of the cured resin from different vendors. The 425 resin mentioned above showed some favorable advantages over the others. The technical specification of the cured pure 425 resin is listed in Table 1.

Table 1. Technical specification of the ultraviolet-curable resin used.

Material Properties	Density (g/mL)	Viscosity cP (20 rpm)	Hardness (Durometer)	Tensile (psi)	Elongation (%)	Modulus of Elasticity (psi)
Dymax 425	1.07	4000	D80	6200	7.3	500,000

2.2. Diamond Abrasive Grain

The abrasive machining tools involved in the machining process of hard and brittle materials mainly utilized with super abrasives as cubic boron nitride (CBN) and artificial synthetic diamond. The CBN tool is mainly used in the machining of the hard metallic material since the diamond is reactive to the ferrous metal at high temperature. Diamond abrasive is superior in hardness, strength, thermal conductivity, and expansion coefficient. It is primarily employed in the machining of hard and brittle material including ceramic, optical glass, and semiconductor material. The abrasive grains selected in this research are the surface textured monocrystalline diamond grains average sized in 15 μm, provided by Engis. Compared with the standard monocrystalline diamond grains, the grain surface of the ones above was textured through a specific etching process, in which the monocrystalline diamond was eroded by oxygen, oxygen compounds, molten metals, and hydrogen at an elevated temperature. As a result, some material on the surface layer of the diamond grains was removed by the erosion and a rough surface was generated. According to our previous study [9,10], the rougher surface of the diamond grain increased the contact surface area between the grains and the bonding agent and consequently improved the bond retention of the abrasive tools. Scanning electronic micrograph comparison of the regular diamond grains and surface textured ones are shown in the figures below, where the surface patterns and pits can be seen in Figure 1b.

(a)

(b)

Figure 1. Scanning electronic micrographs of (a) the regular monocrystalline diamond grains and (b) the surface textured monocrystalline diamond grains.

2.3. Ultraviolet Curing Systems

The ultraviolet curing system for laboratory use mainly consists of an ultraviolet lamp with focalization setup and a power supply. In this study, a customized ultraviolet curing system based on the Innovative Machine UV-100 was primarily employed in the manufacturing experiments of the ultraviolet-curable resin bond abrasive tool, and the Dymax 5000 flood ultraviolet curing system from DYMAX was also used for initial material properties test of the cured ultraviolet-curable resin composites. The curing systems mentioned above are shown in Figure 2. According to the supplier of the resin used in this series of experiments, the recommended wavelengths of the ultraviolet light to initiate the photoreaction is around 365 nm. Hence, the ultraviolet light source of the curing system was professionally optimized to the range shown in Figure 2c. Meanwhile, the detailed technical specification of the curing system is provided in Table 2 below.

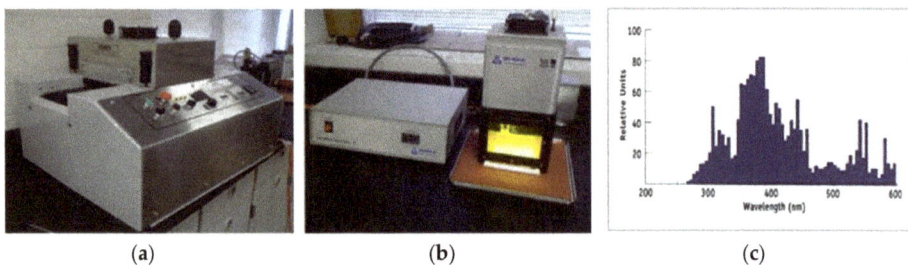

Figure 2. (a) Innovative Machine UV-100 ultraviolet curing system; (b) Dymax 5000 flood ultraviolet curing system; and (c) wavelength distribution of the curing system after optimization.

Table 2. Technique specifications of the ultraviolet curing system.

Output Power	Ultraviolet Light Source	Typical Intensity (320–390 nm, 7.62 cm from the Bottom)	Illumination Area
400 Watts	UVA [1] flood	225 mW/cm^2	161.29 cm^2

[1] Wavelengths of the ultraviolet are classified as UVA, UVB, or UVC, with UVA at 320–400 nm.

2.4. Experimental Fabrication Method

In the experimental fabrication of the ultraviolet-curable resin bond abrasive tool, we developed various laboratory methods to realize the manufacturing. Firstly, the ultraviolet-curable resin and diamond abrasives were uniformly mixed through a stirring machine in the darkroom to prevent any unexpected photoreaction. The composite mixture stood in vacuum conditions for 30 min to release the air bubbles generated in the stirring step. After that, the mixture in liquid form can be either spin-coated on top of the base plate or injected into the separated fan-shape mold for the curing process. For the latter method, the cured fan-shaped pieces were going to be arranged and adhered to the base plate to assemble the desired tool. The schematic diagram of the curing processes of the ultraviolet-curable resin bond abrasive tool is illustrated in Figure 3. The spin-coating method was efficient in curing time and it also ensured the integrity of the cured resin plate. However, since the spin-coated layer of resin and abrasives on top of the base plate was dimensionally large, it was difficult to set up an evenly-distributed ultraviolet exposure in practice. As a result, the photosensitive reaction completion differed by the distance from the light source to the surface of the layer, therefore, the cure depth of the layer could be varied to generate a waviness in the resin plate. The dimensional accuracy and machining efficiency can be principally influenced by this fabrication failure. Hence, the fan-shape molding fabrication method was utilized in the curing process in order to assure the ultraviolet energy absorbed and the corresponding photoreaction within a small area is uniform.

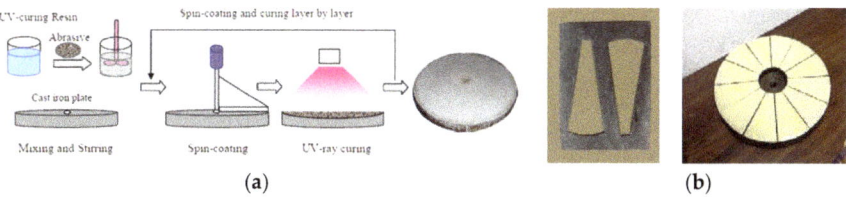

Figure 3. Schematic diagram of the curing processes of the ultraviolet-curable resin bond abrasive tool through (**a**) spin-coating method; and (**b**) fan-shaped pieces assembly method.

In the manufacturing industry of diamond abrasive tools, the diamond concentration is particularly used as a standard to evaluate the weight of diamond in a unit volume of the tool matrix, and it is defined that where each cubic centimeter contains 0.88 g of diamond, the concentration is 100%. In precision machining of hard and brittle materials, a relatively lower abrasive concentration is preferred. On the other hand, the amount of ultraviolet energy passing through the resin and diamond composite strongly depends on the abrasive grain size and concentration, the increase in diamond concentration would decrease the energy absorbed by the sub-surface and bottom layer of the resin composite, and thereby decrease the cure depth of the resin plate [11,12]. For this reason, the diamond abrasive concentration selected in this study is 12.5% to satisfy the requirement of the cure depth, which indicates that each cubic centimeter of the cured resin matrix contains 0.11 g of diamond grains.

2.5. Experimental Machining Test

A group of comparative machining performance test was conducted between the conventional slurry-involved lapping process and the process with the ultraviolet-curable resin bond abrasive tool. Technical ceramic ring-shaped samples from Nanjing Co-Energy Optical Crystal Co. Ltd. (Nanjing, Jiangsu, China). were selected as the workpiece with an initial surface roughness of 0.45 µm, some of the material properties of the workpieces can be found in Table 3 below.

Table 3. Material properties of the ceramic workpiece.

Chemical Formula	Density	Hardness	Tensile Strength	Modulus of Elasticity	Poisson's Ratio
96% Al_2O_3	3.65 g/cm^3	85 HRA	160 MPa	300 GPa	0.20

Each of the workpieces was cleaned through ultrasonic washing and left until dry, the weight of each workpiece was examined by an electronic balance. A photograph with the schematic diagram of the machining test setup is shown in Figure 4, on which six ceramic workpieces were machined at one time. During the machining test, the workpieces were removed and ultrasonically cleaned in acetone every 10 min to record the surface roughness and weight loss. The surface roughness was measured by a Zygo optical profiler, where a 10× magnification Mirau interference objective was equipped. In the surface roughness measurement, filtering is used to highlight the roughness (high-frequency, short-wavelength component) or waviness (low-frequency, long-wavelength component) of a test part. The filtering method in the Zygo optical profiler employed in this series of measurements was set to low pass with a specified wavelength of the higher cutoff point at 5.47 µm. The surface topography was examined with an atomic force microscope. To maintain the consistency of the experiments, the roughness parameter R_a was measured from three randomly-selected areas of 0.70 mm by 0.52 mm on the surface of the machined workpiece. The parameter was based on a pre-positioned straight line that crossed the selected area in each measurement. An average was taken on the roughness parameter R_a from the six workpieces periodically, and the weight loss of each workpiece was evaluated on an electronic balance to study the material removal efficiency of the respective machining process. The process parameters of the machining test on the ceramic workpiece are listed in Table 4. The 15 µm diamond grain slurry used in the iron plate lapping was purchased from Engis in standard concentration, that is, 400 g of diamond abrasive per 750 mL, according to their technical specifications.

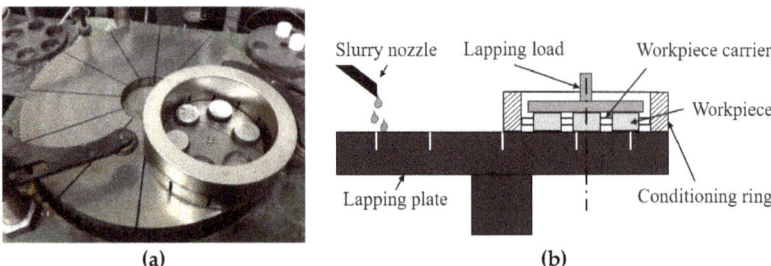

Figure 4. (**a**) The experimental setup and (**b**) schematic diagram of the machining performance test on ceramic workpieces.

Table 4. Process parameters of machining tests on the ceramic workpiece.

Lapping Machine	Diameter of Lapping Plate	Diameter of Workpiece Carrier	Minimum Rotation Speed	Maximum Rotation Speed	Round Count Precision	Lapping Pressure	Lapping Period
Lap Master 12	30.5 cm	10.8 cm	5 RPM	50 RPM	±0.5°	1.77 KPa	10 min

3. Results and Discussion

Figure 5 shows the average roughness parameters periodically recorded from both the machining process of conventional lapping on an iron plate and fixed abrasive lapping on the ultraviolet-curable resin bond tool. The lowest roughness parameter R_a achieved by iron plate lapping and ultraviolet-curable resin plate lapping is 0.201 µm and 0.182 µm, respectively. In most cases, the machined surface roughness is mainly influenced by the grain size of the abrasives. However, considering that the diamond abrasives employed either in the fabrication of resin plate or the slurry mixture were from the same batch in this study, the differences between those two machining processes could be possibly explained with their abrasive wear mechanisms in material removal.

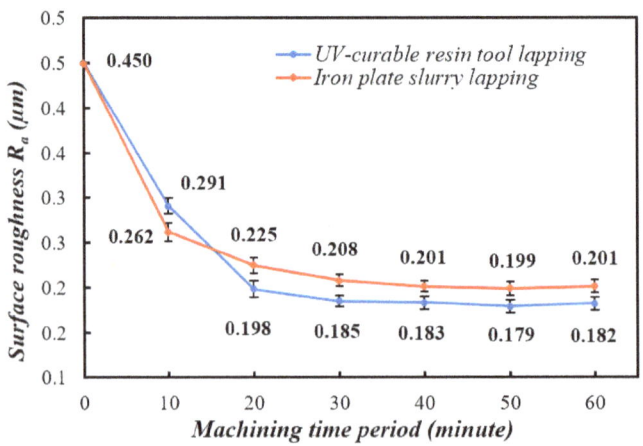

Figure 5. The surface roughness of the ceramic workpiece machined with iron plate lapping and ultraviolet-curable resin tool.

In the beginning 10 min of the machining process, the surface roughness measured from the ceramic workpiece machined with the iron plate is 0.26 μm, which is about 12% lower than 0.29 μm in resin plate lapping. This phenomenon could be explained by the different material properties of the plates. In the case of iron plate lapping, the diamond abrasive grains immediately started to work as long as they were introduced by the slurry and spread in the working zone between workpieces and the iron plate. Slurry-based loose abrasive lapping is a typical process of the three-body abrasion mode, and it is usually considered as a lower-efficiency process to remove material than two-body abrasion lapping. According to Rabinowicz et al. [13], the abrasive grains in three-body abrasion spend 90% of their working time rolling and performing low material removal rate, that can be ten times higher in two-body abrasion [5]. In our ultraviolet-curable resin bond lapping plate, the diamond grains were uniformly distributed and fixed within the tool. Ideally, they should work in two-body abrasion mode as the other fixed abrasive lapping plate, and thereby produced a higher material removal rate. However, because of the gravity settling of the diamond grains in the resin during the curing process, most of the abrasives were buried within the cured resin matrix. At the beginning of the machining process, only a small number of initial protruding diamond grains worked in two-body abrasion mode to remove material from the workpiece. Therefore, the removal rate is relatively lower than that in iron plate lapping. Hence, the volume of material removed is not enough to degrade the surface roughness parameter R_a to the level achieved in iron plate lapping. This assumption also matches the periodical weight loss of the workpieces in the first 10 min, shown in Figure 6.

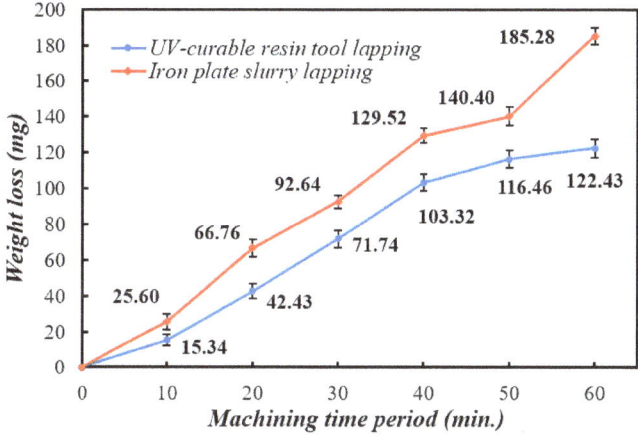

Figure 6. Periodical weight loss of the workpieces.

After 10 min in Figure 5, the surface roughness downward trend is gradually becoming weaker in both processes. The parameter R_a from ultraviolet resin tool lapping surpasses that in iron plate slurry lapping between 10 to 20 min by achieving 0.198 μm. Until the end of the machining test, the periodical measured surface roughness of the resin tool lapping remains at the lower state than the iron plate lapping. As is known, the bonding agent of cured ultraviolet-curable resin is weaker in material properties in terms of hardness, strength, and wear resistance, especially when compared to the metallic or vitrified bond. As a result, some of the protruding grains or fall off grains might be pressed into the resin matrix due to its plastic deformation under the lapping load. This phenomenon would effectively decrease the abrasive grain size and thereby cause disadvantages in machining efficiency. However, from another point of view, the decreased abrasive size and soft bonding matrix would also decrease the machined surface roughness and cause less surface damage. Additionally, the featured viscoelasticity of the resin matrix is considered to help to reduce the surface damage of the workpiece, and improve the process consistency and stability [14]. Thus, in this experiment, a better surface quality could be obtained through the ultraviolet-curable resin plate than the iron plate. This hypothesis reasonably explains the surface roughness difference in Figure 5 and the surface topographies comparison in Figure 7. The relatively rougher surface profile in Figure 7a is a small area of 78.6 μm by 79.3 μm from one of the machined workpieces after 60 min lapping with the iron plate, and the smoother one in Figure 7b is from the resin plate lapping after 60 min.

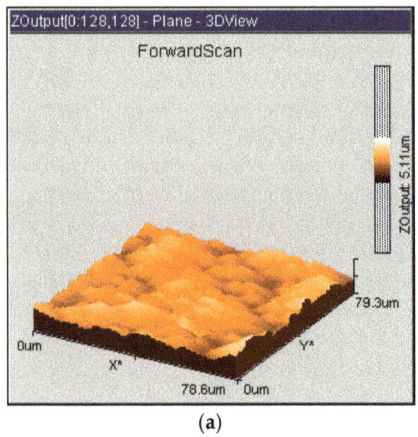 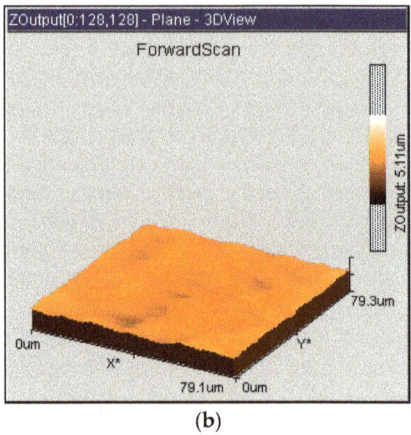

(a)   (b)

Figure 7. Surface topographies of a ceramic workpiece machined with (**a**) iron plate slurry lapping; and (**b**) ultraviolet-curable resin bond tool lapping.

It is noted that, in Figure 6, the material removed in the iron plate slurry lapping is 185.28 mg in total after 60 min, while the number in ultraviolet-curable resin tool lapping is 122.43 mg. It is nearly a 51.34% dropdown in machining efficiency. Additionally, the weight loss trend of the two processes is also quite different. Due to the delivering of lapping slurry, fresh and sharp diamond grains were continuously introduced to the working zone. Hence, the slope rate of the iron plate slurry lapping is constant and varies within a reasonable range. In the case of the resin bond plate, it is assumed that all the diamond grains were embedded in the cured resin matrix of the abrasive tool and worked as a fixed abrasive grain at the beginning. As the machining process continues, the resin matrix started to wear out due to the low hardness and wear resistance. According to the studies in diamond retention of abrasive tools [15,16], the abrasive grains are largely held within the bonding matrix by the mechanical compression generated during the manufacturing process. Therefore, the retention force is predominately determined by the material properties of the solidified bonding agent, which is the ultraviolet-curable resin in this case, considering the disadvantages of cured resin in material properties regarding strength and hardness, within which the diamond abrasive grains tend to be pulled out from the bonding matrix with ease than metallic or vitrified bonding.

Hence, some of the initially-fixed abrasive grains began to fall off from the bonding matrix and turned into loose abrasive grains working in three-body abrasion. Meanwhile, the abrasive grains buried in the underlayer of the resin matrix was continuously revealed and worked as fixed abrasive grain renewedly. This unique mechanism provides the resin tool lapping with fresh diamond grains like the slurry does in iron plate lapping. Additionally, the probability of the falling off diamond grains being pressed into the resin plate exists and converts the loose abrasive grains into fixed ones. Because of this, the lapping process of the ultraviolet-resin bond plate can be summarized as a hybrid process in which the two-body abrasion and three-body abrasion material removal mode corporately contribute to the machining process, while the interconversion between fixed grains and loose grains could be dynamically balanced at a certain period of the process. This presumption explains the stabilization of surface roughness in the resin tool lapping process after 40 min.

According to the data in Figure 6, the weight loss of the machined workpiece in resin tool lapping also becomes stable after 40 min. That means the material removed from the workpiece after 40 min is rapidly reduced. The weight loss of the workpiece in the first 40 min is 103.32 mg, while the number in the period from 40 to 60 min is 19.11 mg. Since the ultraviolet-curable resin bond lapping plate was fabricated in the laboratory, the manufacturing process flaws would cause some of the failures in machining performance. For instance, the absence of diamond grains at certain layers of the tool due to the non-uniform distribution of abrasives, the unstable hardness of the fabricated resin plate because of the incomplete photoreaction during the curing process, and the glazing issue caused by the dull grain and porosity stuck.

An optical image observation on the worn surface of the ultraviolet-curable resin plate after the lapping process is shown in Figure 8. As we presumed, the number of the pull-out holes, which are marked with white circles in Figure 8a, indicate the weaker strength and retention force of the resin matrix to hold the diamond grains from being pulled out. The protruding diamond grains marked in red circles are the fixed abrasives that are primarily employed to remove the materials in the two-body wear mode during machining, while the three-body working abrasives were rolling between the workpiece and the tool and continuously carried away with the machining coolant. Moreover, the nonuniform distribution of the fixed diamond grains and pull-out holes on the worn surface of the tool in Figure 8a reflects the fabrication disadvantage in the laboratory. A 400× magnification on the worn surface in Figure 8b clearly shows the diamond grains that are embedded within the resin matrix after lapping, and it refers to the fact that, as the machining process continues, the fresh diamond grains are not ensured to reveal with the resin matrix wear. In practice, all these potential defects mentioned above possibly lead to the failure of the tool's machining capability. Thus, the improvement in manufacturing process optimization and material selection could be directions for future work.

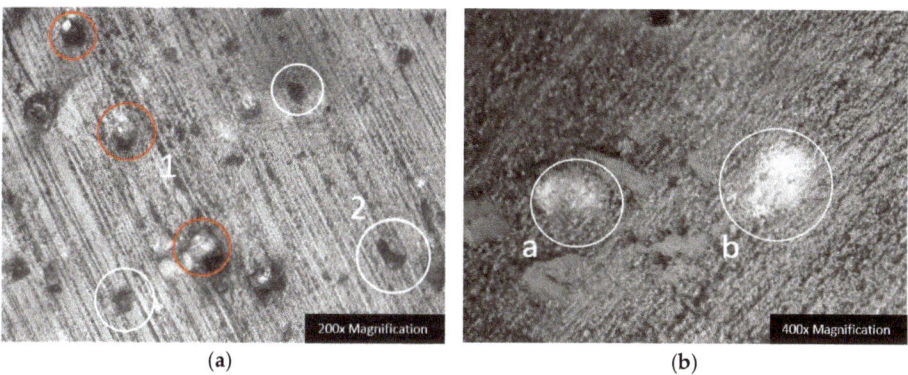

Figure 8. Worn surface of the ultraviolet-curable resin plate after lapping process: (**a**) the diamond grains working condition at 200× magnification with the active protruding diamond grains circled in red and pull-out holes circled in white, and (**b**) the diamond grains embedded within the resin matrix at various underneath heights circled and labelled as *a* and *b*.

4. Conclusions

This study proposed a new type of fixed abrasive lapping plate for precision machining on hard and brittle materials, in which the ultraviolet-curable resin was selected as the bonding agent in the fabrication of the abrasive tool to deliver a rapid, flexible, economical, and environment-friendly manufacturing process. The performance of the ultraviolet-resin bond diamond lapping plate was examined through the comparative experiments with slurry-based iron plate lapping. The conclusions can be summarized as follows:

- The fabrication process of the ultraviolet-curable resin bond plate was completed within a minute light exposure, saving energy costs and labor effort, which occurs in the conventional sintering process.
- In the ceramic workpiece lapping process, the ultraviolet-curable resin bond lapping plate was enabled to achieve an approximately 10% lower surface roughness parameter R_a than that in the iron plate lapping process.
- In the study of the material removal rate evaluated by weight loss of the workpiece, the resin plate lapping showed about 25% less material removed per minute in the stable machining state.
- The machining performance of the resin plate can be explained by the hypothesized discussion that an integrated abrasion mode of two-body and three-body wear was established.

However, the study on the control of the fabrication process parameters that affect the material properties of the cured resin matrix, the methodology to obtain a uniform distribution of the abrasive grains within the bonding agent, and the appropriate technique to evaluate the working condition of the tool is still limited in this research. The effort on these directions should be taken in the future studies regarding related works. It is hoped that this experimental study could inspire the application of the ultraviolet-curable resin bond abrasive tool, and ultimately integrate the two-step precision flattening process of lapping and polishing into one.

Author Contributions: Conceptualization: L.G.; methodology: L.G. and S.C.; investigation: L.G. and X.Z.; resources: J.H.; data curation: L.G.; writing—original draft preparation: L.G.; writing—review and editing: X.Z.; supervision: J.H.

Funding: This research was funded by National Natural Science Foundation of China, grant number 51805044, Natural Science Foundation of Shaanxi Province, grant number 2018JQ5064, and Fundamental Research Funds for the Central Universities, grant numbers 300102258202 and 300102258402.

Conflicts of Interest: The authors declare no conflict of interest.

References

1. Tam, H.Y.; Cheng, H.B.; Wang, Y.W. Removal rate and surface roughness in the lapping and polishing of RB-SiC optical components. *J. Mater. Process. Technol.* **2007**, *192–193*, 276–280. [CrossRef]
2. Kim, H.M.; Manivannan, R.; Moon, D.J.; Xiong, H.; Park, J.G. Evaluation of double sided lapping using a fixed abrasive pad for sapphire substrates. *Wear* **2013**, *302*, 1340–1344. [CrossRef]
3. Shih, A.J.; Denkena, B.; Grove, T.; Curry, D.; Hocheng, H.; Tsai, H.Y.; Ohmori, H.; Katahira, K.; Pei, Z.J. Fixed abrasive machining of non-metallic materials. *CIRP Ann.* **2018**, *67*, 767–790. [CrossRef]
4. Luo, Q.; Lu, J.; Xu, X. Study on the processing characteristics of SiC and sapphire substrates polished by semi-fixed and fixed abrasive tools. *Tribol. Int.* **2016**, *104*, 191–203. [CrossRef]
5. Pyun, H.J.; Purushothaman, M.; Cho, B.J.; Lee, J.H.; Park, J.G. Fabrication of high performance copper-resin lapping plate for sapphire: A combined 2-body and 3-body diamond abrasive wear on sapphire. *Tribol. Int.* **2018**, *120*, 203–209. [CrossRef]
6. Tanaka, T.; Isono, Y. New development of a grinding wheel with resin cured by ultraviolet light. *J. Mater. Process. Technol.* **2001**, *113*, 385–391. [CrossRef]
7. Guo, L.; Huang, Q.; Marinescu, I. Effect of nanosized alumina fillers on manufacturing of UV light-curable-resin bond abrasive tool. *Mach. Sci. Technol.* **2017**, *21*, 223–238. [CrossRef]
8. Guo, L.; Marinescu, I. Al_2O_3 nanoparticle mixed UV-curable resin study in fabrication of the lapping plate. *Trans. N. Am. Manuf. Res. Inst. SME* **2013**, *41*, 154–159.
9. Benea, I.C.; Rosczyk, B.R.; Fitzgerald, L.M. Surface textured diamond particles-properties and applications. *Diam. Technol.* **2011**, *3*, 17–24.
10. Guo, L. Study of the Influence of Nanosized Filler on the UV Light-Curable Resin Bonded Abrasive Tool. Ph.D. Thesis, University of Toledo, Toledo, OH, USA, 2016.
11. Fujita, K.; Ikemi, T.; Nishiyama, N. Effects of particle size of silica filler on polymerization conversion in a light-curing resin composite. *Dent. Mater.* **2011**, *27*, 1079–1085. [CrossRef] [PubMed]

12. Alsharif, S.O.; Bin Md Akil, H.; Abbas Abd El-Aziz, N.; Arifin Bin Ahmad, Z. Effect of alumina particles loading on the mechanical properties of light-cured dental resin composites. *Mater. Des.* **2014**, *54*, 430–435. [CrossRef]
13. Rabinowicz, E.; Dunn, L.A.; Russell, P.G. A study of Abrasvie wear under three-body conditions. *Wear* **1961**, *4*, 345–355. [CrossRef]
14. Deng, Q.F.; Zhou, Z.X.; Ren, Y.G.; Lv, B.H.; Yuan, J.L. Current Research Trends on Resin Bond Used for Abrasive Products. *Adv. Mater. Res.* **2012**, *497*, 105–109. [CrossRef]
15. Luo, S.Y. Effect of fillers of resin-bonded composites on diamond retention and wear behavior. *Wear* **1999**, *236*, 339–349. [CrossRef]
16. Konstanty, J.; Romanski, A. Numerical analysis of diamond retention in cobalt and a copper-base alloy. *Arch. Metall. Mater.* **2014**, *59*, 1457–1462. [CrossRef]

© 2019 by the authors. Licensee MDPI, Basel, Switzerland. This article is an open access article distributed under the terms and conditions of the Creative Commons Attribution (CC BY) license (http://creativecommons.org/licenses/by/4.0/).

Article

Surface Modification of Ti-6Al-4V Alloy by Electrical Discharge Coating Process Using Partially Sintered Ti-Nb Electrode

Chander Prakash [1,*], Sunpreet Singh [1], Catalin Iulian Pruncu [2], Vinod Mishra [3], Grzegorz Królczyk [4], Danil Yurievich Pimenov [5] and Alokesh Pramanik [6]

- [1] School of Mechanical Engineering, Lovely Professional University, Phagwara 144411, India; chander.mechengg@gmail.com
- [2] Mechanical Engineering, Imperial College London, Exhibition Rd., SW7 2AZ London, UK; c.pruncu@imperial.ac.uk
- [3] Optical Devices and Systems, Central Scientific Instruments Organization, Chandigarh 160030, India; vnd.mshr@gmail.com
- [4] Faculty of Mechanical Engineering, Opole University of Technology, 76 Proszkowska St., Opole 45-758, Poland; g.krolczyk@po.opole.pl
- [5] Department of Automated Mechanical Engineering, South Ural State University, Lenin Prosp. 76, Chelyabinsk 454080, Russia; danil_u@rambler.ru
- [6] Department of Mechanical Engineering, Curtin University, Bentley, Perth 6102, WA, Australia; alokesh.pramanik@curtin.edu.au
- * Correspondence: chander.mechengg@gmail.com; Tel.: +91-9878805672

Received: 7 March 2019; Accepted: 24 March 2019; Published: 27 March 2019

Abstract: In the present research, a composite layer of TiO_2-TiC-NbO-NbC was coated on the Ti-64 alloy using two different methods (i.e., the electric discharge coating (EDC) and electric discharge machining processes) while the Nb powder were mixed in dielectric fluid. The effect produced on the machined surfaces by both processes was reported. The influence of Nb-concentration along with the EDC key parameters (Ip and Ton) on the coated surface integrity such as surface topography, micro-cracks, coating layer thickness, coating deposition, micro-hardness has been evaluated as well. It has been noticed that in the EDC process the high peak current and high Nb-powder concentration allow improvement in the material migration, and a crack-free thick layer (215 µm) on the workpiece surface is deposited. The presence of various oxides and carbides on the coated surface further enhanced the mechanical properties, especially, the wear resistance, corrosion resistance and bioactivity. The surface hardness of the coated layer is increased from 365 HV to 1465 HV. Furthermore, the coated layer reveals a higher adhesion strength (~118 N), which permits to enhance the wear resistance of the Ti-64 alloy. This proposed technology allows modification of the mechanical properties and surface characteristics according to an orthopedic implant's requirements.

Keywords: Ti-6Al-4V; alloy; EDC; microcracks; microhardness; adhesion strength

1. Introduction

Among all metallic biomaterials, Ti-6Al-4V (Ti-64) alloy is most widely used material for biomedical applications to fabricate implants and surgical instruments due to its unique feature of great mechanical properties and excellent biocompatibility [1,2]. The Ti-64 alloy has a Ti-oxide bio-inert layer, which has low hardness and poor wear resistance [3]. In order to improve its surface properties and characteristics, a number of surface treatment/modification processes were reported [4]. In the current research scenario, the electrical discharge machining (EDM) process is the only non-conventional machining process, which permits effective manufacturing of Ti-64 alloy that is hard to cut [5].

Moreover, EDM produce a bio-compatible layer on the surface, which promotes a higher surface hardness and better corrosion resistance of Ti-64 alloy [6]. Peng et al. used EDM to machine the Ti-64 alloy and tuned the surface characteristics as required for the osseiointegration process [7]. The nanoporous layer has been synthesized by EDM, which promotes cell adhesion and growth [8]. Bin et al. investigated the effect of EDM to improve the surface hardness, wear resistance, corrosion resistance, and bioactivity of Ti-64 alloy [9]. The application of EDM for coating/deposition/alloying has been extended by altering the tool electrode polarity [10,11], powder metallurgical (P/M) prepared from green compact tool electrode [12–15], and powder mixed dielectric [16,17]. Most researchers use powder mixed dielectric to make deposition into a workpiece surface [18,19]. Prakash et al. investigated the effect of Si-mixed EDM to alter the surface characteristics that allows improving the bio-compatibility, corrosion, and wear resistance properties of a specially designed Ti-35Nb-7Ta-5Zr (β-phase) for orthopaedic applications [20]. Furthermore, multi-objective optimization of Si-mixed EDM by the Non-dominated Sorting Genetic Algorithm-II (NSGA-II) were used to synthesize the bio-mimetic surface [21]. The powder mixed electric discharge machining (PMEDM) was also utilized to enhance the fatigue performance and bone-implant interface [22,23]. Xie et al. reported that the surface hardness of 45-C steel has been increased from 415 to 1420 HV using graphite-mixed EDM [24]. Arun et al. synthesized a hard-layer of Ni-W coating on tool-steel by Ni &W—mixed EDM in order to improve the tribological performance [25]. Ekmekci et al. reported that a hydroxyapatite (HA) enriched bioceramic layer can be successfully deposited on Ti-64 surface using HA-mixed EDM process [26]. Ou and Wang used EDC to deposit a HA-enriched layer on Ti alloy in order to enhance the biocompatibility of base material [27]. Prakash et al. deposited HA-enriched biomimetic porous layer on the Ti-35Nb-7Ta-5Zr (β-phase) surface to enhance the bio-mechanical and corrosion integrity [28]. Recently, Prakash et al. uncovered the ability of HA-mixed EDM process to enhance the mechanical, corrosion, bioactivity of Mg-based biodegradable implants [29,30]. The EDM/PMEDM was utilized for surface modification of biomedical implants [31–33].

There is wide scope in additive mixed EDM process to deposit a bio-compatible layer for orthopaedic applications using a different type of additive mixed in dielectric fluid. To date, no research study has reported the application of niobium (Nb) powder mixed EDM for surface modification of Ti-alloy. As such, the present research investigates the effect of a multi-composite layer (i.e., of TiO_2-TiC-NbC-NbO) deposited on the Ti-64 alloy surface using EDC process. The results gathered in this work prove that the surface of Ti-64 treated by Nb was enhanced in terms of bio-mechanical integrity and wear resistance.

2. Materials and Methods

2.1. Characterization of Sintered Ti-Nb Alloy and Machined Surface by Electric Discharge Coating (EDC) Process

Ti-6Al-4V extra low interstitials alloy (Ti-64 ELI) was used as the workpiece material, provided by Titanium-India, Mumbai. The samples of 10 mm width, 10 mm length, and 5 mm thickness were cut from the as-received cast ingot. The surfaces of the cut samples were polished up to Ra ~ 0.5 μm. The surface modification of Ti64-ELI alloy was carried out by depositing a thin coating layer by the electric discharge coating process while the Nb-powder particles were mixed in a dielectric fluid. Here, were used a commercially pure-Ti tool electrode. Table 1 shows details of experimental conditions. Figure 1a schematically shows the experimental set-up of the EDC process whereas Figure 1b schematically presents the main mechanism of deposition that includes the coating layer of oxides and carbides in the EDC process. A partially sintered tool electrode was used for the coating process.

Table 1. Experimental condition of electric discharge coating (EDC).

Name of Parameter	Range of Parameter
Workpiece	Ti-6Al-4V alloy
Tool electrode	CP-Ti alloy
Polarity	Reverse (−Ve) for EDC and Straight (+Ve) for electrical discharge machining (EDM)
Peak current	5, 10, 15, 20, 25 A
Pulse-on time	50, 100, 200, 400, 800 μs
Duty Cycle	8%
Dielectric medium	Hydrocarbon oil
Machining time	15 min
Powder Concentration	5, 10, 15, 20 g/L

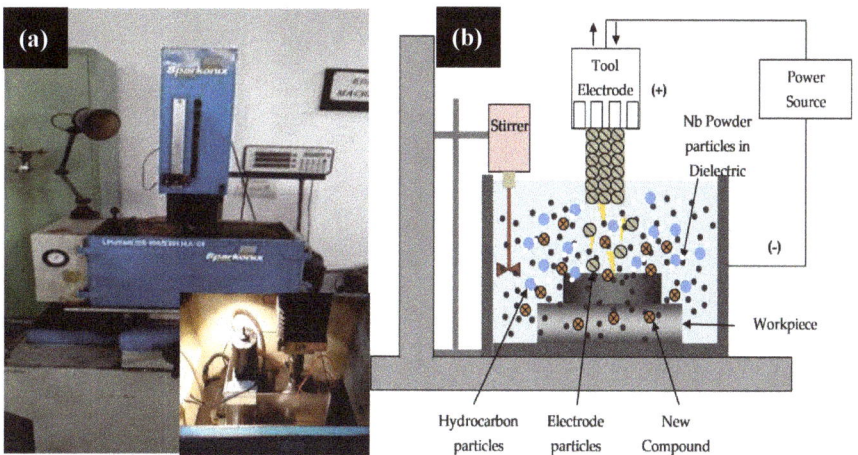

Figure 1. (a) Experimental set-up of EDC process and (b) schematic representation of EDC process.

2.2. Development of Partially Sintered Tool Electrode for EDC Process

The partially sintered Ti-Nb tool electrode was developed by spark plasma sintering process. Titanium (>99.9% purity, 45μm) and niobium (99.9% purity, 45 μm) were procured from N.B. enterprises, Bilaspur, India. Here we present the main steps followed for the fabrication of Ti-Nb alloy: (i) Ti and Nb were mixed 50:50 weight percentages. The mixture of Ti-Nb was developed in a planetary ball mill with tungsten balls (ball to powder ration kept 5:1) at a rotational speed of 200 rpm for 8 h. Figure 2 shows the shape and size of Ti and Nb powders. (ii) The as-blended powder mixture was consolidated via the spark plasma sintering (SPS) method in a graphite die at 800 °C sintering temperature using a heating rate of 100 °C (holding time 15 min) [34]. This was performed under vacuum conditions while the uniaxial pressure was kept constant at 50 MPa. Figure 3 shows a photograph of the SPS machine available at the Indian Institute of Technology, Roorkee, and the red hot sample under vacuum condition during the process.

2.3. Characterization of Sintered Ti-Nb Alloy and Machined Surface by EDC Process

The topology and morphology of coated surface were analyzed with a field emission scanning electron microscopy (FE-SEM; JEOL 7600F; JEOL Inc., Peabody, MA, USA). The phase and element constituents in the as-synthesized composites were determined by the X-ray diffraction (XRD) technique (XRD; X'pert-PRO, PANalytical, Almelo Inc., Almelo, Netherlands) and energy dispersive spectroscopy (EDS) (FE-SEM; JEOL 7600F; JEOL Inc., Peabody, MA, USA), respectively. XRD pattern peaks were analyzed according to the database of the Joint Committee on Powder Diffraction Standards (JCPDS), to identify the phases formed in the coated surface. The thickness of the coating layer (CLT) was measured by investigating the cross-section of the modified specimens [19,20]. For the CLT measurements, the test surfaces were well polished and prepared using adequate grinding and polishing methods according to the ASTM standard (ASTM-E384-11) [19]. The surface hardness of the coated layer was measured with a Shimadzu HMV-G21 (Vickers) micro-hardness tester (HMV-G21ST, SHIMADZU, Japan), using an indenting load of 0.49 N during 15 s dwell time. The cross-sectional surface of the samples was used to measure the microhardness values. The adhesive strength between the coating and the substrate was measured with a scratch tester (TR-102, DUCOM, Bengaluru, India). The loading force was increased up to 150 N by using a loading rate of 25 N/mm and the scratch length was 4 mm.

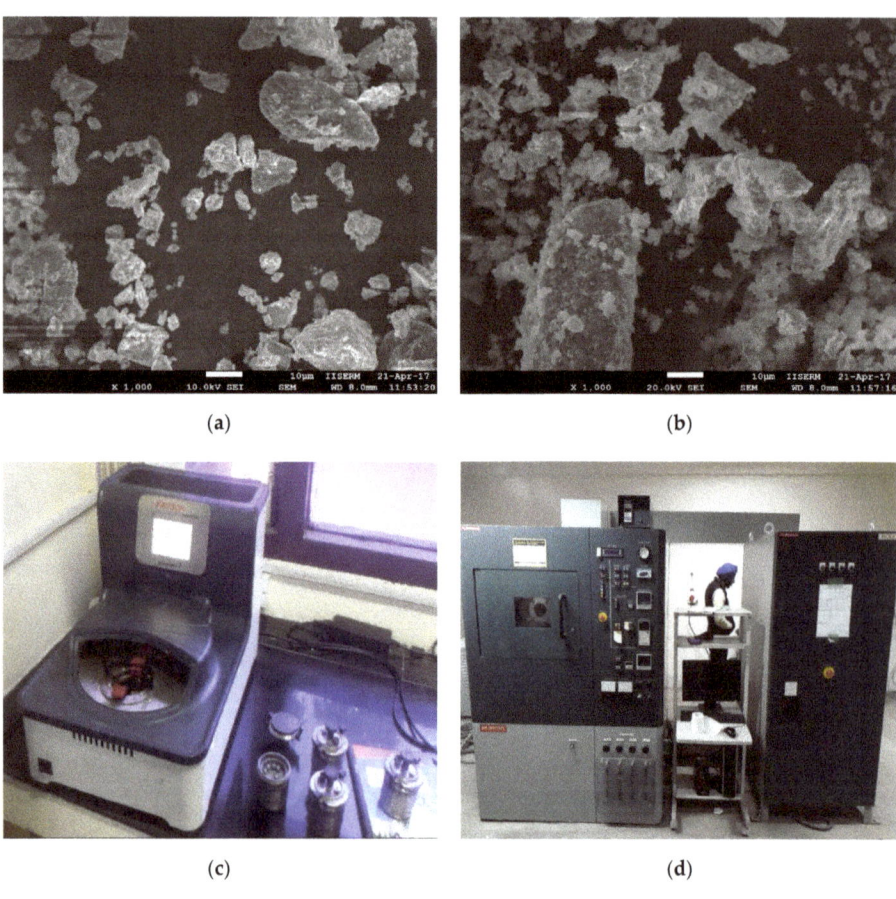

Figure 2. (**a**,**b**) Scanning electron microscope (SEM) micrograph showing the morphology of titanium and niobium powders; (**c**) high energy planetary ball mill; and (**d**) spark plasma sintering (SPS) machine.

3. Results and Discussion

This section is divided into two subsections. The first section provides a concise and precise description of the results from the fabrication of Ti-Nb alloy using the SPS method. The second section provides the experimental results obtained by EDC technique, their interpretation as well as the experimental conclusions formulated from the analysis.

3.1. Microstructure, Morphology, and Chemical Composition of Sintered Ti-Nb Electrode

Figure 3 shows details of the sample produced, its microstructure and phase composition of a partially sintered Ti-Nb alloy obtained with the SPS technique, which was further used in the EDC process. Figure 3a shows the photograph of sintered Ti-Nb alloy that is a compacted sample of 20 mm diameter and thickness of 6 mm. Figure 3b shows the SEM micrographs of the sintered sample obtained at a sintered temperature of 700 °C. These micrograph indicates the presence of porosities in the sample which is a sign of a sample partially sintered. The percentage of porosities in the sintred Ti-Nb alloy is was found ~25%. The porosities are uniformly distributed having the pore size of 5–15 µm. Furthermore, the chemical reaction between the alloy elements generated large amounts of gases that leads to the creation of porosity in the structure. At higher magnification (5000×), the presence of α-Ti and β-Nb phases can be identified clearly; the α-Ti is a dark phase and β-Nb is the brighter phase. The dissolution of Nb is incomplete, and the equiaxed microstructure of β-Nb uniformly distributed and surrounded by α-Ti matrix. Figure 3c shows the associated EDS spectrum of the sintered Ti-Nb alloy. The EDS spectrum confirms the presence of Ti and Nb elements. Apart from the presence of base metal elements, the element O (possible oxides) was noticed as well on the surface, which is a common observation in the SPS-treated surface [35–40]. Figure 3d shows the associated XRD pattern of the sintered Ti-Nb alloy and endorses the phase composition of the sample at the different sintering temperature. The XRD pattern of sintered Ti-Nb alloy revealed the presence of α' phase (Ti), together with β bcc phase (Nb) weak peaks, partially overlapped on α' peaks. It can be seen that the β-phase is the major phase in the sintered samples and the intensity of peaks of β-phase increases as the sintering temperature increases. This is because at lower sintering temperature the sample consists alpha phase, whereas, at higher sintering temperature only the beta phase is produced with a minor amount of alpha phase. Therefore, the peaks of sintered sample are high at higher temperatures associated with the presence of the beta phase. The partially sintered porous Ti-Nb alloy was further utilized as an electrode material for the surface modification of the Ti-6Al-4V alloy by depositing a layer of Ti-Nb by electric discharge coating process to enhance its surface and mechanical properties. The results of the EDC using partially sintered porous Ti-Nb alloy were reported in detail within the next section. The surface topography, elemental composition, coating thickness, material deposition, surface microhardness and adhesion of the coated surface has been simulated with the use of partially sintered porous Ti-Nb alloy as the tool electrode.

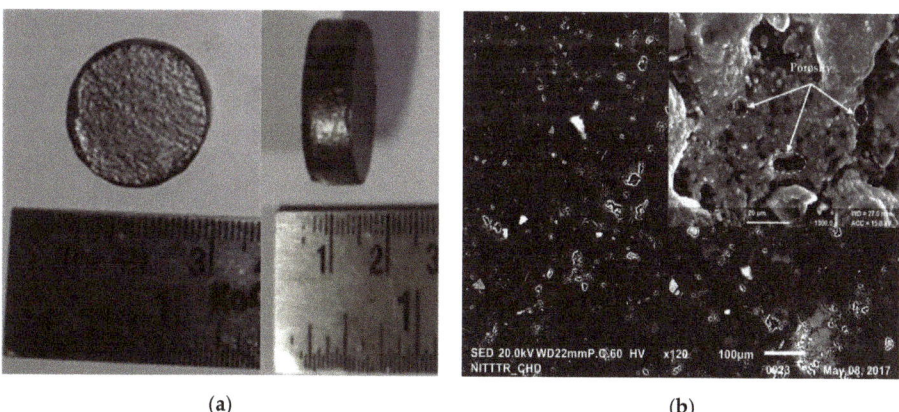

(a) (b)

Figure 3. *Cont.*

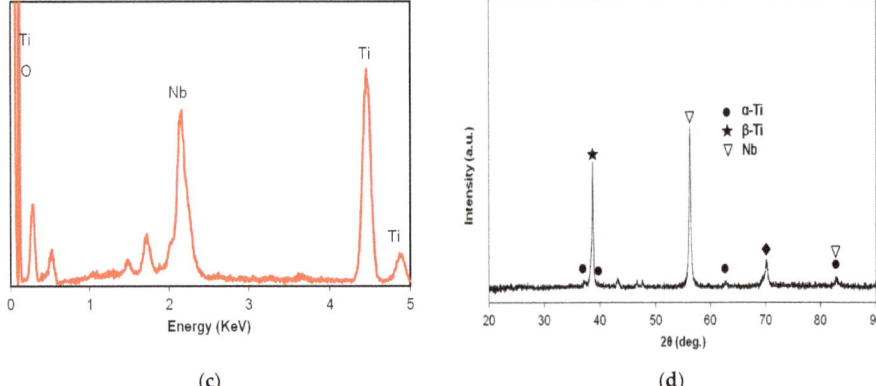

Figure 3. (**a**) Sintered Ti-Nb alloy; (**b**) SEM micrograph showing the morphology; (**c**) EDS spectrum; (**d**) X-ray diffraction (XRD) pattern of Ti-Nb alloy.

3.2. Morphology, Chemical Composition, and Microhardness of EDC-Treated Surface

The effect of process parameters on the Nb deposition, micro-cracks, coating thickness and surface chemistry has been investigated by varying the key parameters of the EDC process (Ip, Ton, and Pc). The amount of Nb deposition on the workpiece surface was measured by EDS analysis. In the present study, the effect of input process parameters on the deposition of weight percentage of Nb powder has been evaluated. Figure 4 shows the variation of process parameters on the deposition of weight percentage of Nb powder in EDM and EDC. As the peak current increases the weight percentage of Nb powder deposition on the workpiece surface first increases up to 20 A. If the peak current increases further, the weight percentage of Nb powder deposition on the workpiece surface start decreases (Figure 4a). The trend for the variation of deposition of Nb on the Ti-64 surface with respect to peak current is the same in both cases, but much less Nb has been deposited on Ti-64 surface as compared to EDC. The maximum Nb (0.42 gm) was deposited at 25A peak current. This is attributed to the increases of peak current and the discharge energy generated which resulted in large exploratory pressure on the dielectric fluid causing migration and deposition of Nb powder towards workpiece surface. Figure 4b shows the variation of weight percentage of Nb powder deposition on the workpiece surface in respect to the pulse duration. The weight percentage of Nb deposition on the workpiece surface first increases with pulse-duration (up to 400 µs) and then start decreasing once the pulse-duration increases further. The explanation agrees with the peak current one, because with the increase in pulse duration the duration of discharge energy in the machining area increases which maintains pressure on the dielectric fluid for the continuous migration and deposition of Nb powder on the workpiece surface. On the other hand, at high pulse duration, the discharge energy generated in a very large exploratory pressure results in a spattering of the molten pool; thus less weight percentage of Nb powder deposition on the workpiece surface. The trend for the variation of deposition of Nb on the Ti-64 surface with respect to pulse duration is the same for both cases, but much less Nb has been deposited compared to EDC. The maximum Nb was deposited such as 0.22 gm at 400 µs of pulse duration. Figure 4c shows the variation of weight percentage of Nb powder deposition on the workpiece surface in respect to Nb powder concentration. As the Nb powder concentration increases, the weight percentage of Nb powder deposition on the workpiece surface increases. This is because, with the increase in Nb-concentration, the Nb-powder particles and eroded-debris are unable to flush out, and are charged due to the ionization of dielectric fluid. As a result, these spark products migrate toward the workpiece surface. The weight percentage of Nb powder deposition on the workpiece surface increases with the increases of the coating thickness. The maximum Nb was deposited 0.34 gm at 20 g/L of Nb powder concentration.

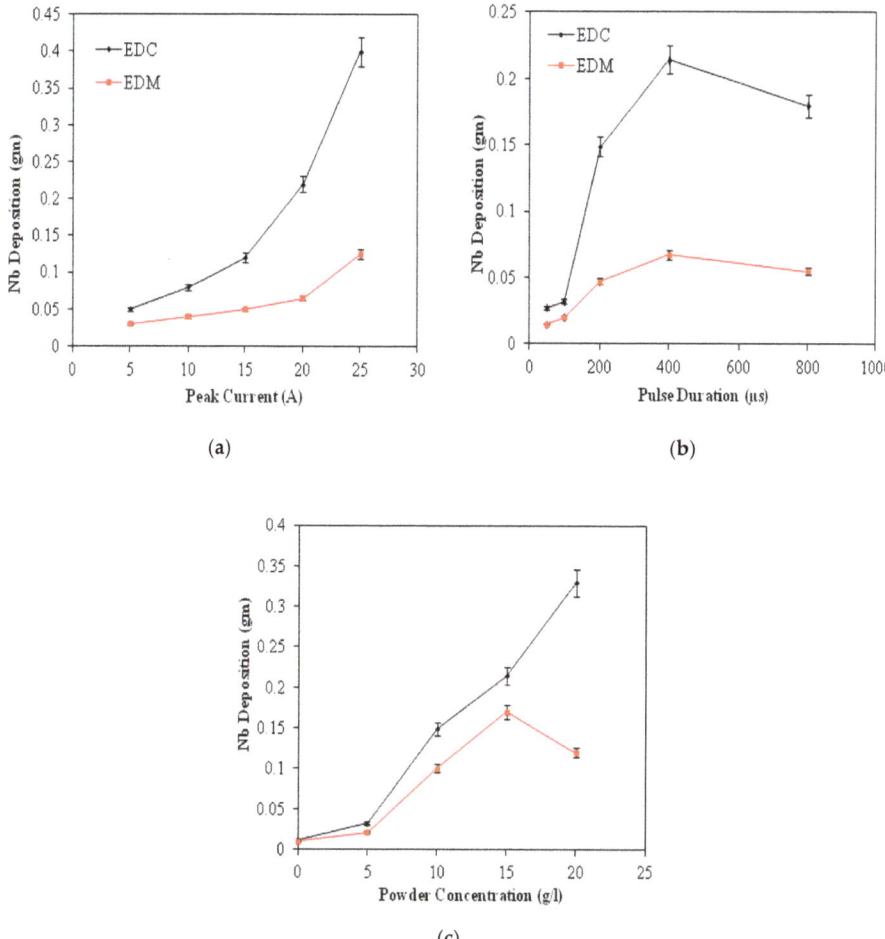

Figure 4. Variation of Nb deposition coating with (**a**) peak current, (**b**) pulse duration, and (**c**) Nb Powder concentration.

Figure 5 shows the surface morphology of the Ti-6Al-4V alloy surface after EDM (tool electrode positive polarity) and EDC (tool electrode negative polarity) setting the peak current = 10 A, pulse duration = 100 μs, duty cycle = 8%, and machining time = 15 min. Many craters resulted from electoral sparks were found on EDM-and EDC-treated surface, but both surfaces have different morphology, which can be observed in Figure 5a,c, respectively. At higher magnification (1700×), a higher density of surface micro-cracks along with high ridges of redeposited molten metal, globules, and pock marks have been identified on the EDM-treated surface, which results in poor surface quality (Figure 5b). On the other hand, the EDC-treated surface shows the micro-cracks but smooth surface as compared to the EDM-treated surface. This is because, in the EDM process, the heat energy is high at the workpiece surface which leads to the removal of workpiece material in the form of deep and wide craters and results in high ridges of redeposited material [18,22,23]. On the other hand, in the case of EDC, the heat energy is high at tool electrode side and low at workpiece side. As a consequence less melting of the workpiece material as compared to the EDM process and results in flat ridges of redeposited material. The results agree with the findings of the previous results reported [25,26]. The EDC-treated surface has less intensity of crack, free from pock marks, globules and Ti and Nb particles in the form of a coated layer can be clearly seen (Figure 5d). This is because during the EDC process, the tool electrode material and degraded carbon form the dielectric fluid, and fill the spaces of micro-cracks, micro-sized craters, and voids. As a result, a layer of oxides and carbides on the machined surface is produced that allows the elimination of micro-cracks.

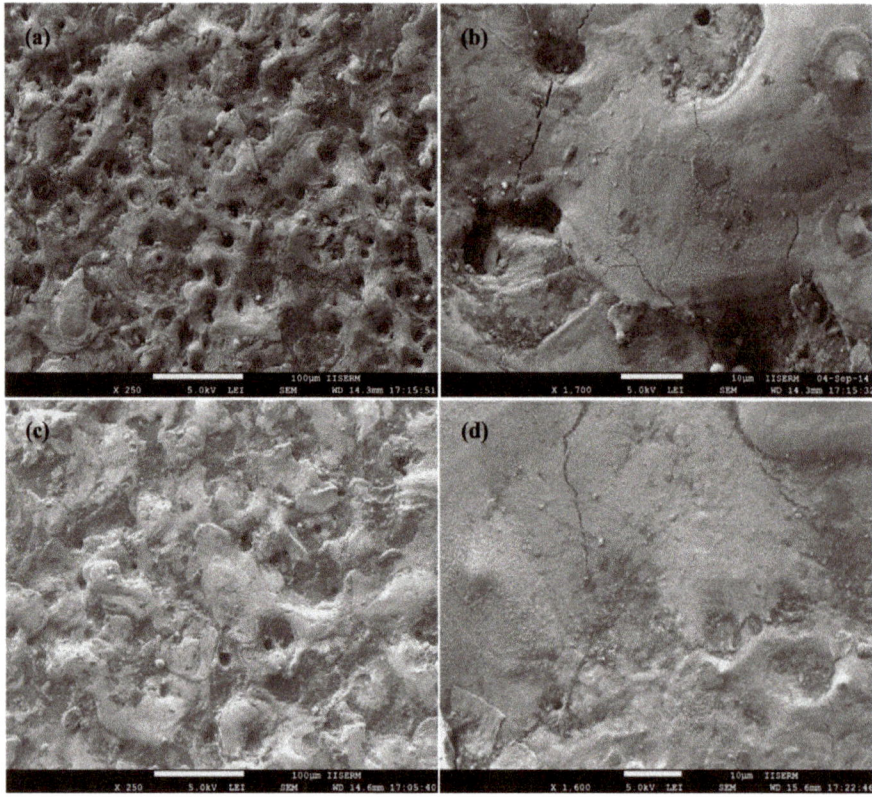

Figure 5. SEM morphology of Ti-6Al-4V alloy after (**a,b**) EDM-treatment and (**c,d**) EDC-treatment.

Figure 6 shows the surface morphology of the Ti-6Al-4V alloy surface after EDC treatment setting the machining condition of peak current = 10 A, pulse duration = 400 µs, duty cycle = 8%, and machining time = 15 min with Nb powder concentration of 10, 15 g/L and 20 g/L, respectively. Relative to the EDM-treated surface, a very flat and smooth surface is observed but only few micro-cracks and micro-pits still exist on the surface at 10 g/L Nb powder concentration, as can be seen in Figure 6a,b. When Nb concentration increased to 15 g/L, a crack and pit-free smooth surface was observed. Apart from this, a layer of oxides and carbides in a higher proportion has been identified, as can be seen in Figure 6c,d. When Nb concentration increased to 20 g/L, a crack and pit-free surface together with a layer of oxides and carbides in a higher proportion was identified that is depicted in Figure 6e,f. This is because, the Nb powder concentration increases the formation of compacted surface cracks while a layer of oxides and carbides (Ti-C, Ti-O, Nb-C, and Nb-O) is formed. Janmanee et al. reported the similar findings in their research for the reduction of micro-cracks on the EDC-treated surface and demonstrated how a layer of WC-Co and Ti-C formed on the machined surface allows the cracks and voids to be filled; causing a reduction in surface cracks [10]. In previous studies, it has been reported that the powder addition in the dielectric fluid significantly reduced the formation of surface cracks [17–20].

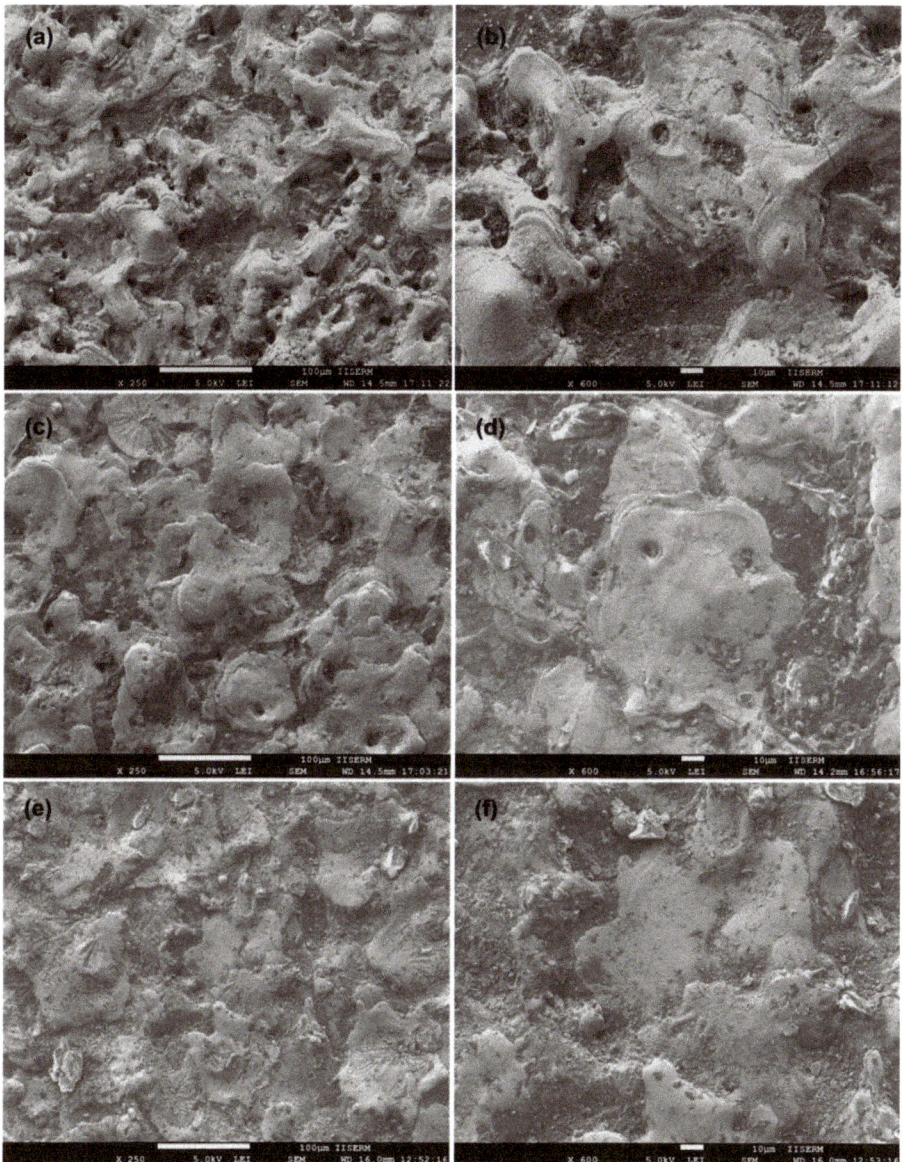

Figure 6. SEM morphology of Ti-6Al-4V alloy after EDC treatment at Nb concentration of (**a,b**) 10 g/L; (**c,d**) 15 g/L and (**e,f**) 20 g/L.

Figure 7 shows the cross section of a micrograph of the coating layer on Ti-6Al-4V alloy after EDM and EDC treatment setting the condition of peak current = 10 A, long pulse duration = 400 µs, duty cycle = 8%, and machining time = 15 min. It can be observed that the thickness of recast layer on Ti-6Al-4V alloy after being EDM-treated was about 149 µm with many surface defects.

Surface micro-cracks and resolidified drops of the molten material can be seen on the EDM-treated surface. The re-solidified material has poor bonding and loosely connected thus there is a risk of their loosening and particles may cause considerable danger since they can penetrate between articulating parts of the joint and damage them. On the other hand, a very smooth and thick layer of coating was observed on the EDC-treated surface when Nb concentration was 10 g/L. The thickness of the coated layer was measured ~195 μm which was thicker than that of the EDM-treated surface, but was defect free. When Nb concentration is 20 g/L. The thickness of the coated layer was measured ~215 μm which was thicker than that of the EDM-treated surface, but still defect free. The loose surface particles are not observed on the EDC-treated surface and excellent metallurgical bonding of re-solidified material with base material were seen, which permits the properties of the modified surface to be enhanced. The deposition of the resolidified material on the workpiece surface has a direct relation with peak current, pulse duration, and powder concentration. The higher peak current and pulse duration generates higher discharge energy which results in repelled both debris and suspended powder particles towards the workpiece surface.

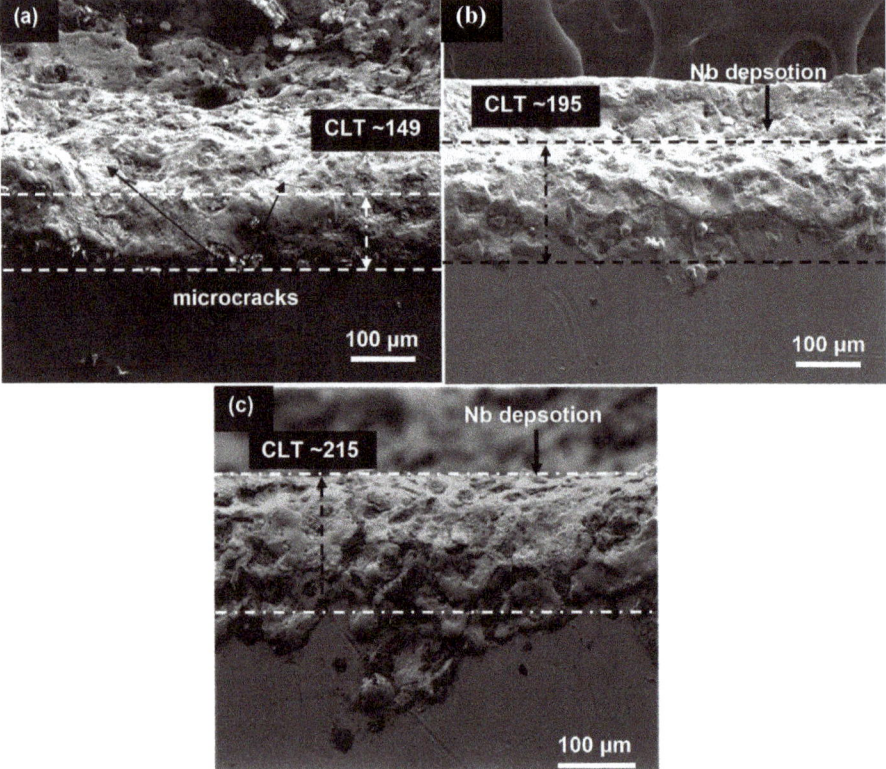

Figure 7. Cross section morphology of coating layer on Ti-6Al-4V alloy: (**a**) 149 μm thickness after EDM-treatment, (**b**) 195 μm thickness after EDC-treatment, and (**c**) 215 μm thickness after EDC-treatment.

Figure 8 shows the EDS spectrum of EDM-treated and EDC-treated of Ti-6Al-4V alloy setting the working condition of peak current = 20 A, pulse duration = 250 μs, duty cycle = 8%, and machining time = 15 min with Nb powder concentration of 0, 10, 15 and 20 g/L. The EDS results show that the EDM-treated surface has no significant presence of Nb powder particles (Figure 8a). On the other hand, the EDC-treated surface has found a significant amount of Nb powder with the weight percentages of 5.98%, 14.87%, and 21.47% as shown in Figure 8b–d. Furthermore, in addition to the presence of Nb, other elements like Ti, Al and, V along with O and C were also present on the EDC-treated surface. The micro-level mapping of the EDC-treated surface at the working condition of peak current = 20 A, pulse duration = 250 µs, duty cycle = 8%, and machining time = 60 min with Nb powder concentration of 20 g/L at a magnification of 250× is shown in Figure 9. The area selected for mapping is shown in Figure 9a. Figure 9b shows the maps of element Nb, Ti, O, and C, changes the surface composition and formed various oxides and carbides, which further increased the surface hardness of the coated layer. Figure 9c shows the composition of the coated surface with carbon of 12.97%, oxygen of 6.97%, niobium 21.67%, and titanium of 52.84%, respectively. The presence of elements O and C indicates the possible formation of oxides and carbides (Ti-C, Ti-O, Nb-C, and Nb-O). The peaks of the coated surface were allocated JCPDS reference nos: 03-065-5714 (TiO2), 00-002-0943 (TiC), JCPDS: 00-002-1031 (NbC), 00-017-0127 (NbC). Figure 9d shows the XRD pattern of un-treated, EDM-treated, and EDC-treated Ti-64 alloy.

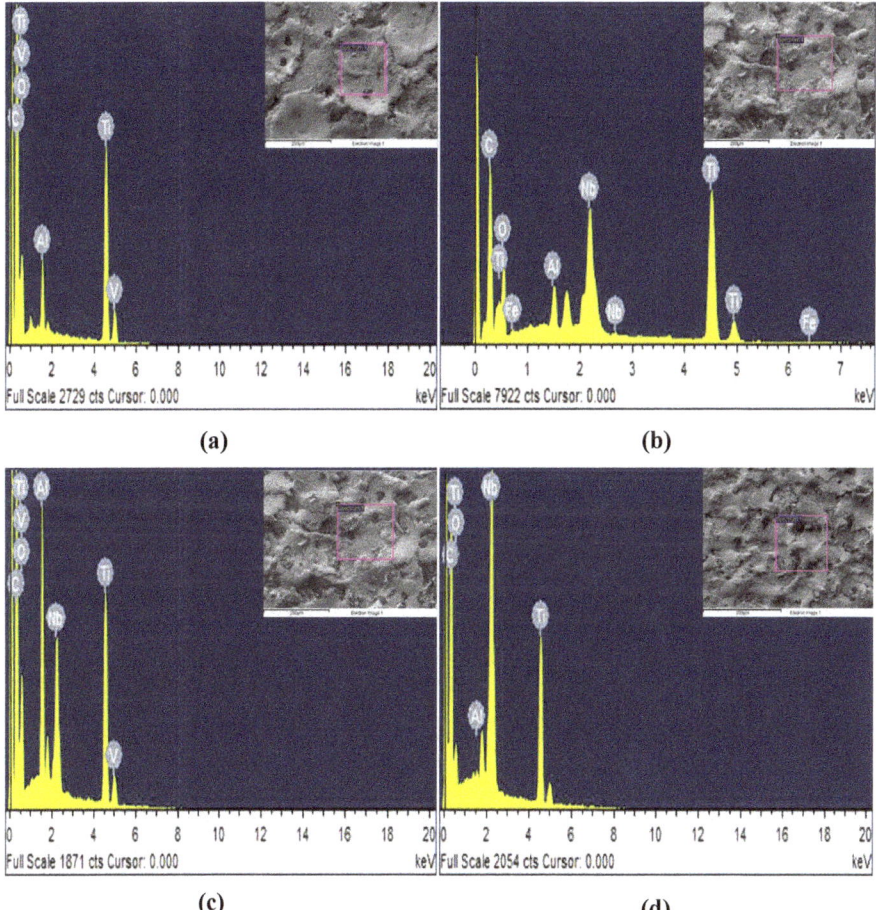

Figure 8. Energy dispersive X-ray spectra of EDM-treated (**a**), EDC-treated Nb coated surface at Nb concentration of (**b**) 10 g/L, (**c**) 15 g/L and (**d**) 20 g/L.

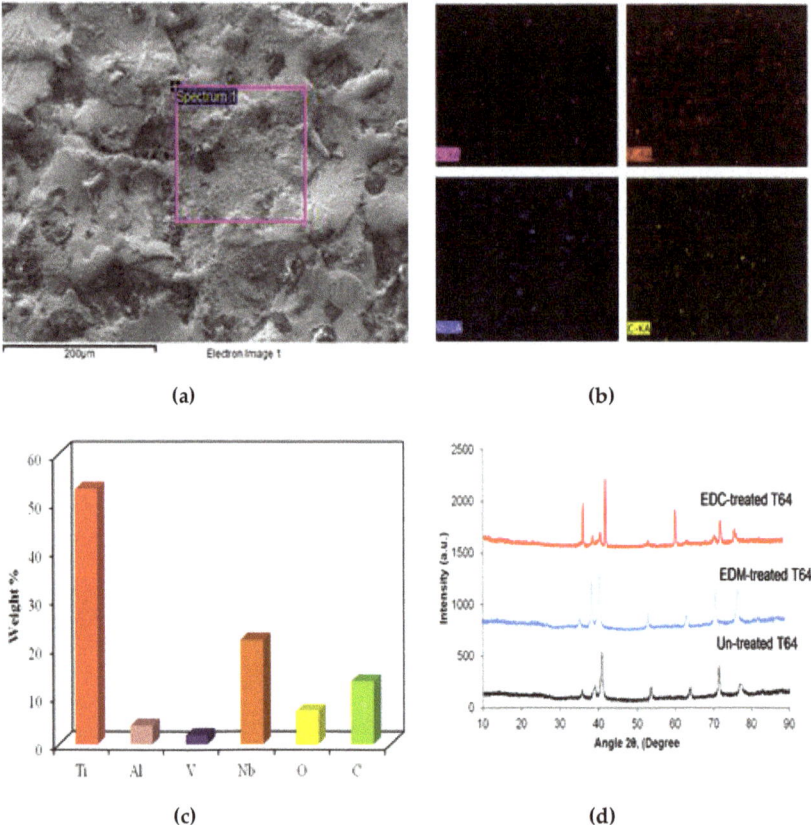

Figure 9. Mapping of EDC-treated Ti-6Al-4V alloy at peak current = 20 A, pulse duration = 250 µs, Duty cycle = 8%, and machining time = 60 min with Nb powder concentration of 20 g/L. (**a**) Spectrum area (**b**) Mapping of Ti, Nb, O and C (**c**) Weight Percentage of elements present in coated layer (**d**) XRD pattern of un-treated, EDM-teraed, and EDC-treated Ti-64.

Figure 10 shows the distribution of micro-hardness along the cross section of ED-coated and EDM-treated surface. The micro-hardness decreases gradually from the top surface to the base surface. The highest micro-hardness of 1465 HV appears at few microns away from the top surface in the case of EDC. The improvement in the microhardness is mainly attributed to the reinforcement effect of Nb/Ti-based oxides and carbides during the manufacturing process, which is transferred to the electrode and dielectric fluid. In the EDC process the resolidification of molten pool is very rapid and changes in the microstructure are evident. The EDC-modified layer exhibited various types of oxides and carbides (Ti-O, TiC, Nb-C, and Nb-O) with excellent metallurgical bond, which enhances the microhardness of the surface. Whereas, the highest microhardness of 1175 HV appears at few microns away from the top surface in the case of EDM. The microhardness of the EDM-treated surface is low as compared to the EDC-treated surface. This is because, during the EDM process at positive polarity, the high temperature leads to the phase transformation being impeded, producing a structure with micro-cracks with the low metallurgical bond. As a result of this, the mechanical properties weaken and the hardness value drops. The metallurgical bond establishes the adhesion strength. A stronger metallurgical bond promote a higher adhesion strength. Figure 10b shows the adhesion strength of the coating. It is clear that, the coated layer has excellent metallurgical bonding to the substrate surface.

Therefore, no delamination of the coated layer was observed in both cases, but the EDM-treated surface is slightly affected in respect to the EDC-treated surface that perform well even at higher load. Figure 10c,d shows the EDM-treated and EDC-treated surfaces, respectively. The Nb-deposited surface possessed high surface hardness and offered excellent mechanical interlocking with the substrate. Benefitting from this mechanical interlocking, the critical adhesion failure of EDM-treated surface and Nb-deposited surface was detected at 82 N and 118 N respectively, as can be seen in Figure 10. This observation permits us to indicate the EDC-treated surface as suitable candidate counterpart for tribological properties.

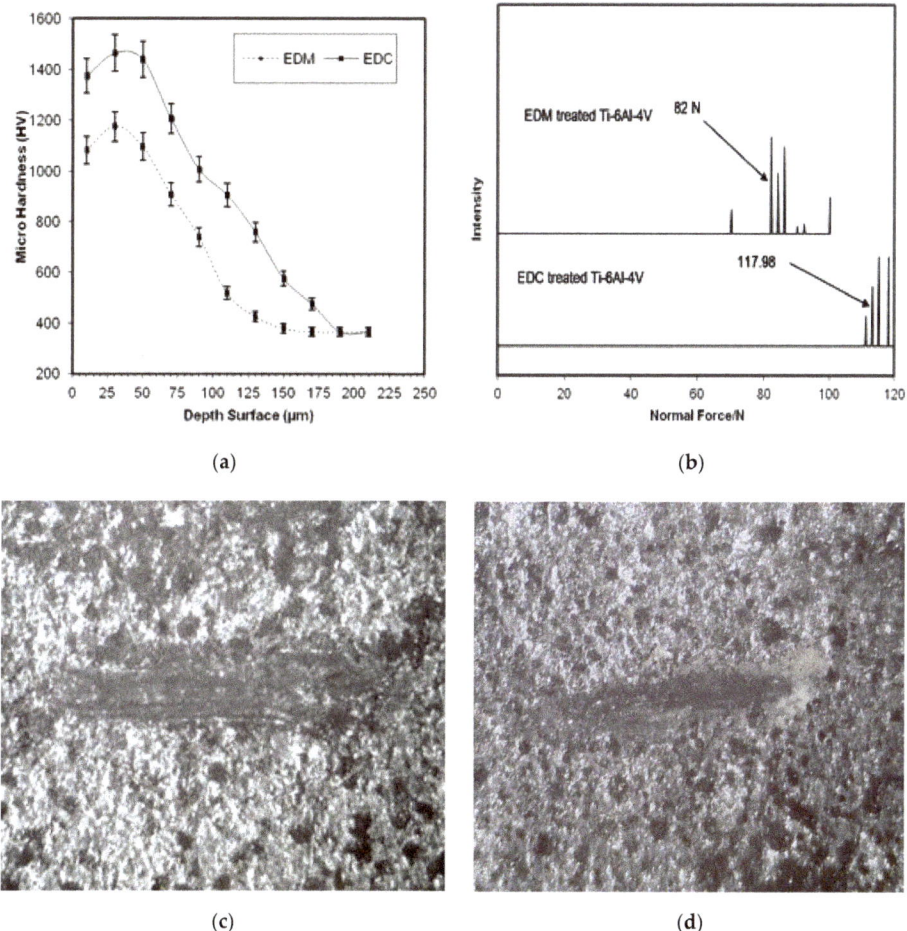

Figure 10. (a) Distribution of micro-hardness along the cross section, (b) adhesion strength of electric. discharge (ED)-coated and EDM-treated surface, and (c,d) scratch images on EDM-treated and EDC-treated surfaces.

4. Conclusions

This study focused on the application of spark plasma sintered Ti-Nb alloy under electric discharge coating in order to improve the surface charctristics of Ti-6Al-4V alloy using the Nb powder mixed dielectric and permits us to drawn the following conclusions:

1. A partially sintered Ti-Nb alloy has been successfully fabricated by mechanical alloying of Ti and Nb from powders, alloy modified later by spark plasma sintering technique, which was further used as tool electrode for the EDC process.
2. The surface of Ti-6Al-4V alloy has been modified by EDM (positive Polarity) and EDC (negative Polarity). The EDC-treated surface contains only few cracks and smooth geometry as compared to the EDM-treated surface.
3. A coating layer of Ti-O, TiC, Nb-C, and Nb-O have been successfully prepared by using partially sintered Ti-Nb alloy and Nb powder mixed in the dielectric fluid of electric discharge machine. The mass deposition of coating layer have almost linear relation and is significantly affected by the concentration of peak current, pulse duration, and Nb powder concentration.
4. The thickness of coating layer was significantly affected by the concentration of peak current, pulse duration, and Nb powder concentration. Using the 0 g/L Nb concentration (EDM), the coating layer thickness ~ 149 μm was obtained for a longer pulse duration (400 μs). On the other hand, when the concentration of Nb was increased to 10 g/L and 20 g/L, the thickness of the recast layer increased to 195 μm and 215 μm, respectively, in the same working conditions.
5. The EDS elemental map and XRD pattern analysis of the EDC-treated surface confirm the process in which the material migrate from the tool electrode to workpiece surface. Here, the suspended powder particles generated in the dielectric fluid which promote adhesion to the workpiece surface are playing the main role in the improvement of surface properties by generating favorable surface chemistry of oxides and carbides. When a higher concentration of Nb powder (20 g/L) is used, the EDC-treated surface is expected to form the Ti-O, Nb-O, Ti-C, and Nb-C like phases.
6. The surface coating layer permits an increase in the microhardness of the workpiece surface from 365 HV to 1465 HV and demonstrates an excellent metallurgical bonding with the base workpiece surface. The adhesion strength is as high as 118 N for the EDC-treated surface when compared to EDM treated surface at 82 N, respectively; thus indicating that the EDC-treated surface may have good tribological properties.
7. In summary, the EDC can be considered a great technique in order to improve surface characteristics and surface properties. The coating obtained by EDC process is more reliable and suitable for its purposes because the surfaces produced by this process demonstrated higher surface hardness that can be associated with better wear resistance of the implant. The Nb content in the coated layer provides superior corrosion resistance and allows improvements in the bioactivity of the implant substrate.

Author Contributions: Conceptualization, C.P., S.S., G.K., D.Y.P.; Data curation, G.K., D.Y.P.; Formal analysis, G.K., D.Y.P., A.P., V.M.; Funding acquisition, G.K., C.I.P.; Investigation, C.P., S.S., A.P., V.M.; Writing—original draft, C.P., S.S., G.K., A.P., V.M.; Writing—review & editing, C.P., G.K., D.Y.P., and C.I.P.

Funding: The APC was funded by Grzegorz Królczyk, Faculty of Mechanical Engineering, Opole University of Technology, 76 Proszkowska St., Opole 45-758, Poland and Catalin Iulian Pruncu, Mechanical Engineering, Imperial College London, Exhibition Rd., SW7 2AZ London, UK.

Conflicts of Interest: The authors declare no conflict of interest.

References

1. Zaman, H.A.; Safian, S.; Idris, M.H.; Kamarudin, A. Metallic biomaterial for medical implant applications: A review. *Appl. Mech. Mater.* **2015**, *735*, 19–25. [CrossRef]
2. Li, Y.; Yang, C.; Zhao, H.; Qu, S.; Li, X.; Li, Y. New developments of Ti-based alloys for biomedical applications. *Materials* **2014**, *7*, 1709–1800. [CrossRef] [PubMed]
3. Geetha, M.; Singh, A.K.; Asokamani, R.; Gogia, A.K. Ti based biomaterials, the ultimate choice for orthopaedic implants—A review. *Prog. Mater. Sci.* **2009**, *54*, 397–425. [CrossRef]
4. Liua, X.; Chu, P.K.; Ding, C. Surface modification of titanium, titanium alloys, and related materials for biomedical applications. *Mater. Sci. Eng. R Rep.* **2004**, *47*, 49–121. [CrossRef]

5. Prakash, C.; Kansal, H.K.; Pabla, B.S.; Puri, S.; Aggarwal, A. Electric discharge machining a potential choice for surface modification of metallic implants for orthopedics applications: A review. *Proc. Inst. Mech. Eng. Part B J. Eng. Manuf.* **2016**, *230*, 331–353. [CrossRef]
6. Peng, P.W.; Ou, K.L.; Lin, H.C.; Pan, Y.N.; Wang, C.H. Effect of electrical-discharging on formation of nanoporous biocompatible layer on titanium. *J. Alloys Compd.* **2010**, *492*, 625–630. [CrossRef]
7. Yang, T.S.; Huang, M.S.; Wang, M.S.; Lin, M.H.; Tsai, M.Y.; Wang, P.Y. Effect of electrical discharging on formation of nanoporous biocompatible layer on Ti-6Al-4V alloys. *Implant Dent.* **2013**, *22*, 374–379. [CrossRef]
8. Lee, W.F.; Yang, T.S.; Wu, Y.C.; Peng, P.W. Nanoporous biocompatible layer on Ti–6Al–4V alloys enhanced osteoblast-like cell response. *J. Exp. Clin. Med.* **2013**, *5*, 92–96. [CrossRef]
9. Bin, T.C.; Xin, L.D.; Zhan, W.; Yang, G. Electro-spark alloying using graphite electrode on titanium alloy surface for biomedical applications. *Appl. Surf. Sci.* **2011**, *257*, 6364–6371. [CrossRef]
10. Janmanee, P.; Muttamara, A. Surface modification of tungsten carbide by electrical discharge coating (EDC) using a titanium powder suspension. *Appl. Surf. Sci.* **2012**, *258*, 7255–7265. [CrossRef]
11. Furutani, K.; Saneto, A.; Takezawa, H.; Mohria, N.; Miyakeb, H. Accretion of titanium carbide by electrical discharge machining with powder suspended in working fluid. *Precis. Eng.* **2001**, *25*, 138–144. [CrossRef]
12. Krishna, M.E.; Patowari, P.K. Parametric study of electric discharge coating using powder metallurgical green compact electrodes. *Mater. Manuf. Process.* **2014**, *29*, 1131–1138. [CrossRef]
13. Krishna, M.E.; Patowari, P.K. Post processing of the layer deposited by electricdischarge coating. *Mater. Manuf. Process.* **2017**, *32*, 442–449. [CrossRef]
14. Patowari, P.K.; Mishra, U.K.; Saha, P.; Mishra, P.K. Surface Integrity of C-40 Steel Processed with WC-Cu Powder Metallurgy Green Compact Tools in EDM. *Mater. Manuf. Process.* **2011**, *25*, 668–676. [CrossRef]
15. Ahmed, A. Deposition and Analysis of Composite Coating on Aluminum Using Ti–B4C Powder Metallurgy Tools in EDM. *Mater. Manuf. Process.* **2016**, *31*, 467–474. [CrossRef]
16. Chen, H.J.; Wu, K.L.; Yan, B.H. Dry Electrical Discharge Coating Process on Aluminum by Using Titanium Powder Compact Electrode. *Mater. Manuf. Process.* **2013**, *28*, 1286–1293. [CrossRef]
17. Wang, Z.L.; Fang, Y.; Wu, P.N.; Zhao, W.S.; Cheng, K. Surface modification process by electrical discharge machining with a Ti powder green compact electrode. *J. Mater. Process. Technol.* **2002**, *129*, 139–142. [CrossRef]
18. Prakash, C.; Kansal, H.K.; Pabla, B.S.; Puri, S. Potential of powder mixed electric discharge machining to enhance the wear and tribological performance of β-Ti implant for orthopedic applications. *J. Nanoeng. Nanomanuf.* **2015**, *5*, 261–269. [CrossRef]
19. Prakash, C.; Kansal, H.K.; Pabla, B.S.; Puri, S. Experimental Investigations in Powder Mixed Electrical Discharge Machining of Ti-35Nb-7Ta-5Zr β-Ti Alloy. *Mater. Manuf. Process* **2017**, *32*, 274–285. [CrossRef]
20. Prakash, C.; Kansal, H.K.; Pabla, B.S.; Puri, S. Processing and characterization of novel biomimetic nanoporous bioceramic surface on β-Ti implant by powder mixed electric discharge machining. *J. Mater. Eng. Perform.* **2015**, *24*, 3622–3633. [CrossRef]
21. Prakash, C.; Kansal, H.K.; Pabla, B.S.; Puri, S. Multi-objective optimization of powder mixed electric discharge machining parameters for fabrication of biocompatible layer on β-Ti alloy using NSGA-II coupled with Taguchi based response surface methodology. *J. Mech. Sci. Technol.* **2016**, *30*, 4195–4204. [CrossRef]
22. Prakash, C.; Kansal, H.K.; Pabla, B.S.; Puri, S. Powder Mixed Electric Discharge Machining an Innovative Surface Modification Technique to Enhance Fatigue Performance and Bioactivity of β-Ti Implant for Orthopaedics Application. *J. Comput. Inf. Sci. Eng.* **2015**, *14*, 041006. [CrossRef]
23. Prakash, C.; Kansal, H.K.; Pabla, B.S.; Puri, S. Effect of Surface Nano-Porosities Fabricated by Powder Mixed Electric Discharge Machining on Bone-Implant Interface: An Experimental and Finite Element Study. *Nanosci. Nanotechnol. Lett.* **2016**, *8*, 815–826. [CrossRef]
24. Xie, Z.J.; Mai, Y.J.; Lian, W.Q.; He, S.L.; Jie, X.H. Titaniumcarbide coating with enhanced tribological properties obtained by EDC using partially sintered titanium electrodes and graphite powder mixed dielectric. *Surf. Coat. Technol.* **2016**, *300*, 50–57. [CrossRef]
25. Arun, M.; Duraiselvam, V.; Senthilkumar, R. Synthesis of Electric Discharge Alloyed Nickel–Tungsten Coating on Tool Steel and its Tribological Studies. *Mater. Des.* **2014**, *63*, 257–262. [CrossRef]
26. Ekmekci, N.; Ekmekci, B. Electrical Discharge Machining of Ti6Al4V in Hydroxyapatite Powder Mixed Dielectric Liquid. *Mater. Manuf. Process.* **2016**, *31*, 1663–1670. [CrossRef]
27. Ou, S.F.; Wang, C.Y. Fabrication of a hydroxyapatite-containing coating on Ti–Ta alloy by electrical discharge coating and hydrothermal treatment. *Surf. Coat. Technol.* **2016**, *302*, 238–243. [CrossRef]

28. Prakash, C.; Uddin, M.S. Surface modification of β-phase Ti implant by hydroaxyapatite mixed electric discharge machining to enhance the corrosion resistance and in-vitro bioactivity. *Surf. Coat. Technol.* **2017**, *326 Pt A*, 134–145. [CrossRef]
29. Prakash, C.; Singh, S.; Pabla, B.S.; Uddin, M.S. Synthesis, characterization, corrosion and bioactivity investigation of nano-HA coating deposited on biodegradable Mg-Zn-Mn alloy. *Surf. Coat. Technol.* **2018**, *346*, 9–18. [CrossRef]
30. Prakash, C.; Singh, S.; Singh, M.; Verma, K.; Chaudhary, B.; Singh, S. Multi-objective particle swarm optimization of EDM parameters to deposit HA-coating on biodegradable Mg-alloy. *Vacuum* **2018**, *158*, 180–190. [CrossRef]
31. Aliyu, A.A.A.; Abdul-Rani, A.M.; Ginta, T.L.; Prakash, C.; Axinte, E.; Razak, M.A.; Ali, S. A review of additive mixed-electric discharge machining: Current status and future perspectives for surface modification of biomedical implants. *Adv. Mater. Sci. Eng.* **2017**, *2017*, 8723239. [CrossRef]
32. Prakash, C.; Kansal, H.K.; Pabla, B.S.; Puri, S. Potential of Silicon Powder-Mixed Electro Spark Alloying for Surface Modification of β-Phase Titanium Alloy for Orthopedic Applications. *Mates. Today Proc.* **2017**, *4*, 10080–10083. [CrossRef]
33. Prakash, C.; Kansal, H.K.; Pabla, B.S.; Puri, S. To optimize the surface roughness and microhardness of β-Ti alloy in PMEDM process using Non-dominated Sorting Genetic Algorithm-II. In Proceedings of the 2015 2nd International Conference on Recent Advances in Engineering & Computational Sciences (RAECS), Chandigarh, India, 21–22 December 2015; pp. 1–5.
34. Ning, C.; Zhou, Y. In Vitro Bioactivity of a Biocomposite Fabricated from HA and Ti Powders by Powder Metallurgy Method. *Acta Biomater.* **2002**, *23*, 2909–2915. [CrossRef]
35. Prakash, C.; Singh, S.; Pabla, B.S.; Sidhu, S.S.; Uddin, M.S. Bio-inspired low elastic biodegradable Mg-Zn-Mn-Si-HA alloy fabricated by spark plasma sintering. *Mater. Manuf. Process.* **2019**, *34*, 357–368. [CrossRef]
36. Bhushan, B.; Singh, A.; Singh, R.; Mehta, J.S.; Gupta, A.; Prakash, C. Fabrication and Characterization of a New Range of β-type Ti-Nb-Ta-Zr-xHaP (x = 0, 10) Alloy by Mechanical Alloying and Spark Plasma Sintering for Biomedical Applications. *Mater. Today Proc.* **2018**, *5*, 27749–27756. [CrossRef]
37. Singh, R.; Singh, B.P.; Gupta, A.; Prakash, C. Fabrication and characterization of Ti-Nb-HA alloy by mechanical alloying and spark plasma sintering for hard tissue replacements. In *IOP Conference Series: Materials Science and Engineering*; IOP Publishing: Bristol, UK, 2017; Volume 225, p. 012051.
38. Prakash, C.; Singh, S.; Gupta, M.; Mia, M.; Królczyk, G.; Khanna, N. Synthesis, Characterization, Corrosion Resistance and In-Vitro Bioactivity Behavior of Biodegradable Mg–Zn–Mn–(Si–HA) Composite for Orthopaedic Applications. *Materials* **2018**, *11*, 1602. [CrossRef] [PubMed]
39. Prakash, C.; Singh, S.; Verma, K.; Sidhu, S.S.; Singh, S. Synthesis and characterization of Mg-Zn-Mn-HA composite by spark plasma sintering process for orthopedic applications. *Vacuum* **2018**, *155*, 578–584. [CrossRef]
40. Singh, H.; Singh, S.; Prakash, C. Current Trends in Biomaterials and Bio-manufacturing. In *Biomanufacturing*; Springer: Cham, Switzerland, 2019; pp. 1–34.

© 2019 by the authors. Licensee MDPI, Basel, Switzerland. This article is an open access article distributed under the terms and conditions of the Creative Commons Attribution (CC BY) license (http://creativecommons.org/licenses/by/4.0/).

Article

Optimization of Power Consumption Associated with Surface Roughness in Ultrasonic Assisted Turning of Nimonic-90 Using Hybrid Particle Swarm-Simplex Method

Navneet Khanna [1], Jay Airao [1], Munish Kumar Gupta [2], Qinghua Song [2,3], Zhanqiang Liu [2,3], Mozammel Mia [4], Radoslaw Maruda [5] and Grzegorz Krolczyk [6,*]

1. Institute of Infrastructure, Technology, Research, and Management, Ahmedabad 380026, India; navneetkhanna@iitram.ac.in (N.K.); jay.airao.16mm@iitram.ac.in (J.A.)
2. Key Laboratory of High Efficiency and Clean Mechanical Manufacture, Ministry of Education, School of Mechanical Engineering, Shandong University, Jinan 250061, China; munishguptanit@gmail.com (M.K.G.); ssinghua@sdu.edu.cn (Q.S.); melius@sdu.edu.cn (Z.L.)
3. National Demonstration Center for Experimental Mechanical Engineering Education, Shandong University, Jinan 250061, China
4. Department of Mechanical Engineering, Imperial College of London, South Kensington, London SW7 2AZ, UK; mozammelmiaipe@gmail.com
5. Faculty of Mechanical Engineering, University of Zielona Gora, 4 Szafrana St, 65-516 Zielona Gora, Poland; r.maruda@ibem.uz.zgora.pl
6. Faculty of Mechanical Engineering, Opole University of Technology, 76 Proszkowska St, 45-758 Opole, Poland
* Correspondence: g.krolczyk@po.edu.pl; Tel./Fax: +48-77-449-8429

Received: 25 August 2019; Accepted: 14 October 2019; Published: 18 October 2019

Abstract: These days, power consumption and energy related issues are very hot topics of research especially for machine tooling process industries because of the strict environmental regulations and policies. Hence, the present paper discusses the application of such an advanced machining process i.e., ultrasonic assisted turning (UAT) process with the collaboration of nature inspired algorithms to determine the ideal solution. The cutting speed, feed rate, depth of cut and frequency of cutting tool were considered as input variables and the machining performance of Nimonic-90 alloy in terms of surface roughness and power consumption has been investigated. Then, the experimentation was conducted as per the Taguchi L9 orthogonal array and the mono as well as bi-objective optimizations were performed with standard particle swarm and hybrid particle swarm with simplex methods (PSO-SM). Further, the statistical analysis was performed with well-known analysis of variance (ANOVA) test. After that, the regression equation along with selected boundary conditions was used for creation of fitness function in the subjected algorithms. The results showed that the UAT process was more preferable for the Nimconic-90 alloy as compared with conventional turning process. In addition, the hybrid PSO-SM gave the best results for obtaining the minimized values of selected responses.

Keywords: ultrasonically assisted turning; Nimonic-90; surface roughness; power consumption; optimization; nature inspired hybrid algorithm

---

1. Introduction

In this growing industrial world, the trend of modern materials, especially nickel based alloys, are prevalent in various sectors such as automobile, aerospace, marine etc. [1]. They are altogether expected in these manufacturing sectors because of their eminent characteristics such as high resistance to corrosion and excellent mechanical properties etc. [2]. These characteristics, however, result in enormous challenges in terms of high tool wear, low finishing, excessive forces etc. in machining of advanced materials [2,3]. Furthermore, the strict environmental policies and concerns are other challenges which must be addressed during the machining of advanced materials. For instance, Japan has established the basic "Energy Policy" which primarily focuses on energy related issues in manufacturing sectors. Similarly, the USA have introduced the special program on "Superior Energy Performance (SEP)" that provides a track in the field of sustainability development for manufacturing sectors. Likewise, the European countries have developed the ISO standards i.e., 5001 for regularization of energy standards in manufacturing sectors [4].

In order to follow the environmental concerns and ISO 5001 standards, the new technologies i.e., hybrid machining processes are considered as main drivers to support the working aspects of sustainability i.e., social, economic and environmental [5,6]. In a hybrid machining process, the material removal mechanism is totally different as compared with conventional machining processes. For instance, in the hybrid machining process, the material is removed with the main machining process while a secondary technique "assists" the material removal by improving the conditions of machining. In recent years, the ultrasonic assisted turning (UAT) process has been termed as one such hybrid process that uses ultrasonic vibrations for the cutting action [5]. In this hybrid process, the interaction of the cutting tool and the workpiece directly takes place and the material is removed under the action of micro chipping [6]. Furthermore, the vibrations of the tool produce some surface texturing effect on the workpiece [7] and thereby good surface finishing, dimensional accuracy and low tool wear are obtained during machining [8]. The efficiency of UAT has been noted by various former researchers. Some of their works are presented here. In the first study of Maurotto et al., it has been seen that the cutting forces produced in the UAT process are significantly less as compared with the conventional dry turning (CT) of Ti-15333 and Ni-625 alloys [9]. In another similar work, the cutting forces were analyzed by Ahmed et al. during machining of Inconel-718 [10]. It was found that the cutting forces induced during CT were 130–140 N whereas; in UAT process were 60–95 N. In the same work, Maurotto et al. showed that the tangential as well as radial cutting forces were reduced up to 70%–80% while machining of Ti-15333 and Ni-625 which was claimed to be possibly due to ultrasonic softening in the base alloys in UAT [11]. It was also determined by the same authors that the cutting speed is the major factor in UAT process that effects the cutting forces when compared to CT [12]. Silberschmidt et al. analyzed the surface roughness values in machining of Inconel-718 and Ti-15333. The comparison was also made between UAT and CT process [13]. Similarly, Zhong and Lin found that the surface roughness improves by 15% with high amplitude as compared to lower amplitude with UAT because of the ironing effect in aluminum metal matrix composites [14]. Moreover, Nath and Rahman [15] studied the effect of frequencies, amplitude and cutting speed on cutting forces values. They concluded that the cutting force generated during the UAT is dependent on tool–workpiece contact ratio (TWCR). In the same context, Vivekananda et al. [16] implemented the Taguchi design of experiment process to optimize the cutting force and surface roughness values in the UAT process.

From the comprehensive state of art review, it has been interestingly noticed that UAT is a very tremendous technology in the modern arena of manufacturing sectors. However, its application is only limited to the cutting forces and tool life while machining of other nickel-based alloys i.e., Inconel 718, whilst surface roughness and power consumption are overlooked. Moreover, the study of UAT of Nimonic-90 has never been performed to the best of the authors' knowledge. Although previous investigations have shown that UAT improves the machinability of Inconel 718 alloy, these results cannot be directly extended to Nimonic-90 alloy. Thus, to bridge this gap, a series of UAT experiments were conducted and the surface roughness and power consumption were investigated and analyzed for Nimonic-90 alloy. In addition, the literature reveals that the performance of UAT is highly dependent upon its process parameters because a large number of process parameters are involved in the ultrasonic assisted turning process. Therefore, the best parameters settings are required to enhance the machining performance of the UAT process. Various types of optimization methods i.e., conventional and advanced methods, are currently available in the literature that improve the process efficiency by changing the input and output settings [17–19]. In the conventional methods Taguchi, signal to noise (S/N) ratio, analysis of variance, regression, desirability analysis etc. have been introduced to solve the optimization issues [20–22]. However, the conventional methods are subjected to some issues, such as the lack of targeting the global optimal solution with these methods, which may result in low accuracy and non-robustness of the results. With these limitations, they are still used in the machining of different materials. For instance, Shokrani et al. [23] used the Taguchi method to optimize the process parameter in cryogenic milling of Ti64 titanium alloy. Islam et al. [24] compared the traditional and Taguchi method in terms of efficiency to analyze the surface roughness values. Ezilarasan et al. [25] used the Taguchi method to discuss the effect of input variables on surface roughness values while machining Nimonic C-263 alloy. Makadia and Ashvin [26] minimized the surface roughness values in machining AISI 410 steel by using the response surface methodology (RSM) method. Bhushan [27] optimized the parameters for minimum power consumption and improve the tool life during machining of 7075 Al alloy SiC particle composites with the help of response surface methodology.

Apart from these optimization methods, the various types of advanced methods such as evolutionary algorithms, nature inspired hybrid algorithms and intelligent methods are well implemented in literature to solve optimization problems [17–19]. The major benefits of these advanced methods are that they accurately achieve the global optimal solution with a small interval of time [17,28]. They are generally presented in the MATLAB code and the objective or fitness function is required to run the program. Moreover, the single or multi-objective problems can be easily tackled with these advanced optimization methods. Numerous studies have been available in the literature that clearly represent the application of advanced optimization algorithms in the machining sector. In the first study, Singh et al. used the two algorithms i.e., particle swarm and bacterial foraging for optimization of cutting parameters in minimum quantity lubrication (MQL) assisted milling of commercially available Inconel-718 alloy [19]. Likewise, Sahu and Andhare used the three advanced algorithms i.e., teaching learning based optimization (TLBO), Jaya algorithm and genetic algorithm (GA) and one conventional method (RSM) to solve the optimization problem of Ti–6Al–4V alloy [29]. They suggested that the performance of the TLBO and Jaya algorithms is better than GA. In another optimization study, Sathish applied the hybrid bee colony cuckoo search (BCCS) and RSM approach in non-conventional machining of Nimonic-263 alloy [30]. In similar work, Gupta et al. implemented the particle swarm optimization (PSO) and bacterial foraging optimization (BFO) while turning titanium (grade-2) alloy under nano-fluid cutting conditions [17]. The performance was also compared with the traditional optimization method i.e., RSM. It has been noted that the PSO and BFO work more efficiently than the RSM methods and significantly enhance the process performance. Furthermore, Rao and Venkaiah used the PSO and RSM to optimize the machining parameters of Nimonic-263 alloy [2].

Thus, as per the availability of current research survey, it has been clearly noted that the machining performance of any material is highly improved with the implementation of advanced algorithms. For instance, Sahu and Andhare, Sathish, Singh et al. and Gupta et al also presented similar findings in machining of Inconel, titanium and Nimonic-based alloys [15,16,26,27]. Still, with all these hard efforts, no research work is available in the literature which shows the application of advanced algorithms in ultrasonic assisted turning of Nimonic-90 alloy for power optimization. Therefore, this research work firstly reported the application of nature inspired hybrid algorithm i.e., particle swarm optimization (PSO) method hybridized with the simplex method (SM) during ultrasonic assisted turning of Nimonic-90 alloy. The input parameters considered were cutting speed, feed rate, depth-of-cut and frequency of the cutting tool used. A Taguchi L9 orthogonal array with three repetitions were used as an experimental design and the power consumption and surface roughness values were optimized with these implemented algorithms. The complete detail of this experimental work complemented with the optimization details are presented in the subsequent sections.

2. Materials and Methods

2.1. Ultrasonic Assisted Turning (UAT) Process

The simplified view of the UAT process is exhibited in Figure 1. The main components of the UAT machine were the frequency generator, piezoelectric transducer and the horn. The frequency generator created the electrical signal which was then converted into a mechanical signal by a piezoelectric transducer. Then, these mechanical signals propagated through the ultrasonic horn to the cutting tool.

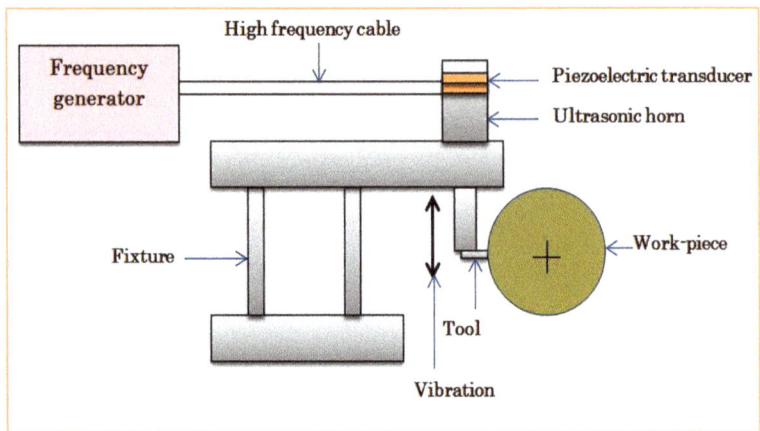

Figure 1. Schematic of ultrasonically assisted turning set-up.

The major aim of this ultrasonic horn was to amplify the vibrations to reasonable magnitudes. Well-known analytical relations exist which are used to facilitate the horn design. For example, the length of stepped horn (L) is determined using $L = \frac{c}{2f}$. Here, f is frequency and c is the speed of sound in the medium (horn material) which depends on the modulus of elasticity (E) and density (ρ) of the material as shown in $c = \sqrt{\frac{E}{\rho}}$. Titanium, aluminum, mild-steel etc. are popular choices for horn material. In our study, mild steel was used to manufacture the horn [31].

2.2. Workpiece and Tool Material

The workpiece materials used were Nimonic-90 alloy. They were precipitation strengthened nickel base super alloys of extra high mechanical properties with corrosion resistance. Nimonic-90 is typically used in extreme stress applications such as turbine blades, hot working tools, exhaust re-heater, disc and high-temperature springs. The chemical composition of Nimonic-90 alloy is shown in Table 1.

Table 1. Chemical composition of Nimonic-90.

Elements	C	Si	Mg	Cr	Ni	Ti	Al	Co	Fe
% Weight	0.08	0.13	0.018	18.1	58	2.4	1.09	18.5	0.82

Similarly, for performing the turning experiments, chemical vapour deposition (CVD) coated carbide inserts with a layer of TiC, Al_2O_3 and TiN were used. The technical specification of the tool is presented in Table 2. Note that the length of cut used was 50 mm and for each cut a fresh cutting edge was used.

Table 2. Cutting insert specifications.

Insert Part Number	CNMG 120408CQ
Rake angle	5°
Relief angle	0°
Nose radius	0.8 mm
Lead angle	45°
Point angle	80°

2.3. Process Parameters

The selection of input parameters was based on the experience of local small and medium-sized enterprises (SMEs), specially involved in machining of Nimonic-90. The selected reposes i.e., cutting speed (V), feed rate (F), depth of cut (DOC) and frequencies (f) chosen for the experimental study are shown in Table 3. Note that for the UAT process, $V_c = 2\pi a f > V = \pi D N$ should be satisfied where a is amplitude, f is frequency, D is diameter of the workpiece, and N is rotating speed (rpm) of the spindle. If the cutting velocity V, exceeds the critical cutting velocity, V_c, the UAT process effectively reduces to a conventional machining process.

Table 3. Range and levels of process parameters.

Parameters	Range		
	Level 1	Level 2	Level 3
Cutting speed (m/min)	27.14	40.77	61.14
Feed rate (mm/rev)	0.11	0.22	0.33
Depth of cut (mm)	0.1	0.2	0.3
Frequency (kHz)	20	18	0 (conventional)
Amplitude (µm)		10	

2.4. Design of Experiment

The turning tests were carried out by considering a Taguchi L9 orthogonal array (as presented in Table 4). According to this design, a total of nine experiments with three repetitions were conducted. Then after, the analysis of variance (ANOVA) test (using the Minitab 18 software, State College, PA, USA) was implemented on the experimental results. The experimental procedure with complete details is exhibited in Figure 2.

Table 4. Design and experimental results of the L$_9$ orthogonal array.

Sr. No.	Control Variables				Average Responses	
	V (m/min)	F (mm/rev)	DOC (mm)	f (kHz)	R$_a$ (µm)	P (W)
1	27.14	0.11	0.1	20	0.37	288.67
2	27.14	0.22	0.2	18	1.56	337.33
3	27.14	0.33	0.3	0	2.21	308.33
4	40.77	0.11	0.2	0	1.06	302.67
5	40.77	0.22	0.3	20	0.9	335.67
6	40.77	0.33	0.1	18	1.14	312.67
7	61.14	0.11	0.3	18	0.64	413.67
8	61.14	0.22	0.1	0	0.67	349.67
9	61.14	0.33	0.2	20	1.26	335.00

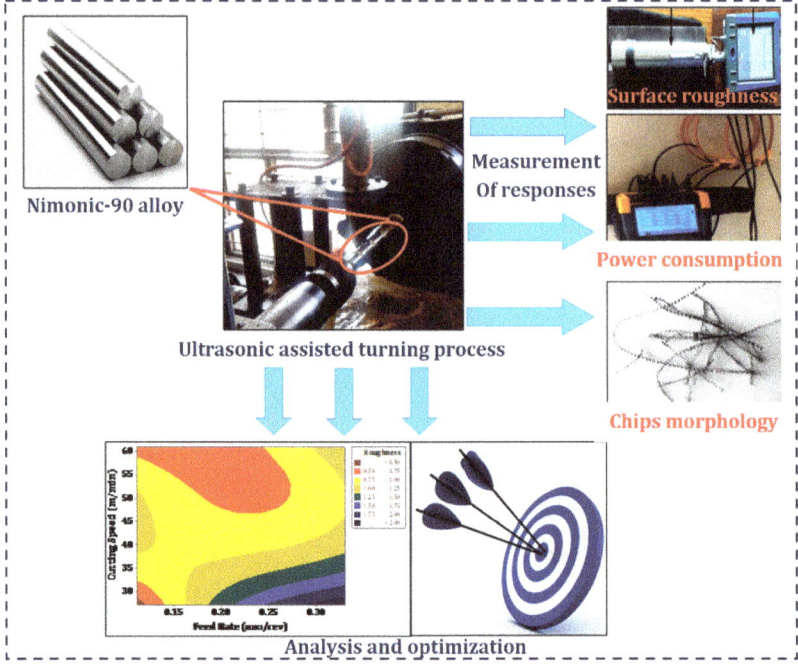

Figure 2. Experimental procedure with complete details.

2.5. Measurement of Responses

In this study, two important machining indices i.e., the average surface roughness (R_a) and power consumption (P) were measured after each experiment. For measurement of surface roughness values, the Taylor Hobson Surface roughness tester (AMETEK, Leicester, UK) was used. The power consumption (P) after each cut was measured with fluke power analyzer 435 series.

3. Nature-Inspired Algorithms

This section describes the overview of implemented algorithms i.e., particle swarm optimization, simplex method and hybrid PSO-SM, respectively. The working principle and procedure are discussed as per the following.

3.1. Particle Swarm Optimization (PSO)

PSO is categorized as the nature inspired-optimization algorithm in which the problem of linear and non-linear programming has been successfully solved [32]. Two paramount terms i.e., particles position as well as velocity has been recognized in the status of PSO method [18]. The *ith* particle position and its velocity in the d-dimensional search space are well described with the following Equations (1) and (2), respectively.

$$X_i = [x_{i,1}, x_{i,2}, \ldots \ldots x_{i,d}], \qquad (1)$$

$$V_i = [v_{i,1}, v_{i,2}, \ldots \ldots v_{i,d}], \qquad (2)$$

where, X_i and x_i up to the *d*th terms are integral values related to the position of particles, V_i, $v_{i,1}$, … $v_{i,d}$ are the velocity values of particles.

In the PSO method, every particle consists of an ideal position (*pbest*) also known as location with respect to the individual ideal values at particle interval of time, t. The *pbest* (Pb_i) is calculated with the help of Equation (3).

$$Pb_i = [pb_{i,1}, pb_{i,2}, \ldots \ldots \ldots pb_{i,d}]. \tag{3}$$

Similarly, the global ideal value (*gbest*) of each particle is termed by Pbg that generally shows the best or ideal particle at time, t. After that, the Equation (4) is used to evaluate the updated velocity of every particle [33,34].

$$v_{i,j}(t+1) = wv_{i,j}(t) + c_1 r_1 (pb_{i,j} - x_{i,j}(t)) + c_2 r_2 (pb_{g,i} - x_{i,j}(t)), \ j = 1, 2, \ldots d, \tag{4}$$

where, $v_{i,j}(t+1)$, $x_{i,j}(t)$ are function values, c_1 and c_2 represent coefficient values, inertia factor is denoted by w, r_1 and r_2 are termed as random variables having values of (0, 1). Therefore, Equation (5) is used to update the position of every particle.

$$x_{i,j}(t+1) = x_{i,j}(t) + v_{i,j}(t+1), \ j = 1, 2, \ldots d. \tag{5}$$

In general, the $v_{i,j}$ in the Equation (4) of every component is expressed in terms of $-v_{max}$ to v_{max}. These values are used to control the tremendous routing of external particles during the search space. Then, the particles follow Equation (5) and the positions of particles are updated towards a newer position [35]. Hence, the process is worked again and again until a global optimal solution is achieved.

3.2. Simplex Method (SM)

In this paper, the simplex method modified by Nelder and Mead, in 1965, was used to tackle the constrained and unconstrained optimization problems [35]. In this method, firstly the n input values at the polyhedron phase is considered and further the $n + 1$ points with R^n series are applied to establish the mathematical model. After that, the initial simplex changes its position i.e., moves, contracts and expands because of their series of primary geometric transformations, respectively. Then, the lower which also knows as the worst point (X_w) at every iteration is calculated by ordering and classifying the vertices values as $X_1, X_2, \ldots, X_n, X_{n+1}$, so that the solution is $f(X_1) < f(X_2) < \ldots f(X_n) < f(X_{n+1})$.

The value of objective function in the simplex method is decided as per the user requirements i.e., whether to minimize it or maximize it. For minimization, the variable with the largest objective value is used for a new reflection and the ideal point value has been placed approximately in the negative gradient direction [36]. For instance, X_1 represents the ideal point, X_{n+1} is termed as the worst or lowest point, X_n describes another worst point and so on. Moreover, the centroid point (X_c) of the n ideal solutions excluding X_{n+1} is calculated. In the end, the lowest or worst point is reflected in Equation (6) and latest point (X_r) is obtained. In addition, at this point, if the function i.e., $f(X_1) \leq f(X_r) < f(X_n)$ and boundary conditions are not desecrated, then the reflection takes place at an ideal region of search space and the replacement of the lower or worst point X_{n+1} is made with (X_r), hence the iteration stops working. Similarly, the other behavior i.e., expansion, contraction, shrinkage, movement of variables are calculated by Equations (7)–(10). Note that, the objective or fitness function is computed at each point of the method and the complete process is processed again and again until the final solution has been achieved.

$$X_r = X_c + \rho(X_c - X_{n+1}), \tag{6}$$

$$X_e = X_c + \gamma(X_r - X_c), \tag{7}$$

$$X_{cont1} = X_c - \gamma(X_c - X_{n+1}), \tag{8}$$

$$X_{cont2} = X_c + \gamma(X_r - X_c), \tag{9}$$

$$P_i = X_1 + \sigma(X_i - X_1), \ i = 2, \ldots, n+1, \tag{10}$$

where, $(X_i, P_i, \ldots, P_{n+1})$ reflects the new vertices, X_e, X_r, X_{cont1} and X_{cont2} shows the behavior at expansion, contraction and stretching.

3.3. Hybrid PSO-SM

PSO is known as the nature inspired algorithm, whereas the simplex method is referred to as an intelligent strategy that is effectively used to solve linear and non-linear problems [36,37]. The main aim of hybridization is to merge the advantages of both methods [38,39]. In addition, the searching of PSO is performed as per the Equations (1) and (2) and integrating PSO with simplex method may enhance the capacity to search the space towards the global optimal solution [36,37]. For instance, Equations (6)–(10) are used to show the behavior i.e., X_e, X_r, X_{cont1} and X_{cont2} and they are further divided by the swarm characteristics with their vector values i.e., $X_i(N_i, x_i, C_{1i}, C_{2i})$, $upto\ i = 1, \ldots, n+1$, where n is referred to the PSO parameters and N_i is an integer value.

The hybridization is performed in two ways: (1) the staged pipelining type in which each population size of PSO is processed by the stochastic optimization method and the simplex search is used for the improvement. Similarly, (2) the additional-operator type hybrid method in which the simplex search is directly applied to the population values and the probability of improvement is targeted by the user [36,37]. Therefore, in the paper, the hybridization of both methods is made by the staged pipelining method. The complete process is described below:

1. Initialization Step: The ideal positions of initials particles, generations of random N particles are selected and evaluated.
2. Repairing Step: The particles have been repaired that affects the boundary conditions by expressing the worst solution towards the ideal solutions. Moreover, terminate the damaged particles.
3. Searching Step: Equation (2) is used to search the individual position of each particle. The step is to select the better or ideal position and evaluate them.
4. Ranking: The obtained solution has been ranked according to their best fitness values, from the Equations (1) and (2).
5. Selection Step: Equation (2) is used to select the better position of each particle and the generation of ideal solution has been obtained.
6. Generation Step: Further, the $D+1$ points have been selected from the population based ranking solution and the initial simplex is well generated.
7. Simplex Method: It is applied on the highest $N+1$ particles and $(N+1)th$ has been updated.
8. Step 6 is replaced with Step 7 i.e., simplex method, so the best solution has been memorized, until the final solution has been achieved.

4. Results and Discussion

This section represents the prominent part of the paper. The statistical analysis was performed with the ANOVA test followed by influence of process parameters and estimation of optimum quality characteristics. The details of these analyses are discussed below:

4.1. Statistical Testing

In this analysis, the relationship between input variables and responses were made from the experimental results. The individual results of selected responses are shown in Table 4. The present statistical analysis was performed at 95% confidence interval (CI), which means at $\alpha = 0.05$ significance level. Further, the F-tests and p-value tests (less than 0.05) at 95% CI were performed on experimental values and are displayed in Tables 5 and 6, respectively. These tests were used to represent the effect of process parameters on responses. For instance, if the F-tests had high values, the more an effect was shown on the process variable. Moreover, the total effect was calculated by the percentage contribution values in respective tables.

Table 5. Analysis of variance of means for surface roughness.

Source	DF	Adj SS	Adj MS	F-Value	p-Value	%C
Cutting speed	2	1.26423	0.63211	134.6	0.001	17.051
Feed	2	3.26703	1.63351	347.83	0.002	44.065
Depth of cut	2	1.79147	0.89574	190.73	0.002	24.163
Frequency	2	1.00732	0.50366	107.25	0.000	13.586
Error	18	0.08453	0.0047			
Total	26	7.41459				

Table 6. Analysis of variance of means for and power consumption.

Source	DF	Adj SS	Adj MS	F-Value	p-Value	%C
Cutting speed	2	15,407	7703.4	20.86	0.000	39.798
Feed	2	2733	1366.3	3.7	0.045	7.0596
Depth of cut	2	6445	3222.3	8.73	0.002	16.648
Frequency	2	7483	3741.6	10.13	0.001	19.32
Error	18	6646	369.2			
Total	26	38,713				

Further, Table 5 shows the experimental results during ultra-sonic assisted turning of Nimonic-90 alloy under different cutting conditions. From the surface roughness analysis, i.e., the F-test showed that the maximum value i.e., 347.83 was for feed rate which meant the feed rate had the highest effect or highest contribution of 44.065% on surface roughness values followed by depth of cut (24.163%), cutting speed (17.051%) and frequency values (13.586%). A similar trend is observed by Reference [15] in machining of titanium alloy. Similarly, from power consumption analysis, the cutting speed (39.798%) had highest effect on power consumption followed by frequency (19.32%), depth of cut (16.648%) and feed rate (7.05%), respectively. In addition, the p-value test showed that the developed models were statistically significant for selected responses.

4.2. Influence of Process Parameters

Surface roughness: The contour effect plots were drawn to demonstrate the influence of different machining conditions on surface roughness values. From the previous statistical analysis, it was clearly noticed that the feed rate highly affected the surface roughness values. This statement is purely justified with the following Equation (11) which shows that the surface roughness is directly proportional to the square of feed rate as per the basic relation.

$$R_a = \frac{f^2}{8r} \qquad (11)$$

where, R_a is arithmetic roughness, f defines as a feed rate in mm/rev, r represents the nose radius in mm.

Therefore, the contour effect plots showing maximum effect of feed rate on surface roughness values were used in this work (as depicted in Figure 3a–c). Figure 3a claims that the surface roughness was minimum at lower values of cutting speed and feed rate. However, it swelled with the rise in feed rate, whereas it dwindled with the change in cutting speed values. The trends of these results were verified with the mechanism given in the Equation (11). Further, the increase in cutting speed lowered the formation of built-up edges at the tool surface. As a consequence, the low surface roughness values were achieved at higher values of cutting speed. Practically, these results may not fulfill the favorable conditions because high surface roughness values are not recommended to achieve the sound machining characteristics. The similar findings were reported by Reference [6]. Similarly, Figure 3b demonstrates the contour effect plot of depth of cut vs. feed rate. The observation results of these plots claim that the lower values of surface roughness were achieved at low depth of cut values and once the depth of cut was changed i.e., from minimum to maximum, undesirable machining surface characteristics were achieved. This is a very interesting fact as the tool area had a higher amount of contact with the subjected workpiece at higher depth of cut values and thereby more frictional heat was produced at the cutting zone. Besides, the heat was not dissipated in the proper manner from the cutting zone because of the intrinsic characteristics i.e., poor thermal conductivity of Nimonic-90 alloy. This high temperature resulted in high affinity to tool materials which may cause the welding of micro-particles of the workpiece to the cutting tool and consequently, reduces the surface finishing values [5].

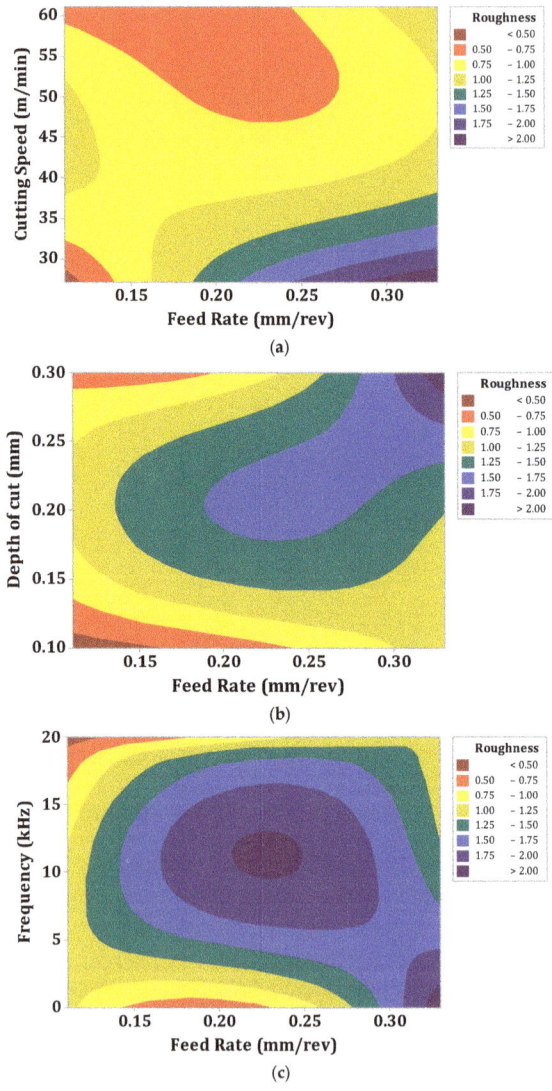

Figure 3. Influence of machining parameters on surface roughness values (**a**) Cutting speed vs. feed rate, (**b**) depth of cut vs. feed rate, and (**c**) frequency vs. feed rate.

Lastly, Figure 3c depicts the contour effect plot of frequency vs. feed rate. This plot exhibits that the lower value of surface roughness was achieved at the higher frequency of cutting tool i.e., at 20 kHz frequency. Moreover, this plot shows that the conventional machining process produced a higher value of surface roughness and it decreased with the change in frequency of cutting tool i.e., 20 < 18 < 0, respectively. This is generally related with the fact that the chips produced in the UAT process are smooth, thinner and shorter when compared to those obtained from conventional turning process (as shown in Figure 4). These smooth and short chips do not stick to the workpiece material and hence reduce the surface roughness values. Moreover, in the conventional turning process, longer chips are produced and these longer chips are undesirable which lead to entanglement of chips with the cutting tool and produces the rough surface. Further, the concept of smooth, thinner and shorter chips are directly related with shear angle and in the case of UAT it is increased. Hence, this increase in shear angle resulted in the decrease in chip thickness and as a consequence a good surface was produced with the increase in frequency of cutting tool (see Figure 5). In addition, the micrographs of chips during UAT and CT processes are presented in Figure 4. From this micro-graph analysis, it was interestingly seen that chips produced during the UAT process were regular while those produced from CT showed irregularities which manifested the poor surface quality of the machined surface. This is subjected to reason that when high frequency vibrations are exposed on cutting tool inserts, the removal of chips takes place because of the effect of vibrations and impact [40]. Moreover, the velocity of the stress wave, because of vibration of the cutting tool, produced a great impact on cutting velocity and hence, the inner stress broke the chips into small segments, and as a result soft, small and smooth chips were produced in the UAT process. Further, the tool work contact ratio was decreased with the increase in frequency of the tool. As a result, the temperature was reduced in the cutting zone because of the aerodynamic lubrication effect and hence the surface finishing was improved in the UAT process [5].

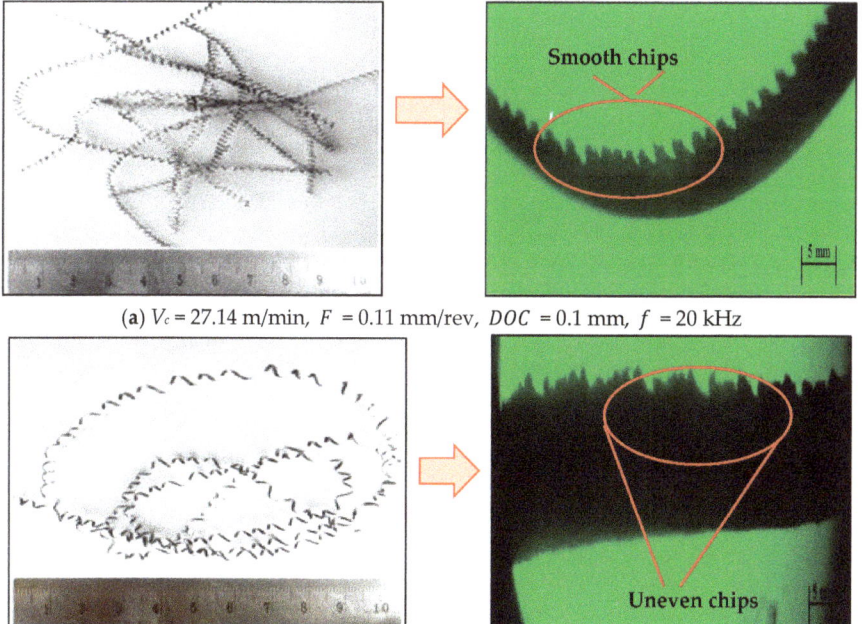

(a) V_c = 27.14 m/min, F = 0.11 mm/rev, DOC = 0.1 mm, f = 20 kHz

(b) V_c = 27.14 m/min, F = 0.33 mm/rev, DOC = 0.3 mm, f = 0 (conventional)

Figure 4. Macrographs and chip formed during machining of Nimonic-90 alloy under different conditions, (**a**) smooth and short chips, (**b**) longer chips.

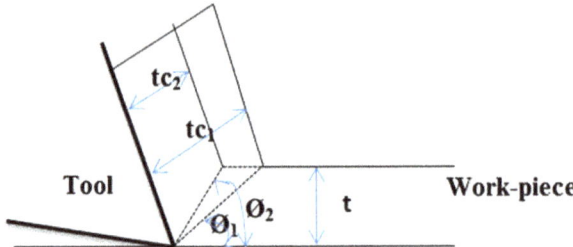

Figure 5. Effect of small ($\varnothing_1$) and large ($\varnothing_2$) shear angle on chip thickness (t_c) and length of shear plane for a given tool and un-deformed chip thickness (t) [26].

Power consumption: The power consumption is a very prominent aspect, especially during machining of hard-to-machine materials. It is also more important from a sustainable or environment point of view as it is directly related to the cutting forces, machine deformation and efficiency etc. Theoretically, it is a multiplication of main cutting force with the cutting speed values. Equation (12) is used to calculate the power consumption during each cut.

$$p = \frac{F_c \times V_c}{60000}, \tag{12}$$

where, p = power consumption in watts, $F_c = k_c \times a_e \times f$ is main cutting force in Newton and V_c is the cutting speed in m/min, a_e is the depth of cut in mm, f is the feed rate in mm/rev and k_c represents as a specific cutting energy coefficient, respectively. Therefore, the power consumption is modified with the following Equation (13). This combination directly states that the power consumption has the direct effect on cutting speed, feed rate and depth of cut. Hence, all these subjected parameters were considered during the power consumption analysis [40].

$$p = \frac{k_c \times a_e \times f \times V_c}{60000}. \tag{13}$$

Figure 6a–c depict the contour effect plots (a) cutting speed vs. feed rate, (b) cutting speed vs. depth of cut and (c) cutting speed vs. frequency. Figure 6a states that the combination of high cutting speed and low feed rate values are responsible for the high-power consumption. Equation (12) already justifies this statement.

This is a true fact that describes the enhancement of power consumption values with the increase in cutting speed because the values of power consumptions were totally dependent on the spindle's rotation per minute (RPM) and the increased in spindle speed consumed more power from the motor. Similarly, the increase in the feed rate value demonstrated the lower power consumption values. This is the general machining fact that high feed rate values lead to low machining time and with this low machining time the tool engagement time is reduced with the workpiece. Hence, low power is consumed during higher feed rates as compared with lower feed rate values.

After that the effect of depth of cut along with the cutting speed is presented in Figure 6b. From this analysis, it has been interestingly noted that the values of power consumption were slightly increased with the depth of cut. In addition, the value of power consumption decreased with the increase in frequency of cutting tool, as depicted in Figure 6c. In the UAT process, as the frequency increased the tool vibration period decreased and tool vibration period for the higher frequency was lower than tool vibration period for lower frequency. Consequently, the tool workpiece contact ratio for higher frequency was lower than tool workpiece contact ratio for lower frequencies and with this a low tool workpiece contact ratio the tool engagement time was reduced with the workpiece. Hence, the slightly low power was consumed during higher frequency as compared with lower frequency values. Another possible reason for this behavior was the increase in shear angle during machining with higher frequency which led to a decrease in cutting forces and consequently power consumption.

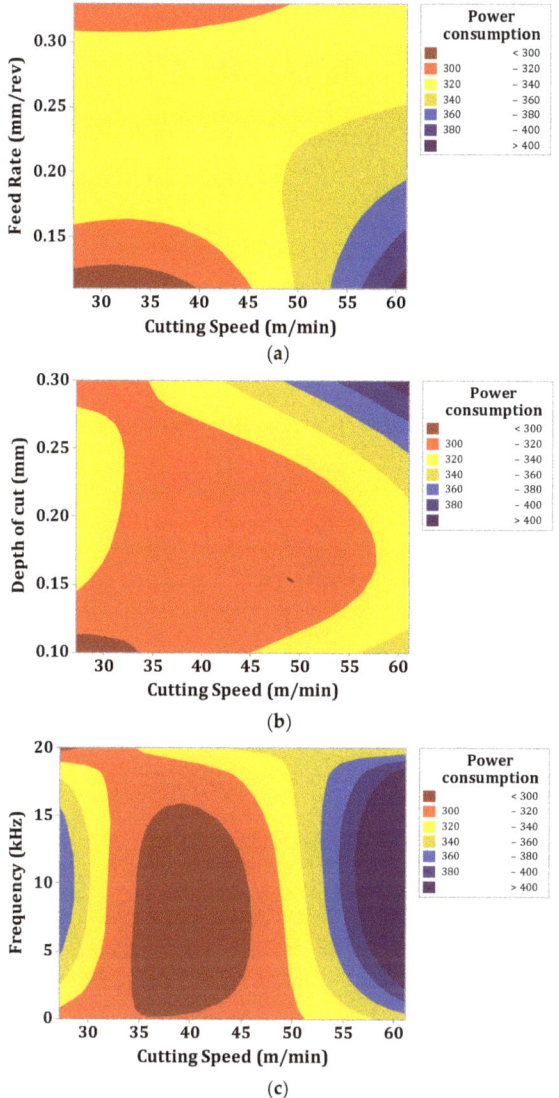

Figure 6. Influence of machining parameters on power consumption values (**a**) Feed rate vs. cutting speed, (**b**) depth of cut vs. cutting speed and (**c**) frequency vs. cutting speed.

4.3. Estimation of Optimum Quality Characteristics for Mono and Bi-Objective Optimization

The implemented algorithms were applied in two ways: (1) mono-objective (2) bi-objective. In mono-objective optimization, the process variables area was individually optimized in terms of input variables as well as responses. For this, the regression equations were directly used in the fitness function of algorithms. Whereas, in bi-objective optimization, one compromised or combined solution was derived for optimization of process parameters. The details of parameters initialization step followed by mono, bi-objective and algorithms confirmation are discussed below:

4.3.1. Basic Parameters: Learning Parametric Setting for PSO and hybrid particle swarm-simplex (HPSO-SM)

The nature-inspired algorithms have some specific parameters i.e., maximum and minimum weight, constants (W_{max}, W_{min}, C_1, C_2 and H) that explore its performance up to certain extent. In general, the role of these algorithm parameters is to decide the effectiveness of algorithm. The basic parameters used for PSO algorithm are shown in Tables 7 and 8. These parameters are selected based upon the user's experience and literature survey. For instance, in previously published work [41], the value of x is introduced in the range of 0–1.4, C_1 and C_2 are 2 and H is in the range of 5–10. Therefore, to effectively preserve the balance between local and global solution, the value of H is selected as 5. Besides, we have noticed that the selected parameters worked in a very efficient manner and significantly improved the efficiency of PSO. Moreover, the simplex method was also coupled with the PSO method and with this integration the performance characteristics with respect to the searching capability of these initial parameters were improved.

Table 7. Initial parameters of PSO.

Input Parameters	Value of Parameters
S, number of agent particles	50
Number of iterations	100
Maximum permissible inertia weight	1.4
Minimum permissible inertia weight	0.5
Maximum defined learning rate, $C_{1max} = C_{2max}$	2
Minimum defined learning rate, $C_{1min} = C_{2min}$	1.5
H	5

Table 8. Initial parameters of HPSO-SM.

Input Parameters	Value of Parameters
S, number of agent particles	50
Number of iterations	100
Maximum permissible inertia weight	1.156
Minimum permissible inertia weight	1.143
Maximum defined learning rate, $C_{1max} = C_{2max}$	1.345
Minimum defined learning rate, $C_{1min} = C_{2min}$	1.845
H	5

4.3.2. Mono-Objective Optimization

In this section, the main aim was to determine the individual optimum parametric setting which showed the minimum values of responses. For this, initially the regression equations for individual parameter in terms of variables were developed and these equations were further used as a fitness function in the MATLAB code of algorithms. The boundary conditions (ranges of input parameters) and objective functions (in terms of regression equations) used in the MATLAB code are discussed below. Boundary conditions: cutting speed: $27.14 \leq V_c \leq 61.14$, feed rate: $0.11 \leq f \leq 0.33$, depth of cut: $0.1 \leq ae \leq 0.3$ and frequency: $0 \leq frequency \leq 20$.

Objective functions:

$$Ra = 0.595 - 0.01487\ Vc + 3.85\ f + 2.62\ ae - 0.0186\ frequency \tag{14}$$

$$P = 231.0 + 1.671\ Vc - 74\ f + 178\ ae + 0.74\ frequency \tag{15}$$

Based upon these boundary conditions and objective functions, the optimization by using the general PSO and hybrid PSO-SM method have been performed. The optimized values selected for surface roughness and power consumption are presented in Table 9. Similarly, the convergence characteristics graph of each factor is shown in Figure 7. From the generated results, the selected values were: $V_{61.14}$, $F_{0.11}$, $DOC_{0.1}$ and f_{20} for minimum surface roughness values (0.35 µm) and $V_{27.14}$, $F_{0.33}$, $DOC_{0.1}$ and f_{20} for minimum power consumption values (270 Watt), where the subscript represents the value of the respective cutting parameter, respectively.

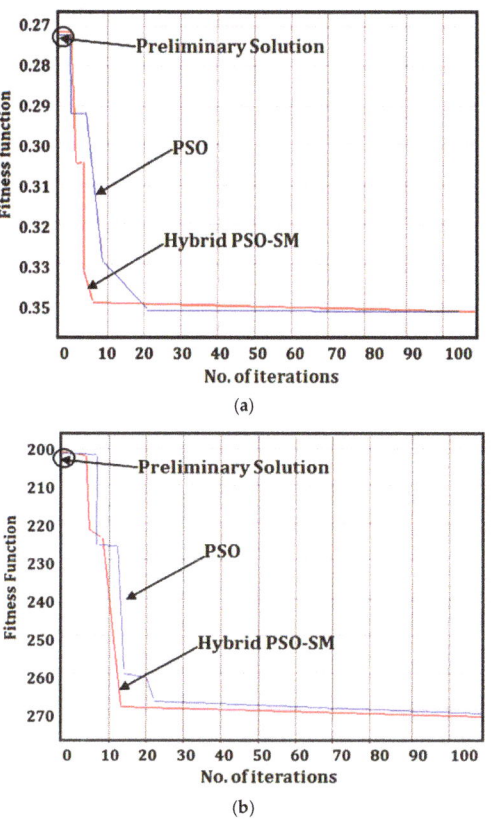

Figure 7. Convergence characteristics graphs for mono-objective optimization (**a**) minimum surface roughness value, (**b**) minimum power consumption value.

Table 9. Control variables and their selected values (for optimal response variables).

Control Variables	Optimal Values for Response Variables					
	Surface Roughness (µm)		Power Consumption (Watts)		Combined Values	
	PSO	HPSO-SM	PSO	HPSO-SM	PSO	HPSO-SM
Cutting speed (m/min)	61.14	61.14	27.14	27.14	40.77	40.77
Feed (mm/rev)	0.11	0.11	0.33	0.33	0.11	0.11
Depth of cut (mm)	0.1	0.1	0.1	0.1	0.2	0.2
Frequency (kHz)	20	20	20	20	20	20
Best solution	<0.35	>0.35	<270	>270	<0.8452	>0.8452
Mean solution	0.353	0.350	272.33	270.52	0.8572	0.8456
Standard deviation	0.458	0.352	0.583	0.383	0.522	0.324
Average time (s)	15	6	15	6	15	6
Success rate	80	90	80	90	80	80
Percentage error	5.34	1.24	6.3	1.5	6.34	1.4

4.3.3. Bi-Objective Optimization

In this section, the multi-objective optimization (in which more than a single factor is involved) with respect to the subjected process parameters was performed. The bi-objective optimization was performed in three manners: (1) maximization of responses, (2) minimization of responses and (3) grouping of minimization and maximization. In this work, the objective was to minimize the surface roughness and power consumption values. Hence, the minimization function was used as a fitness function in this work. The fitness function is initially developed by converting the all responses into single function and then the optimization is performed on this single objective function. The conversions of responses are made by using Equation (16):

$$X_{min} = \frac{W_1 \times X_1}{X_{1min}} + \frac{W_2 \times X_2}{X_{2min}} \tag{16}$$

where, X_{1min} = minimum value of surface roughness, X_{2min} is minimum value of power consumption, W_1 and W_2 are the weights assigned to the responses, i.e., 0.50 for each response. This combined function X_{min} was used as an objective function in MATLAB program and the optimization was performed by considering the same boundary conditions and initial learning parameters, respectively. From the generated results in Table 9, the optimum values selected were: $V_{40.77}$, $F_{0.11}$, $DOC_{0.2}$ and f_{20} for simultaneously minimizing (i.e., 0.8452) both the responses i.e., surface roughness and power consumption values. The convergence characteristics graph is shown in Figure 8.

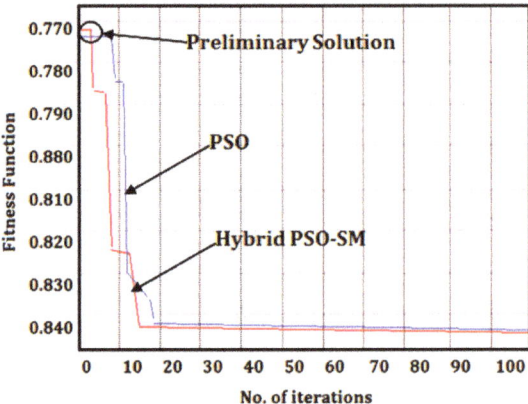

Figure 8. Convergence characteristics graphs for bi-objective optimization of combined objective.

4.3.4. Algorithms Confirmation

To ensure the efficiency of PSO and the hybrid PSO-SM method, comparative analysis in terms of percentage error, standard deviation, success rate and running time etc. was performed. The success rate, ideal values and running time were directly achieved from the MATLAB code, whereas the percentage error and standard deviation were calculated by Equations (17) and (18):

$$\%_{error} = \left| \frac{\#_{Experimental} - \#_{Thoeretical}}{\#_{Thoeretical}} \right| \times 100 \tag{17}$$

$$s = \sqrt{\frac{\sum_{i=1}^{N}(x_i - \bar{x})^2}{N-1}} \tag{18}$$

where, $\{x_1, x_2, \ldots, x_n\}$ are the observed values, $\bar{x}$ is the mean value of these observations, N is the number of observations.

After that, the 100 iterations were run at optimal conditions and the average data were calculated. From the given optimized results and comparative analysis (Table 9), it was noticed that the hybrid PSO-SM method performed better than the standard PSO method in mono as well as bi-objective optimization of the UAT process parameters. The success rate was 90% and running time wass only 6 s in the case of hybrid PSO-SM algorithm. Besides, the low values of percentage error and standard deviation of hybrid PSO-SM proved the high reliability and stability of algorithm towards the global optimal solution. Similarly, the results of the standard PSO showed that the success rate was 80%, running time was 15 s, percentage error and standard deviation were high for achieving the optimal solution. The performance of the hybrid PSO-SM method was high because the initial learning parameters of PSO were improved with the simplex method which was not in the case of the standard PSO. Another relevant aspect is that the independent swarm of PSO method i.e., vector X_i $(N_i, x_i, C_{1i}, C_{2i})$ $i = 1, \ldots, n + 1$, where n is the number of PSO parameters and N_i is an integer number computed with the steps of simplex method i.e., reflection, contraction, expansion and shrinkage and with these integration steps the searching capability of swarms are increased, and as a result swarms rapidly move towards the global optimal solution. Lastly, it is worth noting that the high stability, reliability and confidence of the hybrid PSO-SM method confirmed its effectiveness during optimization of the UAT process.

5. Conclusions

In this work, a robust technique in determining the optimal control parameters in UAT of Nimonic-90 alloy was presented with the goal of obtaining the lowest surface roughness and power consumption values. The optimization was performed in two ways: (1) mono-objective and (2) bi-objective by using a standard PSO and a hybrid PSO-SM, respectively. Further, in-depth analysis of the process mechanism by using contour plots was performed in the Results and Discussion section. From this work, the following conclusions may be drawn:

1. The performance of the hybrid PSO-SM was better in terms of lowering the running time, error and standard deviation as compared with the standard PSO method. The fact is that the initial learning parameters of PSO were improved with the simplex method and they may have increased the performance as compared with the standard PSO.
2. The results of the mono-objective optimization method showed that the cutting speed of 61.14 m/min, feed rate of 0.11 mm/rev, depth of cut of 0.1 mm and frequency of 20 kHz were ideal parameters for surface roughness values. Similarly, the cutting speed of 27.14 m/min, a higher value of feed rate of 0.33 mm/rev, lower value of depth of cut of 0.1 mm and frequency of 20 kHz were the optimum parameters for lowering the power consumption.
3. Likewise, the results of bi-objective optimization show that the medium value of cutting speed of 40.77 m/min, a lower feed rate of 0.11 mm/rev, a medium depth of cut of 0.2 mm and frequency of 20 kHz were the best settings for simultaneously lowering the responses.
4. From the statistical analysis, it has been noticed that the feed rate was the major factor affecting the surface roughness values, whereas the cutting speed claimed the most significant terms for power consumption.
5. The contour effect plots showed that the ultrasonic assisted turning process reduced the surface roughness and power consumption values as compared with the conventional turning process. This was due to the basic reason that the ultrasonic vibration produced the micro-chipping effect and thereby resulted in low surface roughness as well as power consumption values. Besides, the chips formed during the UAT processes were regular and fragmented when compared to those obtained from the CT process.
6. With the ultrasonic assisted machining, the surface roughness was improved by 5%–10% and the power consumption was reduced from 8%–10% when we compared the results with ordinary turning.

Author Contributions: The author contribution is presented here: Conceptualization, N.K.; Data curation, J.A.; Formal analysis, Z.L.; Funding acquisition, Q.S., N.K.; Investigation, M.K.G., G.K., R.M.; Methodology, M.M.; Software; M.K.G.

Funding: The uthors are grateful to the National Natural Science Foundation of China (no. 51875320), the Major projects of National Science and Technology (Grant No. 2019ZX04001031), the National Key Research and Development Program (Grant No. 2018YFB2002201), the Natural Science Outstanding Youth Fund of Shandong Province (Grant No. ZR2019JQ19), and the Key Laboratory of High-Efficiency and Clean Mechanical Manufacture at Shandong University, Ministry of Education.

Acknowledgments: The authors are grateful to the IITRAM for the funding given under M.Tech project fund to develop ultrasonic assisted turning facility at Advanced Manufacturing Laboratory (IITRAM Ahmedabad, India).

Conflicts of Interest: The authors declare no conflict of interest.

References

1. Gupta, M.K.; Sood, P.K. Machining comparison of aerospace materials considering minimum quantity cutting fluid: A clean and green approach. *Proc. Inst. Mech. Eng. Part C J. Mech. Eng. Sci.* **2017**, *231*, 1445–1464. [CrossRef]
2. Rao, M.S.; Venkaiah, N. Parametric Optimization in Machining of Nimonic-263 Alloy using RSM and Particle Swarm Optimization. *Procedia Mater. Sci.* **2015**, *10*, 70–79.
3. Ghoreishi, R.; Roohi, A.H.; Ghadikolaei, A.D. Evaluation of tool wear in high-speed face milling of Al/SiC metal matrix composites. *J. Braz. Soc. Mech. Sci. Eng.* **2019**, *41*, 146. [CrossRef]
4. Jang, D.Y.; Jung, J.; Seok, J. Modeling and parameter optimization for cutting energy reduction in MQL milling process. *Int. J. Precis. Eng. Manuf. - Green Technol.* **2016**, *3*, 5–12. [CrossRef]
5. Sharma, V.; Pandey, P.M. Optimization of machining and vibration parameters for residual stresses minimization in ultrasonic assisted turning of 4340 hardened steel. *Ultrasonics* **2016**, *70*, 172–182. [CrossRef]
6. Sajjady, S.A.; Abadi, H.N.H.; Amini, S.; Nosouhi, R. Analytical and experimental study of topography of surface texture in ultrasonic vibration assisted turning. *Mater. Des.* **2016**, *93*, 311–323. [CrossRef]
7. Sofuoğlu, M.A.; Çakır, F.H.; Gürgen, S.; Orak, S.; Kuşhan, M.C. Numerical investigation of hot ultrasonic assisted turning of aviation alloys. *J. Braz. Soc. Mech. Sci. Eng.* **2018**, *40*, 122. [CrossRef]
8. Puga, H.; Grilo, J.; Oliveira, F.J.; Silva, R.F.; Girão, A.V. Influence of external loading on the resonant frequency shift of ultrasonic assisted turning: Numerical and experimental analysis. *Int. J. Adv. Manuf. Technol.* **2019**, *101*, 2487–2496. [CrossRef]
9. Zhang, C.; Ehmann, K.; Li, Y. Analysis of cutting forces in the ultrasonic elliptical vibration-assisted micro-groove turning process. *Int. J. Adv. Manuf. Technol.* **2015**, *78*, 139–152. [CrossRef]
10. Ahmed, N.; Mitrofanov, A.V.; Babitsky, V.I.; Silberschmidt, V.V. Analysis of forces in ultrasonically assisted turning. *J. Sound Vib.* **2007**, *308*, 845–854. [CrossRef]
11. Maurotto, A.; Roy, A.; Babitsky, V.I.; Silberschmidt, V.V. Analysis of Machinability of Ti- and Ni-Based Alloys. In *Advanced Materials and Structures IV*; Trans Tech Publications ltd.: Zurich, Switzerland, 2012; Volume 188, pp. 330–338.
12. Maurotto, A.; Muhammad, R.; Roy, A.; Silberschmidt, V.V. Enhanced ultrasonically assisted turning of a β-titanium alloy. *Ultrasonics* **2013**, *53*, 1242–1250. [CrossRef] [PubMed]
13. Silberschmidt, V.V.; Mahdy, S.M.; Gouda, M.A.; Naseer, A.; Maurotto, A.; Roy, A. Surface-roughness Improvement in Ultrasonically Assisted Turning. *Procedia Cirp* **2014**, *13*, 49–54. [CrossRef]
14. Zhong, Z.W.; Lin, G. Ultrasonic assisted turning of an aluminium-based metal matrix composite reinforced with SiC particles. *Int. J. Adv. Manuf. Technol.* **2006**, *27*, 1077–1081. [CrossRef]
15. Nath, C.; Rahman, M. Effect of machining parameters in ultrasonic vibration cutting. *Int. J. Mach. Tools Manuf.* **2008**, *48*, 965–974. [CrossRef]
16. Vivekananda, K.; Arka, G.N.; Sahoo, S.K. Finite Element Analysis and Process Parameters Optimization of Ultrasonic Vibration Assisted Turning (UVT). *Procedia Mater. Sci.* **2014**, *6*, 1906–1914. [CrossRef]
17. Gupta, M.K.; Sood, P.K.; Sharma, V.S. Optimization of machining parameters and cutting fluids during nano-fluid based minimum quantity lubrication turning of titanium alloy by using evolutionary techniques. *J. Clean. Prod.* **2016**, *135*, 1276–1288. [CrossRef]

18. Raju, M.; Gupta, M.K.; Bhanot, N.; Sharma, V.S. A hybrid PSO–BFO evolutionary algorithm for optimization of fused deposition modelling process parameters. *J. Intell. Manuf.* **2019**, *30*, 2743–2758. [CrossRef]
19. Singh, G.; Gupta, M.K.; Mia, M.; Sharma, V.S. Modeling and optimization of tool wear in MQL-assisted milling of Inconel 718 superalloy using evolutionary techniques. *Int. J. Adv. Manuf. Technol.* **2018**, *97*, 481–494. [CrossRef]
20. Khanna, N.; Davim, J.P. Design-of-experiments application in machining titanium alloys for aerospace structural components. *Measurement* **2015**, *61*, 280–290. [CrossRef]
21. Khafaji, S.O.W.; Manring, N. Sensitivity analysis and Taguchi optimization procedure for a single-shoe drum brake. *Proc. Inst. Mech. Eng. Part C J. Mech. Eng. Sci.* **2019**, *233*, 3690–3698. [CrossRef]
22. Chen, F.C.; Huang, H.H. Taguchi-fuzzy-based approach for the sensitivity analysis of a four-bar function generator. *Proc. Inst. Mech. Eng. Part C J. Mech. Eng. Sci.* **2006**, *220*, 1413–1421. [CrossRef]
23. Shokrani, A.; Dhokia, V.; Newman, S.T. Investigation of the effects of cryogenic machining on surface integrity in CNC end milling of Ti–6Al–4V titanium alloy. *J. Manuf. Process.* **2016**, *21*, 172–179. [CrossRef]
24. Islam, M.N.; Pramanik, A. Comparison of Design of Experiments via Traditional and Taguchi Method. *J. Adv. Manuf. Syst.* **2016**, *15*, 151–160. [CrossRef]
25. Ezilarasan, C.; Kumar, V.S.; Velayudham, A.; Palanikumar, K. Modeling and analysis of surface roughness on machining of Nimonic C-263 alloy by PVD coated carbide insert. *Trans. Nonferrous Met. Soc. China* **2011**, *21*, 1986–1994. [CrossRef]
26. Makadia, A.J.; Nanavati, J.I. Optimisation of machining parameters for turning operations based on response surface methodology. *Measurement* **2013**, *46*, 1521–1529. [CrossRef]
27. Bhushan, R.K. Optimization of cutting parameters for minimizing power consumption and maximizing tool life during machining of Al alloy SiC particle composites. *J. Clean. Prod.* **2013**, *39*, 242–254. [CrossRef]
28. Gupta, M.K.; Sood, P.K.; Sharma, V.S. Machining Parameters Optimization of Titanium Alloy Using Response Surface Methodology and Particle Swarm Optimization Under Minimum Quantity Lubrication Environment. *Mater. Manuf. Process.* **2015**, *31*, 1671–1682. [CrossRef]
29. Sahu, N.K.; Andhare, A.B. Multiobjective optimization for improving machinability of Ti-6Al-4V using RSM and advanced algorithms. *J. Comput. Des. Eng.* **2018**, *6*, 1–12. [CrossRef]
30. Sathish, T. BCCS Approach for the Parametric Optimization in Machining of Nimonic-263 alloy using RSM. *Mater. Today Proc.* **2018**, *5*, 14416–14422. [CrossRef]
31. Youssef, H.A.; El-Hofy, H. *Machining Technology: Machine Tools and Operations*; Taylor and Francis Group, CRC press: Boca Raton, FL, USA, 2008.
32. Chen, L.; Monteiro, T.; Wang, T.; Marcon, E. Design of shared unit-dose drug distribution network using multi-level particle swarm optimization. *Health Care Manag. Sci.* **2019**, *22*, 304–317. [CrossRef]
33. Ameur, T.; Assas, M. Modified PSO algorithm for multi-objective optimization of the cutting parameters. *Prod. Eng.* **2012**, *6*, 569–576. [CrossRef]
34. Garg, A.; Tai, K.; Lee, C.H.; Savalani, M.M. A hybrid M5 -genetic programming approach for ensuring greater trustworthiness of prediction ability in modelling of FDM process. *J. Intell. Manuf.* **2014**, *25*, 1349–1365. [CrossRef]
35. Gaitonde, V.N.; Karnik, S.R. Minimizing burr size in drilling using artificial neural network (ANN)-particle swarm optimization (PSO) approach. *J. Intell. Manuf.* **2012**, *23*, 1783–1793. [CrossRef]
36. Nie, R.; Yue, J.H.; Deng, S.Q. Hybrid particle swarm optimization-simplex algorithm for inverse problem. In Proceedings of the 2010 Chinese Control and Decision Conference, Xuzhou, China, 26–28 May 2010; pp. 3439–3442.
37. Begambre, O.; Laier, J.E. A hybrid Particle Swarm Optimization—Simplex algorithm (PSOS) for structural damage identification. *Adv. Eng. Softw.* **2009**, *40*, 883–891. [CrossRef]
38. Fan, S.-K.S.; Zahara, E. A hybrid simplex search and particle swarm optimization for unconstrained optimization. *Eur. J. Oper. Res.* **2007**, *181*, 527–548. [CrossRef]
39. Zahara, E.; Kao, Y.-T. Hybrid Nelder–Mead simplex search and particle swarm optimization for constrained engineering design problems. *Expert Syst. Appl.* **2009**, *36*, 3880–3886. [CrossRef]

40. Xu, Y.; Zou, P.; He, Y.; Chen, S.; Tian, Y.; Gao, X. Comparative experimental research in turning of 304 austenitic stainless steel with and without ultrasonic vibration. *Proc. Inst. Mech. Eng. Part C J. Mech. Eng. Sci.* **2016**, *231*, 2885–2901. [CrossRef]
41. Parsopoulos, K.E.; Vrahatis, M.N. Recent approaches to global optimization problems through Particle Swarm Optimization. *Nat. Comput.* **2002**, *1*, 235–306. [CrossRef]

© 2019 by the authors. Licensee MDPI, Basel, Switzerland. This article is an open access article distributed under the terms and conditions of the Creative Commons Attribution (CC BY) license (http://creativecommons.org/licenses/by/4.0/).

Article

Experimental Investigation on Micro-Groove Manufacturing of Ti-6Al-4V Alloy by Using Ultrasonic Elliptical Vibration Assisted Cutting

Rongkai Tan [1], Xuesen Zhao [1], Tao Sun [1,*], Xicong Zou [2] and Zhenjiang Hu [1]

[1] Center for Precision Engineering, Harbin Institute of Technology, Harbin 150001, China; tanrongkai17@gmail.com (R.T.); zhaoxuesen@hit.edu.cn (X.Z.); lyhoo@hit.edu.cn (Z.H.)
[2] School of Mechatronics Engineering, Heilongjiang University, Harbin 150080, China; zouxicong@hlju.edu.cn
* Correspondence: taosun@hit.edu.cn; Tel.: +86-0451-8641-5244

Received: 1 September 2019; Accepted: 20 September 2019; Published: 21 September 2019

Abstract: The micro-groove structure on the planar surface has been widely used in the tribology field for improving the lubrication performance, thereby reducing the friction coefficient and wear. However, in the conventional cutting (CC) process, the high-quality, high-precision machining of the micro-groove on titanium alloy has always been a challenge, because considerable problems including poor surface integrity and a high level of the material swelling and springback remain unresolved. In this study, the ultrasonic elliptical vibration assisted cutting (UEVC) technology was employed, which aimed to minimize the level of the material swelling and springback and improve the machining quality. A series of comparative investigations on the surface defect, surface roughness, and material swelling and springback under the CC and UEVC processes were performed. The experimental results certified that the material swelling and springback significantly reduced and the surface integrity obviously improved in the UEVC process in comparison to that in the CC process. Furthermore, for all the predetermined depths of the cut, when the TSR (the ratio of the nominal cutting speed to the peak horizontal vibration speed) was equal to one of twenty four or one of forty eight, the accuracy of the machined micro-groove depth, width and the profile radius reached satisfactorily to 98%, and the roughness values were approximately 0.1 µm. The experimental results demonstrate that the UEVC technology is a feasible method for the high-quality and high-precision processing of the micro-groove on Ti-6Al-4V alloy.

Keywords: micro-groove; titanium alloy; surface integrity; material swelling and springback; ultrasonic elliptical vibration assisted cutting

1. Introduction

Titanium alloys have been increasingly used in aerospace, aviation, shipbuilding and biomedical fields because of their excellent properties such as high yield stress, high toughness, high strength to weight ratio, high creep and corrosion resistivity and good biocompatibility [1]. However, the surface hardness of titanium alloy is not usually high (approximately 30 HRC), which leads to a poor wear resistance of the titanium alloy part [2,3]. In practical application, the failure of titanium alloy part is often caused by its poor wear resistance [4,5]. Therefore, the studies on the improvement of the wear resistance of titanium alloy hold great significance for improving its reliability and service life. The micro-groove structure has been proven to be useful for improving the lubrication performance during the wet sliding contact condition, thereby reducing the friction coefficient and wear [6]. Numerous fabrication technologies have been proposed for the machining of the micro-groove, including lithographic machining [7], micro electrical discharge machining [8], micro electrochemical machining [9] and micro mechanical machining [10–12]. The lithographic machining technology was more suitable for processing tiny nano-scale structures with a straight sidewall and high aspect ratio due to its high resolution and low removal rate [7]. The micro electrical discharge machining only could be used for the conductive material, and the feature structure was limited by the geometry size of the electrode tool. In addition, the micro electrochemical machining was difficult to obtain the high quality finished surface [9]. Notably, a considerable amount of investigations show that the micro mechanical machining is the most widely used, because it has a large dimension span and allows a high degree of freedom for the structural design as compared with other methods [11,13].

Titanium alloys have been considered to be typical hard-to-cut materials owing to their inherent properties such as the high chemical reactivity, high strain hardening (work hardening), low thermal conductivity and small deformation coefficient [14]. Hence, the high-quality and high-precision machining of the micro-groove on titanium alloy has always been a challenge with the conventional cutting (CC) process [15–17]. During the processing of a micro-groove on titanium alloy using the CC process, the generated cutting heat could not be effectively dissipated through the workpiece or chips because of the low thermal conductivity. Furthermore, the effect of the coolant was greatly limited because the coolant was vaporized before reaching the cutting zone [17]. Thus, the high cutting temperature and thermal stress exist in the narrow cutting area, which are sufficiently high to induce the plastic side flow of melted materials, and leaving the materials behind the cutting edge. Yip et al. [16] reported that the materials expand and their volume increases when the melted materials solidify again, especially at the bottom and side location of cutting edge. Therefore, there is an obvious deviation between the profile shape of the generated micro-groove and the ideal case due to the effect of the material swelling and springback, as shown in Figure 1.

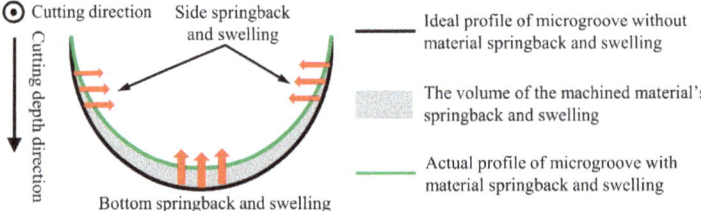

Figure 1. Schematic view of the material swelling and springback in the machining of micro-groove.

More remarkably, the effect of the material swelling and springback of titanium alloy was enhanced due to its low elastic modulus and thermal conductivity [16]. Furthermore, during the CC process, many surface defects such as adhered particles, a welded built-up edge (BUE) and plastic flow grooves appeared on the finished surface [15,18]. Previous studies have suggested that the high cutting temperature was the main factor that caused the formation of surface defect and the serious material swelling and springback [16,19]. Therefore, some coolant technologies, such as the coolant pressurized jet [20], cryogenic cooling [21], minimum quantity lubrication (MQL) [22] have been developed to reduce the cutting temperature. Further high-pressures increased the momentum of the coolant, which led to better heat transfers [17]. However, the environmental pollution and hazards to the operator occurred due to the high use of cutting fluids [23]. In the cryogenic cutting environment, the form accuracy of the machined part was difficult to guarantee and the hardness of workpiece was growing [24]. In MQL machining, a small amount of cooling/lubricating agents, as in the form of an aerosol, enter in the cutting zone for the advantage of effective cooling and lubrication at the tool–chip interface [22]. Krolczyk et al. [25] highlighted that this technology was a practical alternative to drying as well as flood cutting, which could reduce the use of cutting fluid and the manufacturing cost, thus achieving the eco-benign cutting environment. Maruda et al. [26,27] claimed that the MQL technology was performed better when the extreme pressure and anti-wear additives were added, which provided a significant improvement in the cutting tool wear rate. The MQL technology has been applied in many machining techniques, as well as in the wide range of workpieces (especially for Inconel alloys and titanium alloys) [28,29]. Nevertheless, the setup of MQL was complicated and troublesome [25]. Moreover, MQL technology required advanced equipment and caused a higher manufacturing cost, which limited its development [18]. Hence, a better machining method should be adopted to improve the machining performance of titanium alloy and reduce the level of material swelling and springback, thereby resulting in the high-quality and high-precision processing of micro-grooves on titanium alloy.

The ultrasonic elliptical vibration assisted cutting (UEVC) is a promising cutting technique which shows particular advantages over CC, like lower machining forces, higher machining stability, less tool wear and a better surface finish [30–32]. Furthermore, the investigations of the fabrication of micro-groove structures assisted by the UEVC technology have gained more attention from researchers in recent years. Kim et al. [33,34] investigated the machining characteristics of micro-grooves on aluminum and brass by using the UEVC technology. Their results indicated that, in the UEVC process, the machining quality of micro-grooves was improved and the cutting forces were significantly decreased compared with the CC process. Suzuki et al [35,36] performed the micro-groove machining experiments on brittle materials by using the UEVC technology. The results showed that, due to the influence of the UEVC technology, the cutting forces reduced, the critical cutting depth increased and the machining accuracy of the micro-groove improved. Moreover, Zhang et al [11,12] compared and analyzed the machining characteristics of micro-grooves on the stainless steel (0Cr18Ni9) and brass by using the UEVC and CC processes. Their results demonstrated that the machining quality of micro-grooves improved and the cutting forces were reduced in the UEVC process, especially for stainless steel. Similarly, the experimental results obtained by Kurniawan et al. [37] indicated that the UEVC process has shown many advantages in the machining of micro-grooves on steel alloy compared to the CC process.

Therefore, as discussed above, the UEVC process is effective to reduce the cutting forces, lower the surface roughness and attain better machining accuracy in micro-groove machining of easy-to-cut materials (such as copper, brass, and aluminum), brittle materials as well as steel materials. However, the UEVC technology is yet to be used in the machining of micro-grooves on titanium alloy. Moreover, a few studies have been conducted to assess the effect of material swelling and springback in micro-groove machining, which is the dominant factor to deteriorate the form accuracy and surface integrity of micro-grooves, especially for titanium alloy with the high level of material swelling and springback [16,38]. In this work, the UEVC technology is employed to investigate the machining characteristics of micro-grooves on Ti-6Al-4V alloy. In order to clarify the effective mechanisms of the UEVC technology in micro-groove machining of titanium alloy, comparative investigations on the surface defect, surface roughness, and material swelling and springback under the CC and UEVC processes are carried out. Moreover, the effects of different machining parameters on the machining quality of micro-grooves are compared with particular emphasis on the material swelling and springback. It is hoped that this work can provide a feasible method for high-quality and high-precision machining of micro-groove on titanium alloy.

2. Materials and Methods

2.1. The UEVC Principle

Figure 2a shows the schematic illustration of the UEVC process, and Figure 2b presents the schematic diagram of the UEVC device used in this study. The UEVC device worked at the 3rd longitudinal resonant mode and the 6th bending resonant mode. Four groups of longitudinal and bending piezoelectric (PZT) ceramics were stacked between metal blocks. When the excitation signals applied to the longitudinal PZT ceramics and the bending PZT ceramics, the 3rd longitudinal and 6th bending resonant modes of the device were inspired, respectively. An elliptical vibration trajectory was obtained at the tool tip by combining the two resonant vibrations with some phase shift. The detailed working principle of the UEVC device was given in our previous work [39].

As shown in Figure 2a, the cutting tool elliptically vibrates in the *xoz* plane, which is formed by the nominal cutting direction (i.e., *x*-axis) and the chip flow direction (i.e., *z*-axis). Relative to the workpiece, the transient position of the cutting tool can be described as follows:

$$x(t) = a\sin(2\pi ft) - V_C t, \tag{1}$$

$$z(t) = b\sin(2\pi ft + \theta). \tag{2}$$

Thus, the tool velocity relative to workpiece can be written as:

$$x'(t) = 2\pi fa\cos(2\pi ft) - V_C, \tag{3}$$

$$z'(t) = 2\pi fb\cos(2\pi ft + \theta). \tag{4}$$

where a and b are the vibration amplitudes in *x*-direction and *z*-direction, respectively. θ is the phase shift of the 3rd longitudinal resonant and 6th bending resonant, V_C is the nominal cutting speed and f is the vibration frequency. In addition, it should be noted that the ratio of the nominal cutting speed to the peak horizontal vibration speed is an important parameter in the UEVC process. In this study, this ratio is named as TSR. And TSR can be written as:

$$\text{TSR} = \frac{V_C}{2\pi fa}. \tag{5}$$

It should be noted that the intermittent cutting only occurs when TSR < 1. Nath et al. have demonstrated that the machined surface quality improved with the decrease of the value of TSR, and the TSR was usually set to be less than one-twelfth in the ultra-precision machining process [40]. Hence, three different values of TSR, namely one of twelfth, one of twenty-four and one of forty eight, are selected by varying the normal cutting speed for studying the influence of TSR on the machining quality of micro-grooves on titanium alloy. It should be noted that different TSR values in the UEVC process can be implemented by adjusting the radius of the machining area.

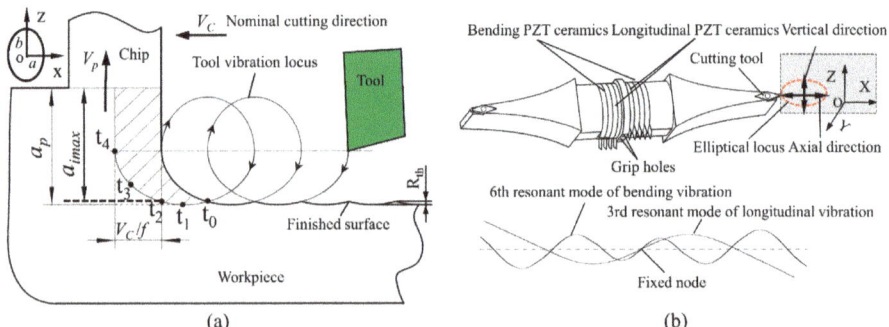

Figure 2. (a) Illustration of the ultrasonic elliptical vibration assisted cutting (UEVC) process, (b) Schematic diagram of the UEVC device.

In the UEVC process, the three most important features are the intermittent cutting, the reduced instantaneous uncut chip thickness and the reversal of friction force between the tool and the chip [13]. As shown in Figure 2a, the cutting motion starts at time t_0, and the cutting tool separates from workpiece at time t_4. Thus, in each cycle of vibration, the contact time between the tool and workpiece is only (t_4–t_0). The cooling medium was more likely to enter the cutting area, and the cooling effect was enhanced [41]. As a result, the cutting temperature was reduced. As presented in Figure 2a, the instantaneous uncut chip thickness continuously varies, and is maximal at time t_2. The maximal instantaneous uncut chip thickness (a_{imax}) is also smaller than the nominal one (a_p). This means that the cutting forces and cutting temperature during the UEVC process are smaller than in CC process. At time t_3, the velocity of the cutting tool in the z direction is equal to the velocity of the chip flow. This can be represented by the following equation:

$$z'(t_3) = 2\pi f b \cos(2\pi f t_3 + \theta) = V_p. \qquad (6)$$

where V_p is the velocity of the chip flow and is considered as a constant. According to the vibration equations of the tool (Equations (3) and (4)), the velocity of the tool in the z direction is increased in time period (t_4–t_3). Therefore, during the time period of (t_4–t_3), the friction force between the chip and tool is reversed. That means the friction force and the velocity of the chip flow have the same direction, which is conducive to break the chip and pull the chip away from the workpiece, and the results in a remarkable decrease in the cutting force, the suppression of the tool chatter, and an increase in the nominal shear angle [40,42]. In the UEVC process, the cutting forces and cutting temperature are reduced, the tool chatter is suppressed, and the chips are smoothly removed. Thus, the UEVC technology maybe a promising machining method to improve the machining performance of titanium alloy and reduce the level of material swelling and springback, thereby resulting in the high-quality processing of micro-grooves on titanium alloy.

2.2. Experimental Setup

In the experimental setup, a typical titanium alloy Ti-6Al-4V alloy, was chosen. The physical properties of Ti-6Al-4V alloy are listed in Table 1. The workpiece is held by the vacuum chuck attached on the spindle of the home-made ultra-precision machine tool, as shown in Figure 3. The machine tool mainly consists of an aerostatic spindle and two horizontal hydrostatic slideways. The UEVC device was positioned on the z-axis, and the high-precision adjustment platform was used to achieve the height adjustment of the UEVC device. It should be noted that the UEVC device could be considered as a traditional tool holder when it was not powered. The samples were round pie with the diameter of 50 mm and the height of 20 mm. The pre-turning was first completed with the CC process. Following this, a series of micro-groove machining experiments under different machining conditions were carried out. The experimental conditions and cutting parameters are listed in Table 2.

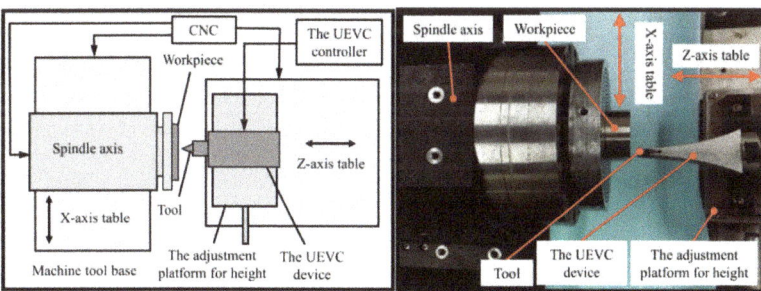

Figure 3. Diagram and picture of experimental setup.

Figure 4 displays the schematic view of the UEVC process in the machining of the micro-groove and the enlarged view of the generated micro-groove. In this study, the cutting tool with a round nose was used. In the ideal situation, the profile shape of the generated micro-groove is anticipated as same as the tool profile. The theoretical value of the micro-groove depth (L_d) is equal to the predetermined depth of cut. The theoretical value of the micro-groove width is L_w, and it can be expressed as:

$$L_w = 2\sqrt{R^2 - (R - L_d)^2} \qquad (7)$$

where R is the nose radius of the tool, and L_d is the theoretical value of the micro-groove depth. Thus, the theoretical value of the micro-groove width can be obtained from Equation (7).

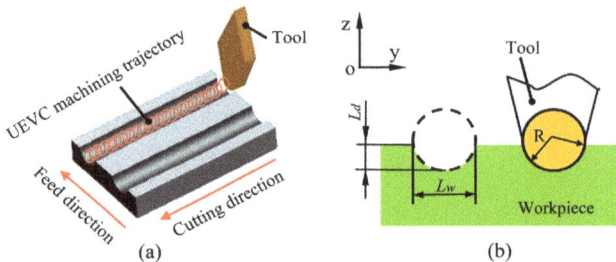

Figure 4. (a) Schematic view of the UEVC process in the machining of the micro-groove; (b) Enlarged view of the machined micro-groove.

Table 1. Physical properties of Ti-6Al-4V alloy.

Tensile Strength (MPa)	Elastic Modulus (GPa)	Hardness (HRC)	Density kg/m^3	Specific Heat Capacity (J·kg^{-1}·°C^{-1})	Thermal Conductivity (W·m^{-1}·°C^{-1})
902	16	32	4500	610	7.6

Table 2. Experimental parameters.

Processing Method		CC Process	UEVC Process
Vibration parameters	Frequency (kHz)	-	29.75
	Amplitude in cutting direction (μm)	-	6
	Amplitude in cutting depth direction (μm)	-	4
	Phase shift (°)	-	150
Cutting parameters	Spindle speed (r/min)	480	20
	Feed rate (μm/r)	500	500
	Depth of cut (μm)	2, 4, 6, 8	2, 4, 6, 8
Cutting Tool	Material	Carbide	
	Rake angle (°)	0	
	Clearance angle (°)	11	
	Nose radius (mm)	1.698	
Workpiece	Material	Ti-6Al-4V alloy	
	Dimension (mm)	Φ50 × L20	
Cutting fluid		No	

3. Results and Discussion

3.1. Material Swelling and Springback

In the micro-groove machining of titanium alloy, the suppression of material swelling and springback is crucial. To evaluate whether the UEVC technology, especially at different TSR, meets to suppress the material swelling and springback, the machining quality assessment of the machined micro-grooves is respectively performed by ultra-depth 3D microscopy system (VHX 1000E; Keyence, Osaka, Japan) and a white light interferometer (NewView 5000; Zygo, Middlefield, CT, USA). The machining quality of the micro-grooves with the predetermined depth of cut of 8 μm produced by different processing methods was first analyzed. Figure 5 shows the micrographs of the generated micro-grooves. As shown in Figure 5a, the clear, obvious and straight swelling marks appeared on the bottom and the side of the micro-groove produced by the CC process. Numerous surface defects were randomly distributed on the finished surface. On the contrary, the machined surfaces of the micro-grooves were clear and smooth, with no surface defect and swelling marks under the UEVC process with different TSR, as shown in Figure 5b–d. These facts implied that the material swelling and springback effect was very obvious in the generated micro-groove produced by CC, and the material swelling and springback effect was obviously suppressed by using the UEVC technology. The underlying reasons could be explained by analyzing the basic function mechanisms of the UEVC technology. As discussed in Section 2.1, the intermittent cutting and the reduction of instantaneous cutting thickness existed in the UEVC process, which caused small cutting forces and resulted in little cutting heat. Furthermore, the reversed friction force led to the increase of the nominal shear angle, so the cutting forces and friction force could be further reduced. On the other hand, the intermittent cutting mechanism provided a more favorable heat dissipation condition. Thus, the cutting temperature in the UEVC process was far lower than the CC process, which resulted in the prominent reduction in the level of material swelling and springback during the processing of micro-grooves on titanium alloy.

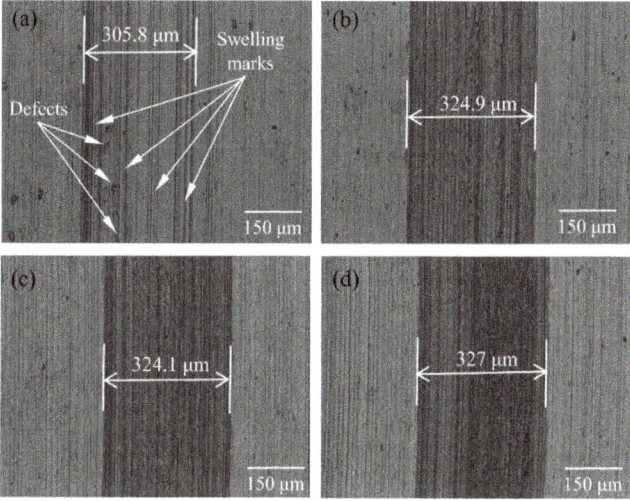

Figure 5. Micrographs of machined micro-groove (the predetermined depth of cut was 8 µm): (**a**) processed by CC, (**b**) processed by UEVC (TSR = 1/12), (**c**) processed by UEVC (TSR = 1/24) and (**d**) processed by UEVC (TSR = 1/48).

The reduction in the level of material swelling and springback was directly reflected in the dimensional parameters of the generated micro-groove. As displayed in Figure 5, the width of the micro-groove machined by the CC process was 305.8 µm, while the width of the micro-grooves machined by the UEVC process with different TSR were 324.9 µm (TSR = 1/12), 324.1 µm (TSR = 1/24) and 327 µm (TSR = 1/48), respectively. These implied that, during the processing of the micro-groove on titanium alloy using the CC process, the side swelling and springback effect was obvious and it introduced the larger volume of the material at the groove side, therefore, the width of generated micro-groove was smaller than the designed one (329.3 µm). During the processing of the micro-groove on titanium alloy using the UEVC process with different TSR values, the deviation values in the micro-groove width were small. The experimental results were all closed to the designed value. In order to further study the details of the generated micro-grooves, the cutting profiles were analyzed, as shown in Figure 6. Due to the effect of material swelling and springback, the ragged profile with wavy and vibration characteristics was obtained, as shown in Figure 6a. The depth of the micro-groove produced by the CC process was 6.4 µm, and it was much less than the designed value (8 µm), which indicated that a significant bottom swelling and springback appeared on the bottom surface of the generated micro-groove. Furthermore, the shape of the cutting profile was even distorted, and deviated largely from the tool shape. Thus, in the CC process, the actual parameters of the generated micro-groove greatly strayed from the designed parameters. In contrast, as shown in Figure 6b–d, the cutting profiles of the generated micro-grooves produced by UEVC process with different TSR displayed a smooth radius curve, and the depths were all close to 8 µm. It is worth noting that as the TSR value decreases, the cutting profile appears smoother. The cutting profile of micro-groove machined by the UEVC process with TSR = 1/48 has a few ripples and vibration marks.

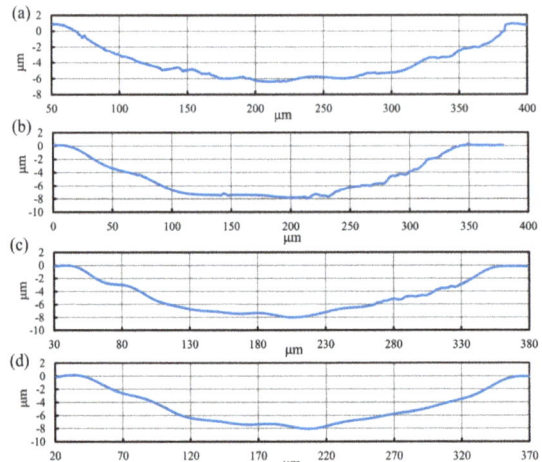

Figure 6. The cutting profiles of machined micro-groove (the predetermined depth of cut was 8 µm): (**a**) processed by CC, (**b**) processed by UEVC (TSR = 1/12), (**c**) processed by UEVC (TSR = 1/24) and (**d**) processed by UEVC (TSR = 1/48).

Table 3. The designed parameters and experimental results of the machined micro-grooves.

Test No.	Parameters of the Micro-Groove	Designed Values	Results of CC Process	Results of UEVC Process (TSR = 1/12)	Results of UEVC Process (TSR = 1/24)	Results of UEVC Process (TSR = 1/48)
1	Micro-groove depth	2	1.74	1.91	1.96	1.96
	Micro-groove width	164.8	152.2	158.9	162.3	163.5
	Profile radius	1698	1709	1653.4	1681	1704.9
2	Micro-groove depth	4	3.32	3.89	3.91	3.94
	Micro-groove width	233	214.9	228.3	229	232.3
	Profile radius	1698	1740.4	1676.8	1678.5	1712.5
3	Micro-groove depth	6	4.91	5.89	5.91	5.97
	Micro-groove width	285.2	264.7	279.6	284.7	284.1
	Profile radius	1698	1786.2	1662	1717.3	1693
4	Micro-groove depth	8	6.4	7.92	7.91	7.98
	Micro-groove width	329.3	305.8	324.9	324.1	327
	Profile radius	1698	1829.7	1670	1663.9	1678.9

For further investigating the effect of the predetermined depth of cut on material swelling and springback under different machining conditions, a series of experiments about micro-groove machining of titanium alloy were carried out. The measured parameters of the generated micro-grooves are presented in Table 3. In addition, according to the experimental data in Table 3, the deviations between the actual parameters of the generated micro-grooves and the theoretical values were analyzed. During the processing of the micro-groove on titanium alloy using the CC process, the deviations of the micro-groove width did not change significantly with the increase of the predetermined depth of cut, as shown in Figure 7a. Noting that, for all the predetermined depths of the cut, the deviations of the micro-groove width in the CC process were almost 7–8%, while the deviations of the micro-groove width in the UEVC process were smaller than 4%. In particular for the UEVC process with TSR = 1/48 and TSR = 1/24, the deviations of the micro-groove width were smaller than 2%. These facts indicate that the UEVC technology can effectively suppress the side swelling and springback during the micro-groove machining of titanium alloy, and in the UEVC process, TSR = 1/24 is recommended because the smaller TSR value means lower processing efficiency.

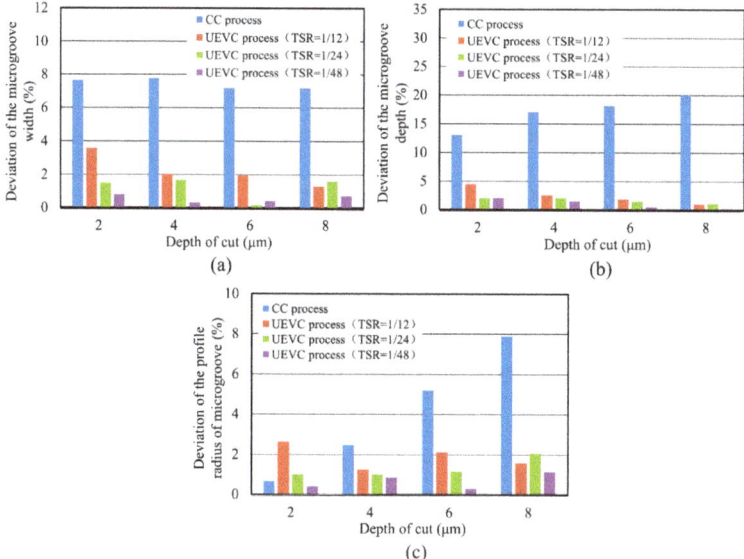

Figure 7. (a) Deviation of the micro-groove width, (b) deviation of the micro-groove depth, and (c) deviation of the profile radius of micro-groove.

Figure 7b presents the deviations of the micro-groove depth under the different processing conditions. In the CC process, the deviation value gradually increases with the increase of the predetermined depth of the cut. The deviation value reached 20%, when the predetermined depth of the cut was 8 μm. This can be explained by the fact that the increase of the predetermined depth of the cut led to an obvious increase in cutting forces and the friction force, thus, the cutting temperature was higher, which greatly increased the level of material swelling and springback. Conversely, the deviations of the micro-groove depth in the UEVC process were smaller than 5% for all predetermined depths of the cut, and were smaller than 2% when TSR = 1/48 or TSR = 1/24. Moreover, the deviations of the micro-groove depth were smaller with the decrease of TSR. This is because the decrease of TSR could further promote the superiority of the UEVC technology, that is, smaller cutting forces and a better heat dissipation condition could be obtained with smaller TSR. Moreover, it is worth noting that, with the increase of the predetermined depth of the cut, the suppression effect of the UEVC technology on material swelling and springback was more obvious. This happened maybe due to the following reasons. In the CC process, the larger depth of the cut means larger cutting forces and a higher cutting temperature. Thus, with the increase of the predetermined depth of the cut, the material swelling and springback becomes the dominant factor in the deviation of the micro-groove depth. However, as discussed above, the effect of material swelling and springback was effectively suppressed by using the UEVC technology. This means the dominant factor in the deviation of the micro-groove depth was eliminated. Therefore, the deviation of the micro-groove depth gradually decreased with the increase of the predetermined depth of the cut.

As can be seen from Figure 7c, it is clear that, in the CC process, the deviation of the profile radius of the machined micro-groove rapidly increased with the increase of the predetermined depth of the cut. However, since the UEVC technology has a good suppression effect on the material swelling and springback in both of the side and bottom during the micro-groove machining of titanium alloy, the deviations of the profile radius of the micro-groove were small for all predetermined depths of the cut. It is worth noting that, for the UEVC process with TSR = 1/48 or TSR = 1/24, the deviations of the profile radius of the micro-groove were smaller than 2%. This indicated that the accuracy of the machined micro-groove could be ensured during the processing of the micro-groove on titanium alloy by using the UEVC process.

3.2. Surface Integrity

Surface integrity is an important indicator for evaluating the machined surface quality and has a large impact on the reliability and service life of the part. Figure 8 shows the surface topography views of the machined micro-grooves with the predetermined depth of cut of 8 μm. It can be observed from Figure 8a,b, that a series of surface defects including irregular ridges, cavities and tearing marks appeared on the micro-groove machined by using the CC process. This can be attributed to the following reasons. During the processing of the micro-groove on titanium alloy using the CC process, the high cutting temperature, large cutting forces and the existence of friction force, which elevated the softening degree of the workpiece material, induced the ruleless vibration of the cutting tool and promoted the formation of the built-up edge (BUE) [18]. The ruleless vibration of the cutting tool and the generated BUE promoted the formation of cavities and tearing marks. In addition, as discussed in Section 3.1, the level of material swelling and springback was very high in the CC process. The irregular ridges were generated by the combined effect of the high level of material swelling and springback and the ruleless vibration of the cutting tool. It should be noted that these defects vastly degraded the surface integrity and the accuracy of the generated micro-groove.

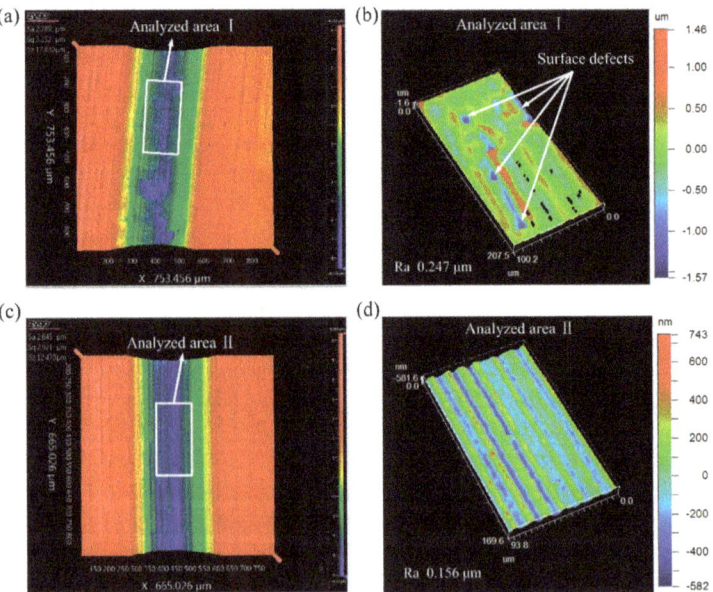

Figure 8. Cont.

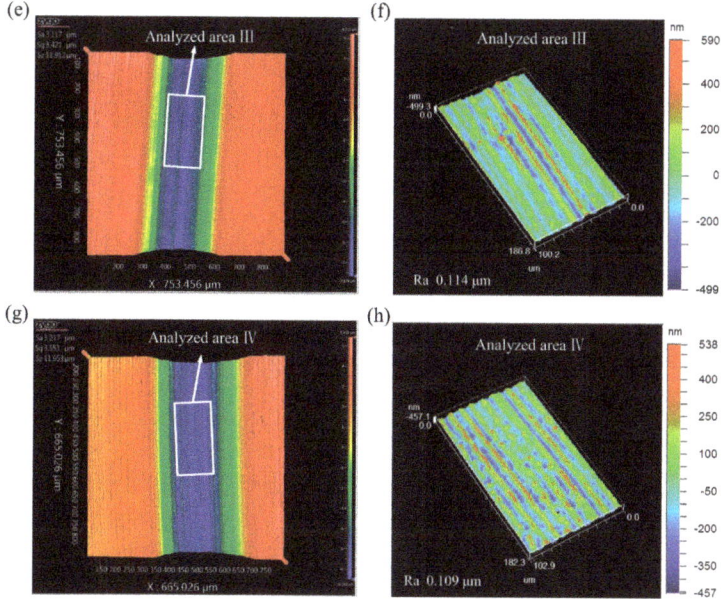

Figure 8. The views of machined micro-grooves (the predetermined depth of cut was 8 μm): (**a**) surface topography of machined micro-groove during CC; (**b**) surface roughness of analyzed area I; (**c**) surface topography of machined micro-groove during UEVC (TSR = 1/12); (**d**) surface roughness of analyzed area II; (**e**) surface topography of machined micro-groove during UEVC (TSR = 1/24); (**f**) surface roughness of analyzed area III; (**g**) surface topography of machined micro-groove during UEVC (TSR = 1/48); (**h**) surface roughness of analyzed area IV.

In contrast, the machined micro-grooves produced by the UEVC process displayed nearly no surface defects on the bottom and side surfaces, as shown in Figure 8c–h. This can be attributed to the following two reasons. The first reason is that, compared to the CC process, the cutting forces and friction force were small due to the features of the UEVC technology such as the intermittent cutting, the reduction of instantaneous cutting thickness and the reversal of the friction force. Additionally, in the UEVC process, a favorable heat dissipation condition was provided because of its intermittent cutting nature. Consequently, the cutting temperature in the UEVC process was far lower than in the CC process, thereby resulting in the effective suppression for the ruleless vibration of the cutting tool and the formation of BUE. The second reason is that, the material swelling and springback was successfully suppressed during the processing of the micro-groove on titanium alloy using the UEVC process, as discussed in Section 3.1. Hence, there were almost no surface defects on the machined surface, as shown in Figure 8d,f,h. Simultaneously, it can be seen that some small and straight ridges also appeared on the machined micro-grooves produced by the UEVC process, especially for TSR = 1/12. This may be caused by the micro notches appearing on the cutting edge of the tool. The effect of the micro notches on the machined surface quality decreased with the decrease of TSR. The reason is that the smaller TSR means the smaller cutting forces and instantaneous uncut chip thickness. As shown in Figure 8f,h, the bottom surface roughness of the machined micro-groove produced by the UEVC process with TSR = 1/24 and TSR = 1/48 were 0.114 μm and 0.109 μm, respectively. It should be noted that the size of the analysis area is approximately 100 μm × 180 μm. However, the bottom surface roughness of the machined micro-groove produced by the CC process was 0.247 μm, which was more than two times than the roughness in the UEVC process with TSR = 1/24 or TSR = 1/48. Therefore, the application of the UEVC technology led to an obvious improvement of the machined micro-groove in surface integrity, which was consistent with the experimental result from the research literature [33–37].

Figure 9 presents the bottom surface roughness values of the machined micro-grooves produced by different processing conditions. For each generated micro-groove, the bottom surface roughness was tested five times and the recorded surface roughness was the average value. The measurement area size was approximately 100 μm × 180 μm. It can be seen that the surface roughness value increased quickly with the increase of the predetermined depth of the cut during the CC process. As discussed above, many surface defects including irregular ridges, cavities and tearing marks appeared on the machined micro-groove under the CC process. It is obvious that a bigger depth of the cut means larger cutting forces and friction force, thereby resulting in the higher cutting temperature. The higher cutting temperature caused more serious ruleless vibration of the tool and the material swelling and springback. Thus, it can be inferred that more machining defects appeared on the generated micro-groove with the increase of the predetermined depth of the cut. The appearance of above-observed machining defects lower the surface quality of the generated micro-groove. However, the surface roughness values did not obviously change with the increase of the predetermined depth of the cut during the UEVC process with TSR = 1/24 or TSR = 1/48. The roughness values were approximately 0.1 μm. These results were consistent with the previous discussion which stated that a significant suppression of the formation of the surface defect was due to the use of the UEVC technology. The experimental results certified that the reduction of material swelling and springback and the improvement of surface integrity could be achieved simultaneously by using the UEVC technology.

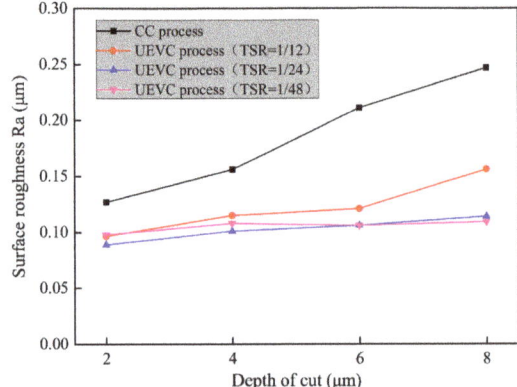

Figure 9. Influence of the predetermined depth of cut on the bottom surfaces roughness of the machined micro-grooves produced by different processing conditions.

4. Conclusions

In this work, the UEVC technology is firstly introduced into the micro-groove machining of Ti-6Al-4V alloy to improve the machining performance and reduce the level of material swelling and springback, thereby realizing the high-quality and high-precision processing of micro-groove. Based on the theoretical analysis and the experiment results, the main conclusions can be drawn as follows:

1. During the processing of the micro-groove on titanium alloy using the CC process, the clear, obvious and straight swelling marks appeared on the bottom and the side of the generated micro-groove. These facts implied that the material swelling and springback effect was very obvious. Further, with the increase of the predetermined depth of the cut, the deviation of the micro-groove width did not change significantly, while the deviation of the micro-groove depth gradually increased. Moreover, the deviation of the profile radius of the micro-groove rapidly increased with the increase of the predetermined depth of the cut. Remarkably, the profile shape of the generated micro-groove was distorted, and deviated largely from the tool shape. Thus,

the actual parameters of the micro-groove machined by the CC process greatly strayed from the designed parameters.
2. During the processing of the micro-groove on titanium alloy using the UEVC process, no surface defect and no swelling mark appeared on the machined surfaces of the generated micro-grooves. The profile shape of the generated micro-grooves was smooth, and the profile became smoother with the decrease of TSR. This indicated that the material swelling and springback effect was obviously suppressed. As the predetermined depth of the cut increased, the deviations between the designed parameters and experimental results of the micro-groove width, depth and the profile radius did not significantly change. It is noticeable that, for all TSR, the percentage errors of the generated micro-groove parameters were smaller than 5%, and were smaller than 2% when TSR = 1/48 or TSR = 1/24. This indicated that the accuracy of the machined micro-groove improved significantly by using the UEVC technology.
3. A series of surface defects including irregular ridges, cavities and tearing marks appeared on the micro-groove machined by using the CC process, and these defects vastly degraded the surface integrity and the accuracy of the generated micro-groove. In contrast, the generated micro-grooves produced by the UEVC process displayed nearly no surface defects. Moreover, the bottom surface roughness of the machined micro-grooves increased quickly with the increase of the predetermined depth of the cut during the CC process. However, the bottom surfaces roughness of the machined micro-grooves did not obviously change with the increase of the predetermined depth of the cut during the UEVC process. The roughness values were approximately 0.1 µm, when TSR = 1/48 or TSR = 1/24. Therefore, the application of the UEVC technology led to an obvious improvement of the machined micro-groove in surface integrity.
4. The experimental results certified that the reduction of the material swelling and springback and the improvement of surface integrity can be achieved simultaneously by using the UEVC technology. The value of TSR has an obvious effect on the action mechanism of the UEVC technology, and the decrease of TSR can further promote the superiority of the UEVC technology. In this study, TSR = 1/24 is recommended. Further investigation should be made to obtain the preferable machined surface by optimizing the combination of processing parameters with the consideration of tool wear.

Author Contributions: Conceptualization, R.T. and X.Z. (Xuesen Zhao); data curation, X.Z. (Xicong Zou); formal analysis, R.T.; funding acquisition, T.S. and Z.H.; project administration, X.Z. (Xuesen Zhao) and T.S.; writing—review & editing, R.T.

Funding: This work was supported by the Science Challenge Project of China (Grant No. TZ2018006-0202-01) and the National Safe Academic Foundation of National Natural Science Foundation of China (Grant No. U1530106).

Conflicts of Interest: The authors declare no conflicts of interest.

Nomenclature

UEVC	ultrasonic elliptical vibration assisted cutting
TSR	the ratio of the nominal cutting speed to the peak horizontal vibration speed
a	the vibration amplitudes in x-direction
b	the vibration amplitudes in z-direction
a_{imax}	the maximal instantaneous uncut chip thickness
V_p	the velocity of the chip flow
L_w	the theoretical value of the micro-groove width
CC	conventional cutting
PZT	piezoelectric
Θ	the phase shift of the 3rd longitudinal resonant and 6th bending resonant
V_C	the nominal cutting speed
F	the vibration frequency

a_p the nominal instantaneous uncut chip thickness
L_d the theoretical value of the micro-groove depth
R   the nose radius of the tool

References

1. Gao, H.; Liu, X. Stability Research Considering Non-Linear Change in the Machining of Titanium Thin-Walled Parts. *Materials* **2019**, *12*, 2083. [CrossRef] [PubMed]
2. Astarita, A.; Durante, M.; Langella, A.; Squillace, A. Elevation of tribological properties of alloy Ti–6%Al–4% upon formation of a rutile layer on the surface. *Met. Sci. Heat Treat.* **2013**, *4*, 662–666. [CrossRef]
3. Yadav, S.; Kumar, A.; Paramesh, T.; Sunita, K. A review on enhancement of wear resistance properties of titanium alloy using nano-composite coating. *IOP Conf. Ser. Mater. Sci. Eng.* **2018**, *455*, 012120. [CrossRef]
4. Martinez, J.M.V.; Pedemonte, F.J.B.; Galvin, M.B.; Gomez, J.S.; Barcena, M.M. Sliding Wear Behavior of UNS R56400 Titanium Alloy Samples Thermally Oxidized by Laser. *Materials* **2017**, *10*, 830. [CrossRef] [PubMed]
5. Yuan, W. Effect of the bionic morphologies on bio-tribological properties of surface-modified layers on ti6al4v with Ni+/N+ implantation. *Ind. Lubr. Tribol.* **2018**, *70*, 325–330. [CrossRef]
6. Gachot, C.; Rosenkranz, A.; Hsu, S.M.; Costa, H.L. A critical assessment of surface texturing for friction and wear improvement. *Wear* **2017**, *372–373*, 21–41. [CrossRef]
7. Dhupal, D.; Doloi, B.; Bhattacharyya, B. Pulsed Nd: YAG laser turning of micro-groove on aluminum oxide ceramic (Al2O3). *Int. J. Mach. Tools Manuf.* **2008**, *48*, 236–248. [CrossRef]
8. Lim, H.S.; Wong, Y.S.; Rahman, M.; Lee, M.K.E. A study on the machining of high-aspect ratio micro-structures using micro-EDM. *J. Mater. Process. Technol.* **2003**, *140*, 318–325. [CrossRef]
9. Zhou, X.; Qu, N.; Hou, Z.; Zhao, G. Electrochemical micromachining of microgroove arrays on phosphor bronze surface for improving the tribological performance. *Chin. J. Aeronaut.* **2017**, *31*, 1609–1618. [CrossRef]
10. Xiaobin, D.; Tianfeng, Z.; Siqin, P.; Zhiqiang, L.; Qian, Y.; Benshuai, R.; Xibin, W. Defect analysis in microgroove machining of nickel-phosphide plating by small cross-angle microgrooving. *J. Nanomater.* **2018**, 1478649. [CrossRef]
11. Zhang, C.; Guo, P.; Ehmann, K.F.; Li, Y. Turning of microgrooves both with and without aid of ultrasonic elliptical vibration. *Mater. Manuf. Process.* **2015**, *30*, 1001–1009. [CrossRef]
12. Zhang, C.; Guo, P.; Ehmann, K.F.; Li, Y. Effects of ultrasonic vibrations in micro-groove turning. *Ultrasonics* **2016**, *67*, 30–40. [CrossRef] [PubMed]
13. Zhang, J.; Cui, T.; Ge, C.; Sui, Y.; Yang, H. Review of micro/nano machining by utilizing elliptical vibration cutting. *Int. J. Mach. Tools Manuf.* **2016**, *106*, 109–126. [CrossRef]
14. Liu, D.; Zhang, Y.; Luo, M.; Zhang, D. Investigation of Tool Wear and Chip Morphology in Dry Trochoidal Milling of Titanium Alloy Ti–6Al–4V. *Materials* **2019**, *12*, 1937. [CrossRef] [PubMed]
15. Tan, R.; Zhao, X.; Zhang, S.; Zou, X.; Guo, S.; Hu, Z.; Sun, T. Study on ultra-precision processing of Ti-6Al-4V with different ultrasonic vibration assisted cutting modes. *Mater. Manuf. Process.* **2019**, in press. [CrossRef]
16. Yip, W.S.; To, S. Reduction of material swelling and recovery of titanium alloys in diamond cutting by magnetic field assistance. *J. Alloys Compd.* **2017**, *722*, 525–531. [CrossRef]
17. Revuru, R.S.; Posinasetti, N.R.; Vsn, V.R.; Amrita, M. Application of cutting fluids in machining of titanium alloys-a review. *Int. J. Adv. Manuf. Technol.* **2017**, *91*, 2477–2498. [CrossRef]
18. Xiaoliang, L.; Zhanqiang, L.; Wentao, L.; Xiaojun, L. Sustainability assessment of dry turning Ti-6Al-4V employing uncoated cemented carbide tools as clean manufacturing process. *J. Clean. Prod.* **2019**, *214*, 279–289. [CrossRef]
19. Sun, F.J.; Qu, S.G.; Pan, Y.X.; Li, X.Q.; Li, F.L. Effects of cutting parameters on dry machining Ti-6Al-4V alloy with ultra-hard tools. *Int. J. Adv. Manuf. Technol.* **2015**, *79*, 351–360. [CrossRef]
20. Pramanik, A. Problems and solutions in machining of titanium alloys. *Int. J. Adv. Manuf. Technol.* **2014**, *70*, 919–928. [CrossRef]
21. Pereira, O.; Rodríguez, A.; Fernández-Abia, A.I.; Barreiro, J.; Lacalle, L.N.L.D. Cryogenic and minimum quantity lubrication for an eco-efficiency turning of AISI 304. *J. Clean. Prod.* **2016**, *139*, 440–449. [CrossRef]
22. Maruda, R.W.; Krolczyk, G.M.; Feldshtein, E.; Pusavec, F.; Szydlowski, M.; Legutko, S.; Sobczak-Kupiec, A. A study on droplets sizes, their distribution and heat exchange for minimum quantity cooling lubrication (MQCL). *Int. J. Mach. Tools Manuf.* **2016**, *100*, 81–92. [CrossRef]

23. Davoodi, B.; Tazehkandi, A.H. Experimental investigation and optimization of cutting parameters in dry and wet machining of aluminum alloy 5083 in order to remove cutting fluid. *J. Clean. Prod.* **2014**, *68*, 234–242. [CrossRef]
24. Klocke, F.; Settineri, L.; Lung, D.; Claudio Priarone, P.; Arft, M. High performance cutting of gamma titanium aluminides: Influence of lubricoolant strategy on tool wear and surface integrity. *Wear* **2013**, *302*, 1136–1144. [CrossRef]
25. Krolczyk, G.M.; Maruda, R.W.; Krolczyk, J.B.; Wojciechowski, S.; Mia, M.; Nieslony, P.; Budzik, G. Ecological trends in machining as a key factor in sustainable production—A review. *J. Clean. Prod.* **2019**, *218*, 601–615. [CrossRef]
26. Maruda, R.W.; Krolczyk, G.M.; Wojciechowski, S.; Zak, K.; Habrat, W.; Nieslony, P. Effects of extreme pressure and anti-wear additives on surface topography and tool wear during MQCL turning of AISI 1045 steel. *J. Mech. Sci. Technol.* **2018**, *32*, 1585–1591. [CrossRef]
27. Maruda, R.W.; Krolczyk, G.M.; Feldshtein, E.; Nieslony, P.; Tyliszczak, B.; Pusavec, F. Tool wear characterizations in finish turning of AISI 1045 carbon steel for MQCL conditions. *Wear* **2017**, *372*, 54–67. [CrossRef]
28. Gupta, M.K.; Jamil, M.; Wang, X.; Song, Q.; Liu, Z.; Mia, M.; Hegab, H.; Khan, A.M.; Collado, A.G.; Pruncu, C.I.; et al. Performance Evaluation of Vegetable Oil-Based Nano-Cutting Fluids in Environmentally Friendly Machining of Inconel-800 Alloy. *Materials* **2019**, *12*, 2792. [CrossRef] [PubMed]
29. Deiab, I.; Raza, S.W.; Pervaiz, S. Analysis of lubrication strategies for sustainable machining during turning of titanium Ti-6Al-4V alloy. *Procedia CIRP* **2014**, *17*, 766–771. [CrossRef]
30. Shamoto, E.; Moriwaki, T. Ultaprecision Diamond Cutting of Hardened Steel by Applying Elliptical Vibration Cutting. *CIRP Ann.-Manuf. Technol.* **1999**, *48*, 441–444. [CrossRef]
31. Xu, W.-X.; Zhang, L.-C. Ultrasonic vibration-assisted machining: Principle, design and application. *Adv. Manuf.* **2015**, *3*, 173–192. [CrossRef]
32. Brehl, D.E.; Dow, T.A. Review of vibration-assisted machining. *Precis. Eng.* **2008**, *32*, 153–172. [CrossRef]
33. Kim, G.D.; Loh, B.G. Characteristics of Elliptical Vibration Cutting in Micro V-grooving with Variations of Elliptical Cutting Locus and Excitation Frequency. *J. Micromech. Microeng.* **2008**, *18*, 025002. [CrossRef]
34. Kim, G.D.; Loh, B.G. An ultrasonic elliptical vibration cutting device for micro V-groove machining: Kinematical analysis and micro V-groove machining characteristics. *J. Mater. Process. Technol.* **2007**, *190*, 181–188. [CrossRef]
35. Suzuki, N.; Masuda, S.; Haritani, M.; Shamoto, E. Ultraprecision micromachining of brittle materials by applying ultrasonic elliptical vibration cutting. In Proceedings of the International Symposium on IEEE Micro-Nanomecha-tronics and Human Science and the 4th Symposium Micro-Nanomecha-tronics for Information-Based Society, Nagoya, Japan, 31 October–3 November 2004; pp. 133–138. [CrossRef]
36. Suzuki, N.; Haritani, M.; Yang, J.; Hino, R.; Shamoto, E. Elliptical Vibration Cutting of Tungsten Alloy Molds for Optical Glass Parts. *CIRP Ann.-Manuf. Technol.* **2007**, *56*, 127–130. [CrossRef]
37. Kurniawan, R.; Kumaran, S.T.; Ali, S.; Nurcahyaningsih, D.A.; Kiswanto, G.; Ko, T.J. Experimental and analytical study of ultrasonic elliptical vibration cutting on AISI 1045 for sustainable machining of round-shaped microgroove pattern. *Int. J. Adv. Manuf. Technol.* **2018**, *98*, 2031–2055. [CrossRef]
38. To, S.; Cheung, C.F.; Lee, W.B. Influence of material swelling on surface roughness in diamond turning of single crystals. *Mater. Sci. Technol.* **2001**, *17*, 102–108. [CrossRef]
39. Tan, R.; Zhao, X.; Zou, X.; Sun, T. A novel ultrasonic elliptical vibration cutting device based on a sandwiched and symmetrical structure. *Int. J. Adv. Manuf. Technol.* **2018**, *97*, 1397–1406. [CrossRef]
40. Nath, C.; Rahman, M.; Neo, K.S. Machinability study of tungsten carbide using PCD tools under ultrasonic elliptical vibration cutting. *Int. J. Mach. Tools Manuf.* **2009**, *49*, 1089–1095. [CrossRef]
41. Zhang, X.Q.; Liu, K.; Kumar, A.S.; Rahman, M. A study of the diamond tool wear suppression mechanism in vibration-assisted machining of steel. *J. Mater. Process. Technol.* **2014**, *214*, 496–506. [CrossRef]
42. Zhang, J. Micro/Nano Machining of Steel and Tungsten Carbide Utilizing Elliptical Vibration Cutting Technology. Ph.D. Thesis, Nagoya University, Nagoya, Japan, 2014.

 © 2019 by the authors. Licensee MDPI, Basel, Switzerland. This article is an open access article distributed under the terms and conditions of the Creative Commons Attribution (CC BY) license (http://creativecommons.org/licenses/by/4.0/).

Article

Performance Evaluation of Vegetable Oil-Based Nano-Cutting Fluids in Environmentally Friendly Machining of Inconel-800 Alloy

Munish Kumar Gupta [1], Muhammad Jamil [2], Xiaojuan Wang [1], Qinghua Song [1,3,*], Zhanqiang Liu [1,3], Mozammel Mia [4], Hussein Hegab [5], Aqib Mashood Khan [2], Alberto Garcia Collado [6], Catalin Iulian Pruncu [7] and G.M. Shah Imran [4]

[1] Key Laboratory of High Efficiency and Clean Mechanical Manufacture, Ministry of Education, School of Mechanical Engineering, Shandong University, Jinan 250000, China
[2] College of Mechanical and Electrical Engineering, Nanjing University of Aeronautics and Astronautics, Nanjing 210016, China
[3] National Demonstration Center for Experimental Mechanical Engineering Education, Shandong University, Jinan 250000, China
[4] Mechanical and Production Engineering, Ahsanullah University of Science and Technology, Dhaka 1208, Bangladesh
[5] Mechanical Design and Production Engineering Department, Cairo University, Giza 12163, Egypt
[6] Department of Mechanical and Mining Engineering, University of Jaen, EPS de Jaen, Campus Las Lagunillas, 23071 Jaen, Spain
[7] Mechanical Engineering, Imperial College London, Exhibition Rd., SW7 2AZ London, UK
* Correspondence: ssinghua@sdu.edu.cn; Tel.: +86-(0)531-88392539

Received: 31 July 2019; Accepted: 27 August 2019; Published: 30 August 2019

Abstract: Recently, the application of nano-cutting fluids has gained much attention in the machining of nickel-based super alloys due their good lubricating/cooling properties including thermal conductivity, viscosity, and tribological characteristics. In this study, a set of turning experiments on new nickel-based alloy i.e., Inconel-800 alloy, was performed to explore the characteristics of different nano-cutting fluids (aluminum oxide (Al_2O_3), molybdenum disulfide (MoS_2), and graphite) under minimum quantity lubrication (MQL) conditions. The performance of each nano-cutting fluid was deliberated in terms of machining characteristics such as surface roughness, cutting forces, and tool wear. Further, the data generated through experiments were statistically examined through Box Cox transformation, normal probability plots, and analysis of variance (ANOVA) tests. Then, an in-depth analysis of each process parameter was conducted through line plots and the results were compared with the existing literature. In the end, the composite desirability approach (CDA) was successfully implemented to determine the ideal machining parameters under different nano-cutting cooling conditions. The results demonstrate that the MoS_2 and graphite-based nanofluids give promising results at high cutting speed values, but the overall performance of graphite-based nanofluids is better in terms of good lubrication and cooling properties. It is worth mentioning that the presence of small quantities of graphite in vegetable oil significantly improves the machining characteristics of Inconel-800 alloy as compared with the two other nanofluids.

Keywords: environmentally friendly; nano-cutting fluids; nickel-based alloys; turning; optimization

1. Introduction

Nickel (Ni)-based alloys have become widely accepted materials for the manufacture of critical parts owing to their exceptional characteristics such as high creep, good rupture strength, and resistance to corrosion and oxidation [1]. Due to excellent fatigue strength and possession of yield strength at high temperature and pressure (600 °C, 1000 MPa), Ni alloys are used in the manufacturing of aero-engines, turbine blades, nuclear reactors, and in chemical industries, where there is a requirement for use of cyclic loads and high temperatures. Thakur and Gangopadhyay have examined an aero-engine consisting of 50% Ni alloy in weight, due its high thermal stability in severe environments [2]. In addition, Ni alloys are ductile materials under cryogenic temperature because of their face-centered cubic (FCC) structure, which is why they are used in cryogenic tanks, as superconducting materials, and in rocket motor casings [3]. Nowadays, Ni-based alloys have several grades, such as Inconel-718, FGH-95, ME-16, IN-100, Inconel-800, and Inconel-825. Among numerous Inconel grades, Inconel-800 is a Fe–Ni–Cr alloy that offers adequate resistance to oxidation, and carburization even at elevated temperatures with moderate strength [4]. It is highly desirable for the manufacturing of high temperature equipment which is resistant to chloride stress corrosion cracking and shows high creep and stress rupture characteristics in the temperature range of 594–983 °C.

Despite all its advantages and applications, machining of such difficult-to-cut materials is also a great challenge due to their poor thermal conductivity, hot hardness, and chemical reaction with tool materials [5]. Such limitations have compelled the manufacturing industries to revise tool failure criteria for turning (ISO-3685) to attain adequate surface quality and tool life. Therefore, considerable attention has been dedicated to researching the manufacture aerospace components without compromising surface quality, in addition to tool edge damage. In the turning process, shearing and friction due to rubbing of chip at the tool rake face produces an elevated temperature in primary and secondary machining zones. This generated heat strongly effects the tool wear and surface quality because, above a certain temperature, tool binding may start losing its strength and accelerate wear.

In order to reduce the temperature and acceleration of tool wear and to improve the surface quality, several lubri-cooling techniques have been practiced in industry. These coolants and lubricants remove chips and reduce temperature and friction due to the rubbing of chip and tool. Water-soluble oils and minerals oils are frequently applied in industry. However, due to their adverse effects on ecology, operator heath, and some restrictions from the EPA (Environmental Protection Agency), advanced industries have started accepting some sustainable cooling/lubrication techniques [6], such as minimum quantity lubrication (MQL) machining [7], cryogenic machining [8], and nanofluid-assisted MQL machining [9] in order to enhance the machinability of Inconel-800 alloy. In MQL machining, small quantities (microlubrication) of pure oil (vegetable oil) are mixed with compressed air to impinge a fine mist (10~100 mL) to attain the advantage of effective cooling and lubrication at the tool–chip interface. Most of research studies have shown better surface quality and tool life under MQL machining compared with dry or flood cooling [10,11]. Similarly, Gurraj et al. have investigated the machining of difficult-to-cut material under MQL to enhance machinability. Turning tests under the MQL lubri-cooling technique were carried out to improve the machinability in terms of surface quality, tool wear, and cutting forces. Findings have depicted a 15% improvement in all the responses under the MQL cooling technique [12]. Also, Joshi et al. [13] have investigated the turning of Incoloy-800 under dry, flood (600 L/h) and MQL (150 mL/h, 230 mL/h) cutting conditions from the perspective of surface quality and flank wear. The findings have depicted less wear and better surface quality under MQL conditions. However, MQL (230 mL/h) provided favorable results compared to MQL (150 mL/h) under all conditions. Maruda et al. [14,15] also studied MQL conditions. They claimed that MQL performs very well during the machining of different materials. From the above findings, it can be understood that although lower flow rates of MQL achieved better performance, they were nevertheless not suitable for machining due to the material being difficult to cut. The key reason behind this problem is due to the lower oil flow rate which fails to limit heat generation at primary and secondary cutting zones and evaporates immediately in the machining of difficult-to-cut materials.

In order to enhance the machinability performance of MQL, specifically for difficult-to-cut materials, several advancements in MQL have been applied in research, i.e., nanofluid-assisted MQL [16,17], hybrid nanofluid MQL [18], Ranque–Hilsch vortex tube [12], ionic liquid-assisted MQL

(IL-MQL) [19], electrostatic MQL [20], vegetable oil-based solid lubricant MQL [21], and time-controlled MQL pulse [22]. Among these advancements of the MQL system, nanofluid-assisted machining of difficult-to-cut materials is a widely accepted alternative. In order to enhance the thermal characteristics of heat transfer in machining, different types nanoparticles are used with vegetable base oil, such as alumina (Al_2O_3), graphite, aluminum nitride (AlN), and molybdenum disulfide. The mentioned nanoparticles provide superior heat transfer, thermal conductivity, surface area, and Brownian motion. Considering the sustainable machining of difficult-to-cut materials, Khan et al. [23] carried out machining under conventional MQL and Al_2O_3 nanofluid-assisted MQL (NFMQL) from the viewpoint of temperature, surface roughness, and energy consumption. Findings have depicted the superiority of NFMQL with a 16%.2~34.5% reduction in temperature and 11.3%~12% reduction in surface roughness for all cutting conditions. It is mentioned that nanoadditives (size < 100 nm) have a biodegradable base oil-enhanced tribological behavior, owing to an amending effect, polishing, tribo-film formation, and ball bearing effect. The existence of nanoadditives in nanofluids enhance thermal conductivity, the heat transfer rate, and the nanoparticles deposited on the machining region behave as small bearings and fins, leading to heat dissipation and lubrication. This was proposed for industrial applications, where nanofluids have provided stability at the tool–chip interface for better surface quality due to a ball bearing effect. Padmni et al. [24] have applied molybdenum disulfide (MoS_2) nanoparticle-based vegetable oil in conventional machining in order to improve the machinability from the perspective of surface roughness and tool wear. Results have underscored a maximum of a 37% reduction in tool wear and 44% reduction in surface roughness with 0.5 vol% nanoparticles in comparison with dry turning. Khan et al. [25] applied copper nano-additives (Cu-np)-based MQL in the conventional machining to improve the surface quality and machinability. Findings have depicted superior surface quality under nanofluid-assisted machining. They reported that the application of Cu-nps in biodegradable oil extended tribological film formation as well as thermal properties. Hence, Cu-np-assisted machining has minimized surface roughness and lowered the environmental impact.

According to the aforementioned state-of-art review, it is worth mentioning that the MQL system along with the nanofluid cooling conditions have been considered as a good alternative and help to significantly improve the machining performance in terms of lower cutting forces, surface roughness, tool wear, etc. Therefore, with this aim, the three type of nanofluids i.e., aluminum oxide (Al_2O_3), molybdenum disulfide (MoS_2), and graphite with MQL system have been firstly implemented in the turning of new Inconel-800 alloy and various important characteristics such as cutting force, tool wear, and surface roughness were evaluated. Further, the process parameters were tested for their statistical significance levels using Box Cox transformation, normally distributed plots, and analysis of variance (ANOVA) methods, respectively. In the end, the optimized parameters were obtained using composite desirability approach (CDA). The paper is organized into the following sections (1) Introduction and Literature Review followed by (2) Materials and Methods, (3) results are presented in the Results and Discussion section and (4) the findings of complete paper are presented here.

2. Materials and Methods

This section discusses the experimental setup used for machining of Inconel-800 alloy under nanofluid-enriched MQL conditions. The complete details of workpiece materials, mechanical properties, tool materials, and equipment used for machinability study are discussed below:

2.1. Workpiece, Cutting tool, and Machine Tool Details

In this work, the turning experiments were performed on new nickel-based alloy, i.e., Inconel-800. This alloy is mainly used in the aerospace, nuclear, and marine sectors. This alloy is used under heat-treated conditions. The chemical composition and heat treatment conditions of Inconel-800 alloy are presented in Tables 1 and 2. The diameter and length of subjected material used was 50 mm × 120 mm, respectively. Further, the cutting tool used for machining the Inconel-800 alloy is cubic boron nitride (CBN) having model no. CCGW 09T304-2 tips and with rhombic shape. The insert was rigidly fixed on the tool holder of lathe tool dynamometer. The details of dynamometer are given in the following sections. Note that no separate tool holder is used for experimentation. This insert contains 50% of CBN content having a grain size of 2 µm, titanium carbide binder, and titanium nitride coating. Moreover, it is highly recommended to use interrupted cutting and heavy operations on high-strength temperature-resistant alloys, i.e., Inconel-800. The complete specifications of cutting tool are presented in Table 3. Further, the CNC lathe is used for performing the turning experiments on Inconel-800 alloy. This machine tool consists of two concurrently controlled axes, namely, Z axis (movement of carriage parallel to spindle axis (longitudinal)), and X-axis (movement of turret slide at right angle to spindle axis (cross)), and equipped with a Siemens control system.

Table 1. Chemical composition of Inconel-800 alloy.

Ni	Cr	Fe	C	Al	Ti	Al + Ti
30.0–35.0	19.0–23.0	39.5 min	0.10 max	0.15–0.60	0.15–0.60	0.30–1.20

Table 2. Heat treatment conditions of Inconel-800 super alloy.

Heat Treatment	Intermediate Treatment	Final Treatment	Rockwell Hardness
1050 °C for 2 h, air-cooling	850 °C for 6 h, air-cooling	700 °C for 2 h, air-cooling	RC

Table 3. Tool geometry of cutting tool.

Inclination Angle	−6°
Orthogonal rake angle	6°
Orthogonal clearance angle	80°
Auxiliary cutting-edge angle	15°
Principal cutting-edge angle	90°
Nose radius	0.4 mm
Shape	Rhombic

2.2. Cooling-Lubrication Conditions

Environmentally friendly cooling conditions, i.e., the minimum quantity lubrication system, have been implemented in this work. The MQL system used in this work was "*NOGA mini cool system'*. The main parts of this system are two pipes, nozzles, control valve, syphon line, and powerful on/off Popeye magnet system. The coolant flow rate, air flow rate and air pressure used in this work were 30 mL/h, 6 L/min, and 5 bar, respectively.

2.3. Preparation of Nanofluids

The current work involves the application of three types of commercially available nanoparticles having an average size of 40 nm. These nanoparticles are aluminum oxide (Al_2O_3), molybdenum disulfide (MoS_2), and graphite respectively. The 3 wt. % nanoparticles were mixed in vegetable base oil as additives, i.e., in sunflower oil having the following physical properties: kinematic viscosity 40 1C (cSt): 40.05; viscosity index: 206; flashpoint (0 °C): 252; and pour point (0 °C): −12.00. For proper mixing of nanoparticles with base oil, the two-step method was adopted. In this method, sonication was carried out with the help of an ultra-sonication bath for about one hour followed by hot magnetic stirring of half an hour. In order to enhance the dispersion, reduce surface tension, and improve wettability and oxidation resistance, sodium lauryl sulfate and the natural antioxidant tocopherol (vitamin E) were used as surfactant at a ratio of 1:10. The properties of different nanofluids are shown in Table 4.

Table 4. Properties of different nanofluids [26].

Properties	Vegetable Base Oil	Al_2O_3 Nanofluid	MoS_2 Nanofluid	Graphite Nanofluid
Appearance	Bright and clear	White	Black	Grayish Black
Viscosity (CP) (at 20 °C)	68.16	120.23	100.56	83.12
Thermal Conductivity (W/mK)	0.1432	0.2085	0.2362	0.2663

2.4. Machining Characteristic Measurements

In this work, three prominent machining indices, namely main cutting forces (Fc), tool wear (VB_{max}), and surface roughness values (Ra) were evaluated under three different nanofluid cutting conditions. The main cutting forces were measured using online mode and the tool wear as well as surface roughness measured using offline mode. For the measurement of cutting forces, the TeLC made lathe tool dynamometer associated with XKM 2000 software was used. In same context, the tool flank wear measurements for the finish turning operation were recorded using a standard Mitutoyo's make toolmaker's microscope (i.e., $VB_{max} \geq 0.60$ mm, as per the ISO 3685 standard). Similarly, the arithmetic roughness values have been measured with the Mitutoyo make SJ301 surface roughness tester. Moreover, these conditions were considered for evaluation of surface roughness values, i.e., standard ISO 1999 profile R cut off length of 0.8 mm, range—auto, and speed of 0.25 mm/s. In the end, the tool wear was analyzed with the help of scanning electron microscopy (SEM).

2.5. Process Parameters and Design Methodology

The three types of machining parameters with three different levels, namely cutting speed, feed rate, and depth of cut have been used in this work. The complete details of parameters and their used levels are detailed in Table 5. The selection of these parameters was based purely on pilot experiments, literature review, and tool manufacturer recommendations. The machining time of 1 min was fixed in all set of turning experiments. Moreover, these experiments were performed by following the Box-Behnken response surface methodology (RSM) design. In this, the machining parameters are considered as a continuous factor and different cooling conditions are termed as a categorical factor. Note that the total 29 experiments was suggested by RSM and for each set of experiments, a fresh cutting edge of CBN tool has been used to accurately study the effect of process parameters on machining responses. In the end, the given parameters were optimized by composite desirability approach (CDA). The main aim of the implementation of CDA is to achieve the most accurate predictions and results in the minimum possible time. The complete methodology of this scientific work is shown in Figure 1.

Table 5. Machining parameters and their levels.

Parameters	Coded Value	Units	Low Level (−1)	Middle Level (0)	High Level (+1)
Cutting Speed (Vc)	A	m/min	200	250	300
Feed Rate (f)	B	mm/rev	0.1	0.15	0.20
Depth of cut (a_p)	C	mm	0.25	0.50	0.75
Cooling condition	D	-	Al_2O_3	MoS_2	Graphite

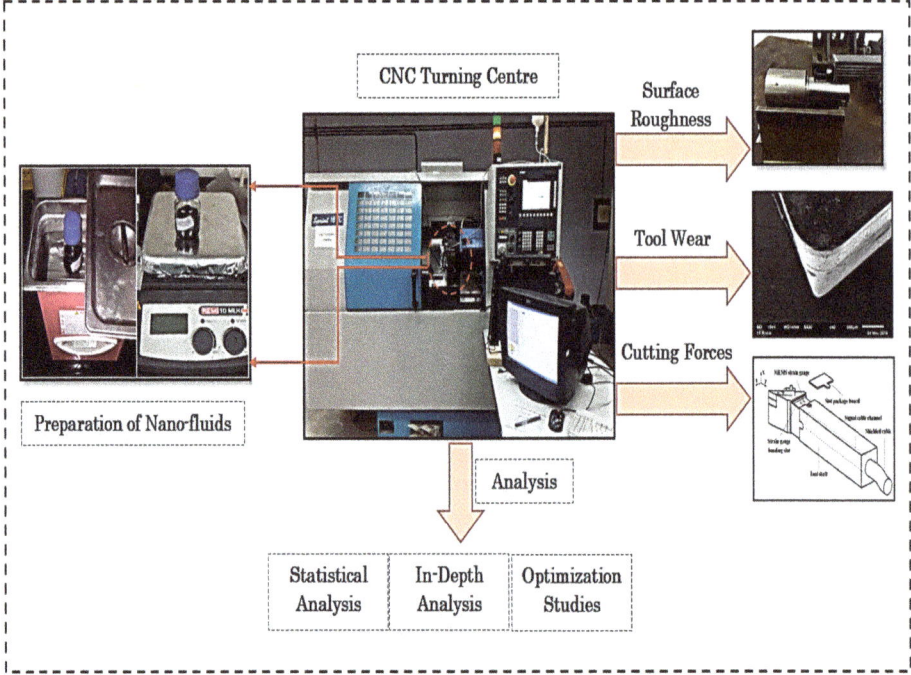

Figure 1. Research methodology of current work.

3. Results and Discussion

This section is divided into three phases: (1) Statistical Analysis, (2) Experimental Investigation of Process Parameters and (3) Optimization Studies. The complete details are given in the following sections.

3.1. Statistical Analysis

Statistical relevance was determined for the tested parameters. Therefore, the Box Cox transformation was developed to train the predictive models. It can help to build a family of transformations that can contains normalized data. They are normally unevenly distributed and linked to an appropriate exponent (lambda, l). By using the lambda value, it is possible to easily control the power in order to modify these data. Initially, the Box and Cox were applied to simultaneously correct the normality, linearity, and homogeneity.

The Box Cox plots associated with the cutting forces, surface roughness, and tool wear was presented in Figure 2a–c. In all responses, the blue line obtained from the cutting forces and surface roughness shows a value for lambda residuals equal to 1, having values outside of the 95% confidence limits. As per the green line observation, the lambda is approx. 0.5. The square root transformation is applied to the responses, which allows for the generation of normally distributed residuals. The Box Cox transformation results for the normal distribution plot of residuals are depicted in Figure 3a–c. We can note that the residuals fall over the straight line conveying the evolution of residuals were distributed as normal. Finally, the ANOVA was introduced for verification.

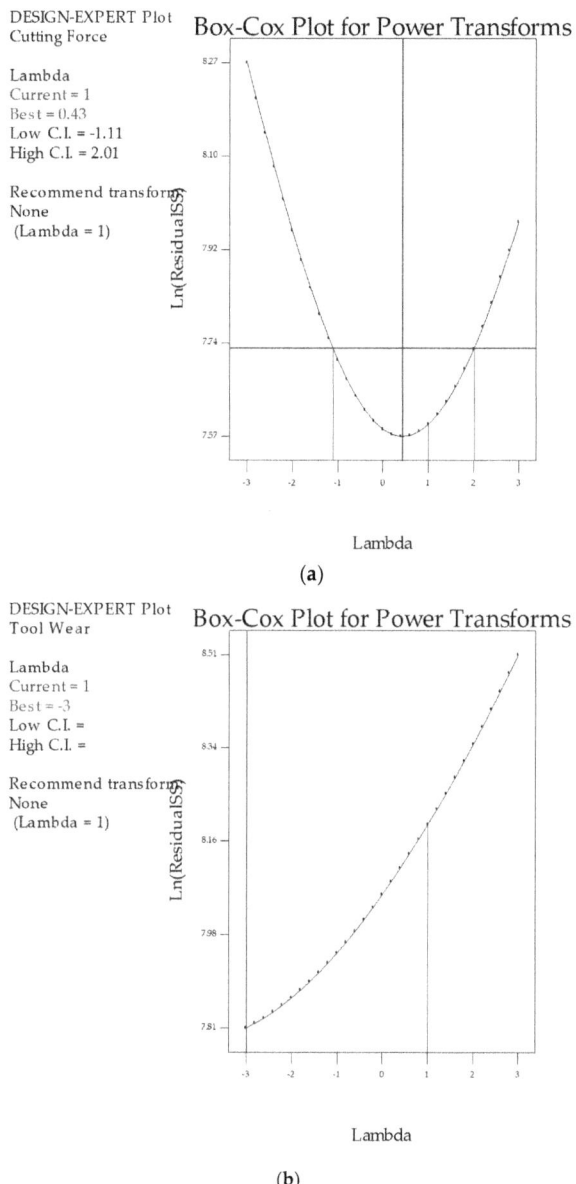

Figure 2. Cont.

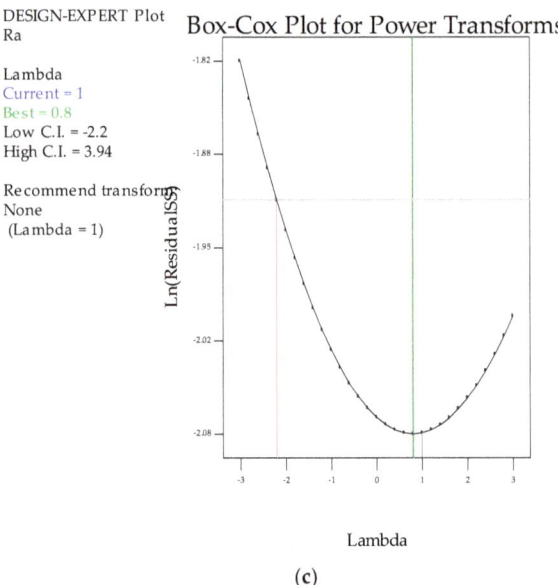

(c)

Figure 2. Box-Cox plots. (**a**) Cutting forces; (**b**) Tool wear; (**c**) Surface roughness.

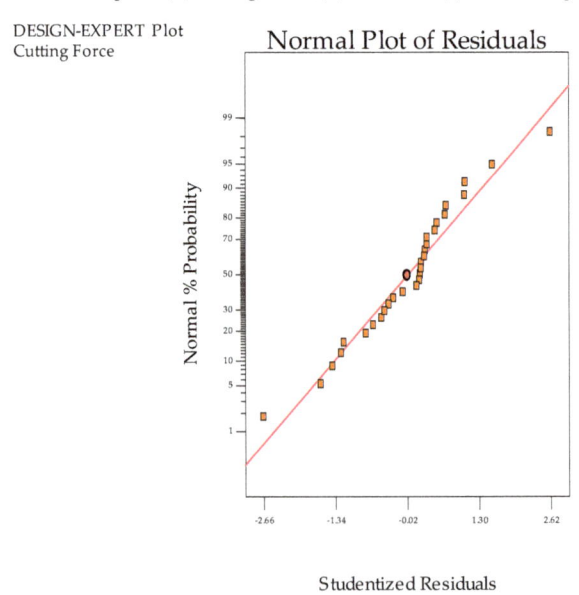

(a)

Figure 3. *Cont.*

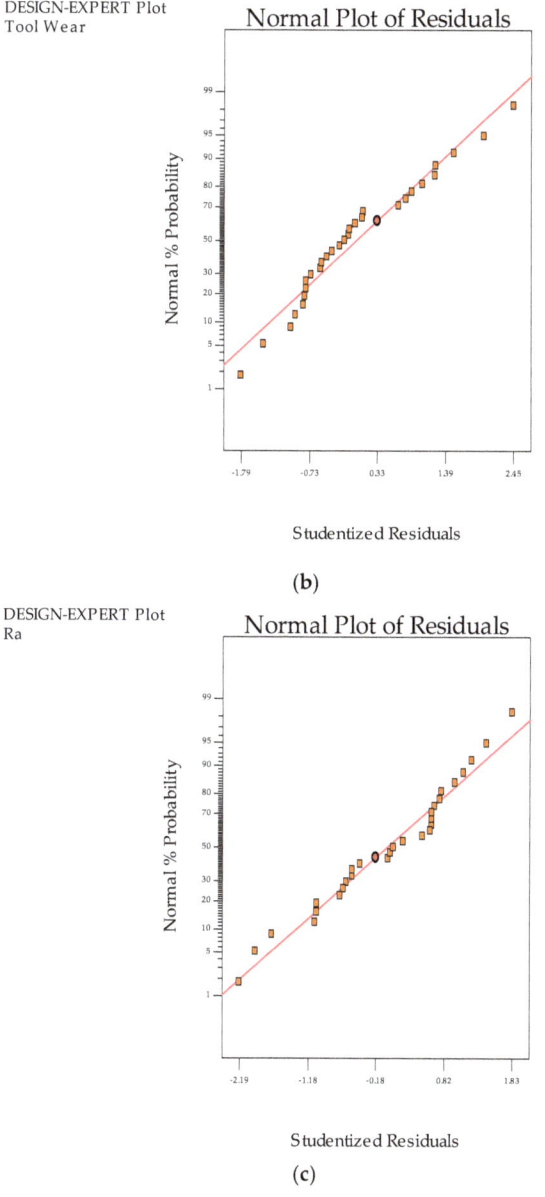

Figure 3. Normal distribution plots. (**a**) Cutting forces; (**b**) Tool wear; (**c**) Surface roughness.

The values of manual elimination procedure for the cutting force, surface roughness, and tool wear determined by ANOVA were introduced. The data from these tables demonstrates that the simulated models for all the responses are significant. There were noted the value of "Probability> F'" that is associated with lack of fit ~0.0124 (cutting forces), 0.0431 (tool wear), and 0.0173 (surface roughness). As it is larger than 0.05, the lack of fit is considered insignificant. The R^2 values are 0.918 (cutting forces), 0.742 (tool wear), and 0.7173 (surface roughness). The measure of proportion for the entire variability, R2, helps to explain the model and was found to equal or be close to 1, as per recommendations [27]. Further, the adjusted R^2 value was found good agreement that aided in comparing the models if they had a different number of terms. Our simulation indicates a close match between the adjusted and predicted R^2 value. Finally, the Equations (1) to (3) were used to determine the final regression model used to determine the output parameters (i.e., cutting forces, tool wear, and surface roughness) when used different nano-cutting fluids as coded factors:

Main Cutting Force

Cutting Fluid 1:
$$\text{Cutting Force} = -58.33 + 0.645 \times \text{Cutting Speed} + 521.66 \times \text{Feed Rate} + 0.356 \times \text{Depth of Cut} \quad (1a)$$

Cutting Fluid 2:
$$\text{Cutting Force} = -70.28 + 0.983 \times \text{Cutting Speed} + 324.61 \times \text{Feed Rate} + 0.74 \times \text{Depth of Cut} \quad (1b)$$

Cutting Fluid 3:
$$\text{Cutting Force} = -79.66 + 0.85 \times \text{Cutting Speed} + 642.74 \times \text{Feed Rate} + 0.42 \times \text{Depth of Cut} \quad (1c)$$

Tool Wear-

Cutting Fluid 1:
$$\text{Tool Wear} = 73.33 + 0.485 \times \text{Cutting Speed} + 291.66 \times \text{Feed Rate} - 0.144 \times \text{Depth of Cut} \quad (2a)$$

Cutting Fluid 2:
$$\text{Tool Wear} = 67.36 + 0.647 \times \text{Cutting Speed} + 472.34 \times \text{Feed Rate} - 0.248 \times \text{Depth of Cut} \quad (2b)$$

Cutting Fluid 3:
$$\text{Tool Wear} = 59 + 0.32 \times \text{Cutting Speed} + 542.62 \times \text{Feed Rate} - 0.376 \times \text{Depth of Cut} \quad (2c)$$

Surface Roughness

Cutting Fluid 1:
$$\text{Surface Roughness} = 0.5 + 1.82 \times 10^{-3} \times \text{Cutting Speed} + 2.466 \times \text{Feed Rate} - 1.94 \times 10^{-3} \times \text{Depth of Cut} \quad (3a)$$

Cutting Fluid 2:
$$\text{Surface Roughness} = 0.42578 + 1.2354 \times 10^{-3} \times \text{Cutting Speed} + 5.42367 \times \text{Feed Rate} - 1.424 \times 10^{-3} \times \text{Depth of Cut} \quad (3b)$$

Cutting Fluid 3:
$$\text{Surface Roughness} = 0.45 + 1.897 \times 10^{-3} \times \text{Cutting Speed} + 7.336 \times \text{Feed Rate} - 1.798 \times 10^{-3} \times \text{Depth of Cut} \quad (3c)$$

3.2. Experimental Investigation

3.2.1. Cutting Forces

The cutting forces play an important role in the deformation of machine tool structure. However, they are directly associated with the machine tool dynamics, and the main factors which affect the cutting forces are cutting speed, feed rate, and depth of cut. Moreover, the application of cooling conditions significantly helps to reduce the cutting forces during turning operation, and this phenomenon becomes more prominent when the nanofluids are used along with the MQL system. Therefore, the main cutting force is considered as the important machining index in this experimental work. The effect of machining parameters and cooling conditions on main cutting force is shown in Figure 4a–c. It is clearly noted that the cutting forces are increased with the increase in cutting speed and feed rate values. However, the change in depth of cut produces very little variations in the main cutting force values. The increase in cutting speed may have increased the temperature which caused local work hardening. Consequently, a higher force is required for material deformation. The increase of force in cutting with increased feed rate is due to an increase in the chip cross-section, i.e., more friction [28]. In other words, the cutting speed and feed rate are the most significant factors that contribute in increasing the main cutting force values as compared with the depth of cut. The same results are also calculated using the ANOVA tests. Similarly, the trend of cutting environment are described in these figures. It is highly visible that the cutting forces are significantly reduced with the application of change in cutting fluid from Al$_2$O$_3$-based nanofluids to graphite-based nanofluids and MoS$_2$-based nano-cutting fluid shows a moderate effect on the main cutting force values. This is justified with the two properties of cutting fluids presented in Table 4: (1) viscosity and (2) thermal conductivity. It should be stated that using graphite-based nanofluid enhances the heat transfer and tribological performance of the cutting process. The applied nano-mist with compressed air shows capabilities in penetrating into the tool–workpiece interface area, and reduces the severity of the induced heat during machining operation. In addition, the employed nano-mist acts like rollers and causes a noticeable effect in the frictional behavior of the cutting operation, as has been discussed in the literature [29–31].

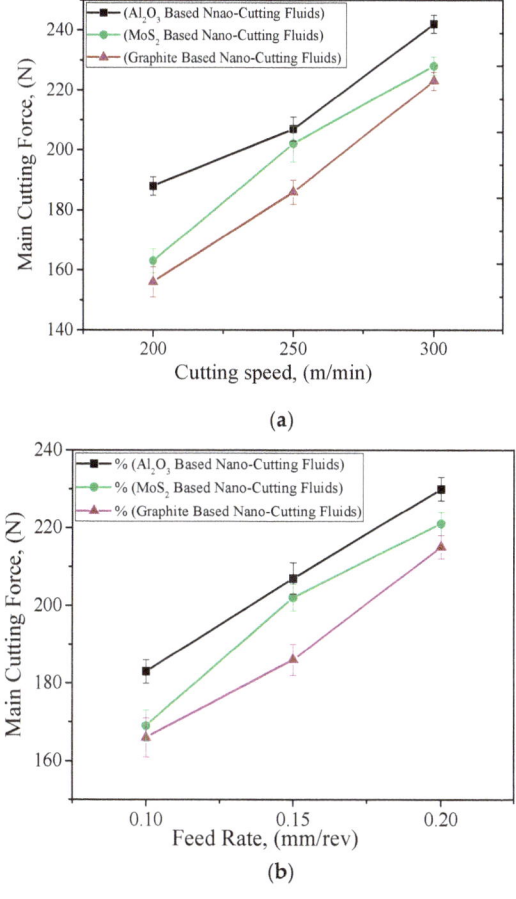

Figure 4. *Cont.*

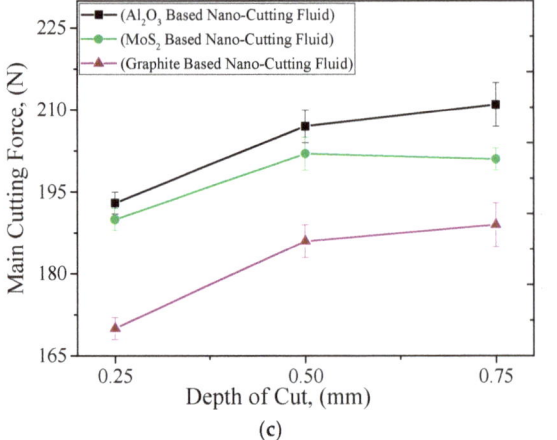

Figure 4. Effect of cutting parameters on main cutting forces. (**a**) Cutting Speed, where $f = 0.15$ mm/rev & $a_p = 0.50$ mm; (**b**) Feed Rate, where $Vc = 250$ mm/rev and $a_p = 0.50$ mm; (**c**) Depth of Cut, where $Vc = 250$ mm/rev and $f = 0.15$ mm/rev.

3.2.2. Tool Wear

The optimum machining cost and minimum energy is achieved with the tool having longer life because the tool wear and tool life are critically damaged with the direct values of machining parameters. Therefore, the careful selection of cutting speed, feed rate, and depth of cut values are required to achieve the maximum tool life values. In addition, the presence of coolant/lubricant has cemented their potential to improve the tool life values. Therefore, in this work, the maximum flank wear criteria according to the ISO 3685 (i.e., $VB_{max} \geq 400$ μm) has been selected to evaluate the results. The influence of cutting speed, feed rate, and depth of cut under the different nanofluids conditions are presented in Figure 5a–c. It is noted that for cutting speed and feed rate values, the tool flank wear follows an increasing pattern. However, the depth of cut shows the opposite trend and very little effect on tool flank wear values as compared with the cutting speed followed by the feed rate values. As mentioned earlier, the increased chip contact area and the hardening effect at local zone may have caused the difficulty in cutting. As such, the wear of tool increased. Furthermore, it is observable that the values of tool flank wear were reduced with the different cutting fluids values and the graphite-based nano-cutting fluids shows promising results. In terms of the tool performance, the nano-mist from all nano-additives showed promising performance in reducing the severity of the tool wear, and this is mainly due to the enhancement in the rubbing level (i.e., at the tool–workpiece interface zone). As mentioned earlier, the nano-mist acts as rollers, and that reduces the induced coefficient of friction. As can be seen in Figure 5, graphite-based nano-mist offers less flank wear than Al_2O_3 and MoS_2 nano-mist. This can be due to the higher thermal conductivity (see Table 4), which means better heat transfer and less rubbing effects. In addition, the giant covalent structure in graphite atomic structure can lead to a better rolling effect, especially when it is compared to the alumina atomic structure (see Figure 6). That can be reflected in the results obtained in Figure 5 as the graphite and MoS_2 nano-mist offer almost the same performance with a slight advantage over the graphite nanofluid. Furthermore, noticeable effects of both graphite and MoS_2 nano-mist clearly appeared at high cutting speeds (see Figure 5a) which mean that a high heat was generated at the cutting zone. Moreover, the SEM images in Figure 7 confirmed the same findings in Figure 5 as all used nano-additives offered better tool performance; however, greater advantages have been observed for both graphite and MoS_2 nano-mist, especially in crater wear.

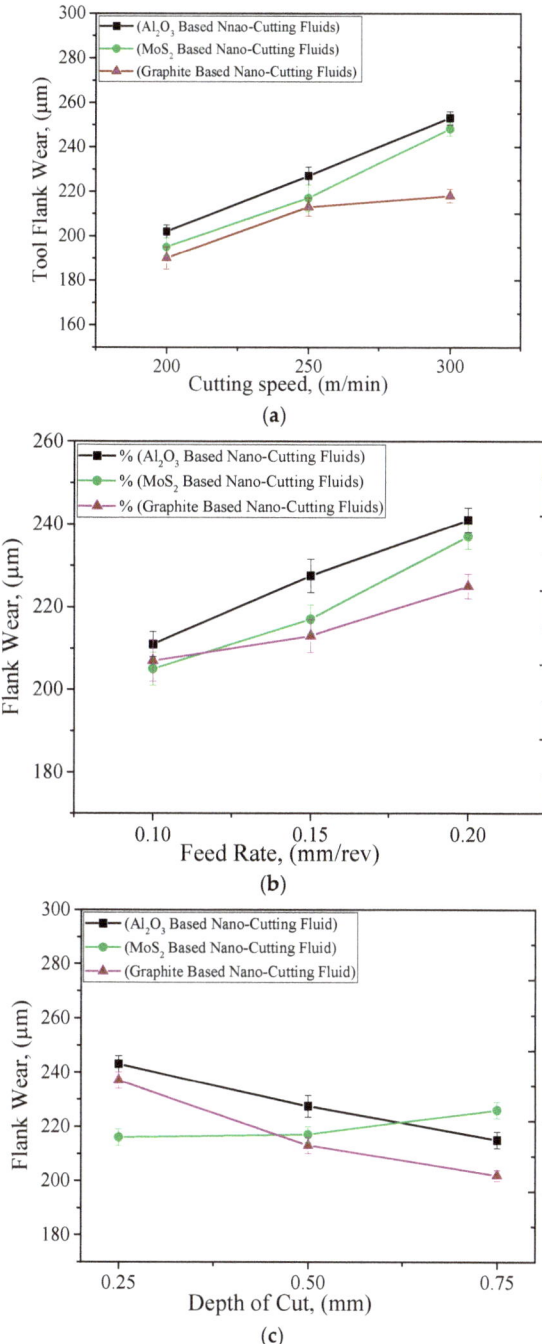

Figure 5. Effect of cutting parameters on tool wear. (**a**) Cutting Speed, where f = 0.15 mm/rev & a_p = 0.50 mm; (**b**) Feed Rate, where Vc = 250 mm/rev and a_p = 0.50 mm; (**c**) Depth of Cut, where Vc = 250 mm/rev and f = 0.15 mm/rev.

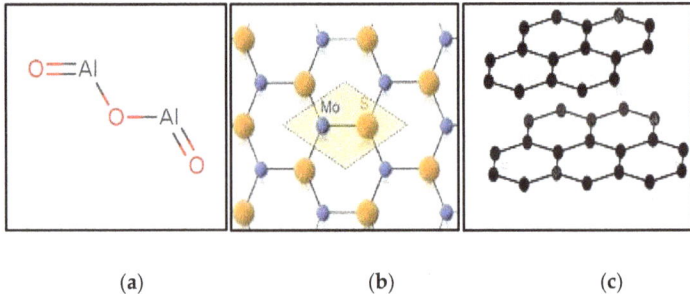

(a)   (b)   (c)

Figure 6. Atomic structure of the used nanoadditives. (**a**) Alumina structure; (**b**) MoS$_2$ structure; (**c**) Graphite structure.

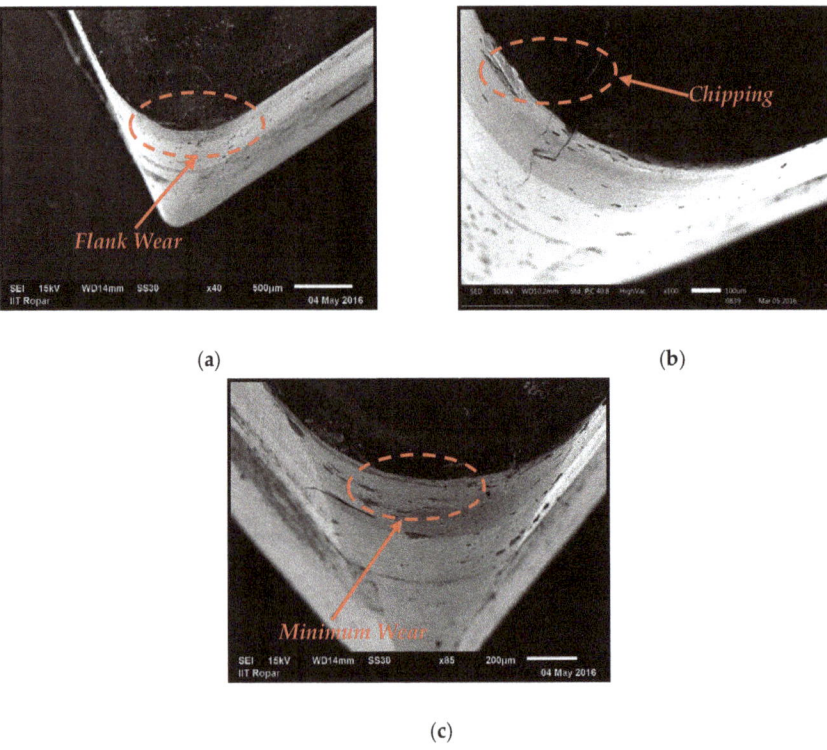

Figure 7. Tool wear images at different working conditions at V_c = 300 m/min, f = 0.15 mm/rev, a_p = 0.50 mm. (**a**) Al$_2$O$_3$ nanofluid; (**b**) MoS$_2$ nanofluid; (**c**) Graphite nanofluid.

3.3. Surface Roughness

In order to evaluate the quality of any product, arithmetic surface roughness (Ra) is considered as a valuable parameter. As evident from Figure 8, the increase in the cutting speed caused an increment in surface roughness. This can be attributed to the increased chatter of the machine tool at such a high cutting speed (300 m/min) used for a superalloy. Similarly, the increase in feed rate caused an increase in the area of tool travel; in other words, more friction was endured. As a result, the tool wear increased. Higher friction and tool wear may have caused the surface roughness to be increased at such a higher feed rate. However, the increase in depth of cut has caused the surface roughness to be lower. Here, it is to be noted that the impingement of cooling agent with nanofluid have caused the surface to be smoother and a lower roughness value was found. It can be noted that the performance of the three used nanoadditives were nearly the same; however, both graphite and MoS$_2$ nano-mist showed better results at the highest cutting speed (see Figure 8). As mentioned earlier, this can be due to the higher thermal conductivity compared to Al$_2$O$_3$ nanofluid. This enhancement results from the promising heat transfer and tribological performance of such nanoadditives, which improve the interactional effect between the tool and workpiece, and reduce the induced coefficient of friction as well as the high generated temperature in the cutting zone. Therefore, better tool performance and surface quality can be clearly observed compared to the classical technique, as previously discussed in some previous studies [32,33].

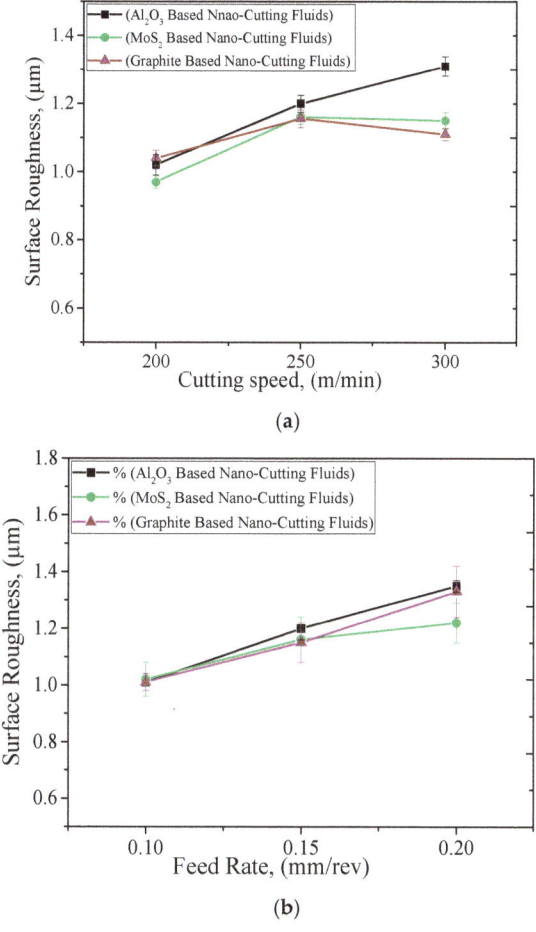

Figure 8. Cont.

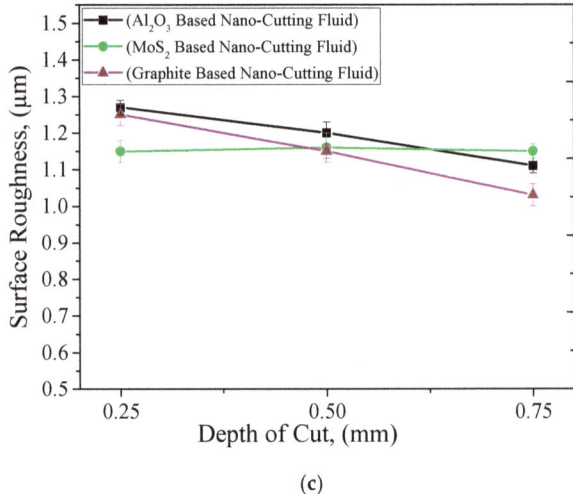

(c)

Figure 8. Effect of cutting parameters on surface roughness. (**a**) Cutting Speed, where f = 0.15 mm/rev & a_p = 0.50 mm; (**b**) Feed Rate, where Vc = 250 mm/rev and a_p = 0.50 mm; (**c**) Depth of Cut, where Vc = 250 mm/rev and f = 0.15 mm/rev.

In order to physically understand the provided results, the nano-mist mechanism during cutting processes should be discussed in terms of the tribological and heat transfer aspects. Due to the effect of the MQL compressed air along with the nanofluids, a very thin layer is formed at the cutting zone as described in the literature [34,35]. This layer includes two main advantages to the overall performance of the cutting process. The first advantage is to absorb the high heat generated during the process as it has a high heat transfer coefficient because of the employed nanoadditives [36]. The second aspect is related to the friction behavior. The applied nano-mist at cutting zone plays as rollers which reduce the induced rubbing between the tool and workpiece [37]. Therefore, lower cutting forces and tool wear can be observed for the cutting tests performed with MQL nanofluid.

3.4. Optimization of Process Parameters: Composite Desirability Approach (CDA)

The main objective was to identify the best possible process parameters that lead to sustainable machining of Inconel-800 alloy. Table 6 presents the results obtained using the typical desirability strategy. This approach is very common and provides an efficient solution with a friendly interface [8,9]. It is developed to adjust the characteristic weight and their importance. Through this approach, it is possible to combine all assigned goals to a unique desirable function in the range $0 \leq d_i \leq 1$. Details of this approach were presented elsewhere [38]. The outputs responses determined through this approach are classified as (1) higher-the-better, (2) smaller-the-better, and (3) nominal-the-better. To ensure the goal of this research, here, we implemented 'smaller-the-better' responses that were evaluated numerically through Equation (4).

$$d_i = \begin{cases} 1 & x_i \leq x_1'' \\ \left[\frac{x_i^* - x_i}{x_1^* - x_1''}\right]^r & x_1'' < x_i < x_i^* \\ 0, & x_i \geq x_i^* \end{cases} \quad (4)$$

Table 6. Optimum results using composite desirability approach (CDA).

Sr. No.	Cutting Speed	Feed Rate	Depth of Cut	Cutting Fluid	Cutting Force	Tool Wear	Surface Roughness	Desirability
1	200	0.10	0.70	3	143	181	0.87	1.00
2	202	0.10	0.64	3	141	183	0.88	0.88
3	201	0.10	0.63	3	140	183	0.88	0.74
4	201	0.10	0.70	3	145	182	0.87	0.72
5	200	0.10	0.62	2	141	182	0.88	0.65

Here, x_1'' represents the smallest value associated to x_i, x_1^* is the largest value associated to x_i while r denotes the shape function.

Further, the cutting speed, feed rate, and depth of cut were selected within the range parameters. Table 6 presents the optimum five solutions determined by applying the CDA. The best solution was considered as the one having the maximum desirability value. Similarly, Figure 9 shows the histogram determined for the ideal solution with following parameters: 200 m/min for the cutting speed, 0.10 mm/rev of feed rate, 0.70 mm of depth of cut, and graphite-based nano-cutting fluids having the maximum desirability value.

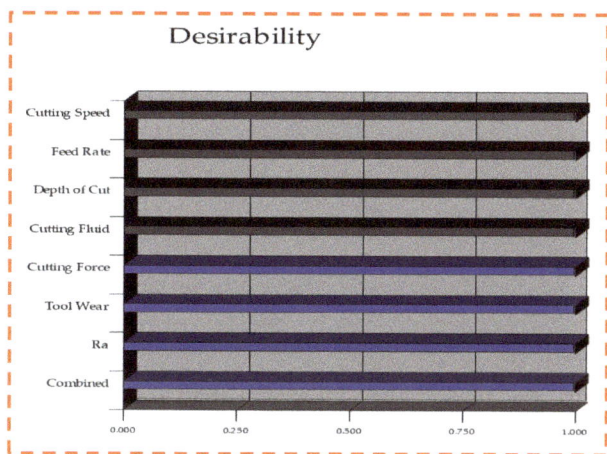

Figure 9. Histogram plot represent the optimum values.

4. Conclusions

This work primarily focuses on the performance of three different nano-cutting fluids during the turning of new nickel-based alloy, i.e., Inconel-800. In the literature, major efforts have been focused on the Inconel-718 alloy. That is why this new alloy has been selected as a subject material in this experimental work. The major conclusions drawn from these experiments are given below:

- Statistical analysis results: The results determined through experiments were statistically significant in terms of Box Cox transformation, R^2 values, and ANOVA tests. Therefore, the prediction models are useful for researchers and academics to determine the values for their reference.
- Experimental investigation: The trend of almost all parameters were found to be the same, i.e., the cutting forces, tool wear, and surface roughness values were significantly affected with small changes in any one of these machining parameters.
- Comparison results: When the comparison was made between all cutting fluids, the overall performance of graphite-based nanofluids was found to be better in improving the machining

characteristics. This is because of the good tribological and cooling properties of graphite-based nano-cutting fluids. Moreover, the chemical structure of graphite is more covalent and this drastically affects its performance as compared to other nanofluids.

- Optimization results: CDA is also a very efficient optimization method for determining the optimal solution, i.e., 200 m/min for the cutting speed, 0.10 mm/rev of feed rate, 0.70 mm of depth of cut, and graphite-based nano-cutting fluids.
- Future recommendations: Even though the results obtained from this study were highly useful for practical applications, some future avenues are still pending to improve the machining performance of Inconel-800 alloy. For instance, the high-pressure cooling (HPC) approach could be integrated with the nano-cutting fluids and the results compared with the MQL technique.

Author Contributions: The author contributions are presented here: Conceptualization, M.J.; Data curation, X.W.; Formal analysis, Z.L.; Funding acquisition, Q.S.; Investigation, M.K.G.; Methodology, M.M.; Software, H.H.; Supervision, A.G.C.; Validation, G.M.S.I.; Writing—review & editing, A.M.K., A.G.C. and C.I.P.

Funding: The authors are grateful to the National Natural Science Foundation of China (no. 51875320), the Major projects of National Science and Technology (Grant No. 2019ZX04001031), the National Key Research and Development Program (Grant No. 2018YFB2002201), the Natural Science Outstanding Youth Fund of Shandong Province (Grant No. ZR2019JQ19), and the Key Laboratory of High-efficiency and Clean Mechanical Manufacture at Shandong University, Ministry of Education. The authors declare that there is no conflict of interest.

Conflicts of Interest: The authors declare no conflict of interest.

References

1. Darshan, C.; Jain, S.; Dogra, M.; Gupta, M.K.; Mia, M. Machinability improvement in Inconel-718 by enhanced tribological and thermal environment using textured tool. *J. Therm. Anal. Calorim.* **2019**, 1–13. [CrossRef]
2. Thakur, A.; Gangopadhyay, S. State-of-the-art in surface integrity in machining of nickel-based super alloys. *Int. J. Mach. Tools Manuf.* **2016**, *100*, 25–54. [CrossRef]
3. Behera, B.C.; Alemayehu, H.; Ghosh, S.; Rao, P.V. A comparative study of recent lubri-coolant strategies for turning of Ni-based superalloy. *J. Manuf. Process.* **2017**, *30*, 541–552. [CrossRef]
4. Gupta, M.; Pruncu, C.; Mia, M.; Singh, G.; Singh, S.; Prakash, C.; Sood, P.; Gill, H. Machinability investigations of inconel-800 super alloy under sustainable cooling conditions. *Materials* **2018**, *11*, 2088. [CrossRef] [PubMed]
5. M'Saoubi, R.; Outeiro, J.C.; Chandrasekaran, H.; Dillon Jr, O.W.; Jawahir, I.S. A review of surface integrity in machining and its impact on functional performance and life of machined products. *Int. J. Sustain. Manuf.* **2008**, *1*, 203–236. [CrossRef]
6. Krolczyk, G.M.; Maruda, R.W.; Krolczyk, J.B.; Wojciechowski, S.; Mia, M.; Nieslony, P.; Budzik, G. Ecological trends in machining as a key factor in sustainable production—A review. *J. Clean. Prod.* **2019**, *218*, 601–615. [CrossRef]
7. Maruda, R.W.; Krolczyk, G.M.; Feldshtein, E.; Pusavec, F.; Szydlowski, M.; Legutko, S.; Sobczak-Kupiec, A. A study on droplets sizes, their distribution and heat exchange for minimum quantity cooling lubrication (MQCL). *Int. J. Mach. Tools Manuf.* **2016**, *100*, 81–92. [CrossRef]
8. Mia, M.; Singh, G.; Gupta, M.K.; Sharma, V.S. Influence of Ranque-Hilsch vortex tube and nitrogen gas assisted MQL in precision turning of Al 6061-T6. *Precis. Eng.* **2018**, *53*, 289–299. [CrossRef]
9. Gupta, M.K.; Sood, P.K.; Sharma, V.S. Optimization of machining parameters and cutting fluids during nano-fluid based minimum quantity lubrication turning of titanium alloy by using evolutionary techniques. *J. Clean. Prod.* **2016**, *135*, 1276–1288. [CrossRef]
10. Yıldırım, Ç.V.; Kıvak, T.; Erzincanlı, F. Tool wear and surface roughness analysis in milling with ceramic tools of Waspaloy: A comparison of machining performance with different cooling methods. *J. Brazilian Soc. Mech. Sci. Eng.* **2019**, *41*, 83. [CrossRef]
11. Sarikaya, M.; Güllü, A. Multi-response optimization of minimum quantity lubrication parameters using Taguchi-based grey relational analysis in turning of difficult-to-cut alloy Haynes 25. *J. Clean. Prod.* **2015**, *91*, 347–357. [CrossRef]

12. Singh, G.; Pruncu, C.I.; Gupta, M.K.; Mia, M.; Khan, A.M.; Jamil, M.; Pimenov, D.Y.; Sen, B.; Sharma, V.S. Investigations of machining characteristics in the upgraded MQL-assisted turning of pure titanium alloys using evolutionary algorithms. *Materials* **2019**, *12*, 999. [CrossRef] [PubMed]
13. Joshi, K.K.; Kumar, R. Anurag An Experimental Investigations in Turning of Incoloy 800 in Dry, MQL and Flood Cooling Conditions. *Procedia Manuf.* **2018**, *20*, 350–357. [CrossRef]
14. Maruda, R.W.; Krolczyk, G.M.; Wojciechowski, S.; Zak, K.; Habrat, W.; Nieslony, P. Effects of extreme pressure and anti-wear additives on surface topography and tool wear during MQCL turning of AISI 1045 steel. *J. Mech. Sci. Technol.* **2018**, *32*, 1585–1591. [CrossRef]
15. Maruda, R.W.; Feldshtein, E.; Legutko, S.; Krolczyk, G.M. Research on emulsion mist generation in the conditions of Minimum Quantity Cooling Lubrication (MQCL). *Teh. Vjestn. Tech. Gaz.* **2015**, *22*, 1213–1218.
16. Sharma, A.K.; Tiwari, A.K.; Dixit, A.R. Effects of Minimum Quantity Lubrication (MQL) in machining processes using conventional and nanofluid based cutting fluids: A review. *J. Clean. Prod.* **2016**, *127*, 1–18. [CrossRef]
17. Sharma, A.K.; Singh, R.K.; Dixit, A.R.; Tiwari, A.K. Characterization and experimental investigation of Al2O3 nanoparticle based cutting fluid in turning of AISI 1040 steel under minimum quantity lubrication (MQL). *Mater. Today Proc.* **2016**, *3*, 1899–1906. [CrossRef]
18. Jamil, M.; Khan, A.M.; Hegab, H.; Gong, L.; Mia, M.; Gupta, M.K.; He, N. Effects of hybrid Al2O3-CNT nanofluids and cryogenic cooling on machining of Ti–6Al–4V. *Int. J. Adv. Manuf. Technol.* **2019**, *102*, 3895–3909. [CrossRef]
19. Abdul Sani, A.S.; Rahim, E.A.; Sharif, S.; Sasahara, H. Machining performance of vegetable oil with phosphonium- and ammonium-based ionic liquids via MQL technique. *J. Clean. Prod.* **2019**, *209*, 947–964. [CrossRef]
20. Huang, S.; Lv, T.; Wang, M.; Xu, X. Effects of machining and oil mist parameters on electrostatic minimum quantity lubrication–EMQL turning process. *Int. J. Precis. Eng. Manuf. Green Technol.* **2018**, *5*, 317–326. [CrossRef]
21. Marques, A.; Paipa Suarez, M.; Falco Sales, W.; Rocha Machado, Á. Turning of Inconel 718 with whisker-reinforced ceramic tools applying vegetable-based cutting fluid mixed with solid lubricants by MQL. *J. Mater. Process. Technol.* **2019**, *266*, 530–543. [CrossRef]
22. Mia, M.; Razi, M.H.; Ahmad, I.; Mostafa, R.; Rahman, S.M.S.; Ahmed, D.H.; Dey, P.R.; Dhar, N.R. Effect of time-controlled MQL pulsing on surface roughness in hard turning by statistical analysis and artificial neural network. *Int. J. Adv. Manuf. Technol.* **2017**, *91*, 3211–3223. [CrossRef]
23. Khan, A.M.; Jamil, M.; Ul Haq, A.; Hussain, S.; Meng, L.; He, N. Sustainable machining. Modeling and optimization of temperature and surface roughness in the milling of AISI D2 steel. *Ind. Lubr. Tribol.* **2019**, *71*, 267–277. [CrossRef]
24. Padmini, R.; Vamsi Krishna, P.; Krishna Mohana Rao, G. Effectiveness of vegetable oil based nanofluids as potential cutting fluids in turning AISI 1040 steel. *Tribol. Int.* **2016**, *94*, 490–501. [CrossRef]
25. Khan, A.M.; Jamil, M.; Salonitis, K.; Sarfraz, S.; Zhao, W.; He, N.; Mia, M.; Zhao, G.L. Multi-objective optimization of energy consumption and surface quality in nanofluid SQCl assisted face milling. *Energies* **2019**, *12*, 710. [CrossRef]
26. Gupta, M.K.; Sood, P.K. *Machining Behavior of High Strength Temperature Resistant Alloys Under MQL Environment*; National Institute of Technology: Hamirpur, India, 2018.
27. Vasantharaja, P.; Vasudevan, M. Optimization of A-TIG welding process parameters for RAFM steel using response surface methodology. *Proc. Inst. Mech. Eng. Part L J. Mater. Des. Appl.* **2015**, *232*, 121–136. [CrossRef]
28. Gutnichenko, O.; Bushlya, V.; Zhou, J.; Ståhl, J.-E. Tool wear and machining dynamics when turning high chromium white cast iron with pcBN tools. *Wear* **2017**, *390*, 253–269. [CrossRef]
29. Hegab, H.; Umer, U.; Soliman, M.; Kishawy, H.A. Effects of nano-cutting fluids on tool performance and chip morphology during machining Inconel 718. *Int. J. Adv. Manuf. Technol.* **2018**, *96*, 3449–3458. [CrossRef]
30. Hegab, H.; Kishawy, H. Towards sustainable machining of inconel 718 using nano-fluid minimum quantity lubrication. *J. Manuf. Mater. Process.* **2018**, *2*, 50. [CrossRef]
31. Eltaggaz, A.; Hegab, H.; Deiab, I.; Kishawy, H.A. Hybrid nano-fluid-minimum quantity lubrication strategy for machining austempered ductile iron (ADI). *Int. J. Interact. Des. Manuf.* **2018**, *12*, 1273–1281. [CrossRef]

32. Hegab, H.; Kishawy, H.A.; Umer, U.; Mohany, A. A model for machining with nano-additives based minimum quantity lubrication. *Int. J. Adv. Manuf. Technol.* **2019**, *102*, 2013–2028. [CrossRef]
33. Hegab, H.; Umer, U.; Deiab, I.; Kishawy, H. Performance evaluation of Ti–6Al–4V machining using nano-cutting fluids under minimum quantity lubrication. *Int. J. Adv. Manuf. Technol.* **2018**, *95*, 4229–4241. [CrossRef]
34. Hegab, H.; Kishawy, H.A.; Gadallah, M.H.; Umer, U.; Deiab, I. On machining of Ti-6Al-4V using multi-walled carbon nanotubes-based nano-fluid under minimum quantity lubrication. *Int. J. Adv. Manuf. Technol.* **2018**, *97*, 1593–1603. [CrossRef]
35. Eltaggaz, A.; Zawada, P.; Hegab, H.A.; Deiab, I.; Kishawy, H.A. Coolant strategy influence on tool life and surface roughness when machining ADI. *Int. J. Adv. Manuf. Technol.* **2018**, *94*, 3875–3887. [CrossRef]
36. Hegab, H.; Darras, B.; Kishawy, H.A. Sustainability assessment of machining with nano-cutting fluids. *Procedia Manuf.* **2018**, *26*, 245–254. [CrossRef]
37. Kishawy, H.A.; Hegab, H.; Deiab, I.; Eltaggaz, A. Sustainability assessment during machining Ti-6Al-4V with Nano-additives-based minimum quantity lubrication. *J. Manuf. Mater. Process.* **2019**, *3*, 61. [CrossRef]
38. Gupta, M.K.; Sood, P.K.; Singh, G.; Sharma, V.S. Sustainable machining of aerospace material—Ti (grade-2) alloy: Modeling and optimization. *J. Clean. Prod.* **2017**, *147*, 614–627. [CrossRef]

 © 2019 by the authors. Licensee MDPI, Basel, Switzerland. This article is an open access article distributed under the terms and conditions of the Creative Commons Attribution (CC BY) license (http://creativecommons.org/licenses/by/4.0/).

Article

Dimensionless Analysis for Investigating the Quality Characteristics of Aluminium Matrix Composites Prepared through Fused Deposition Modelling Assisted Investment Casting

Sunpreet Singh [1], Chander Prakash [1], Parvesh Antil [2], Rupinder Singh [3], Grzegorz Królczyk [4] and Catalin I. Pruncu [5,6,*]

1. Department of Mechanical Engineering, Lovely Professional University, Phagwara, Punjab 144411, India; snprt.singh@gmail.com (S.S.); chander.mechengg@gmail.com (C.P.)
2. Department of Basic Engineering, College of Agricultural Engineering and Technology, CCS HAU, Hisar, Haryana 125004, India; parveshantil.pec@gmail.com
3. Manufacturing Research Laboratory, Production Engineering Department, Guru Nanak Dev Engineering College, Ludhiana 141006, India; rupinderkhalsa@gmail.com
4. Faculty of Mechanical Engineering, Opole University of Technology, 76 Proszkowska St., 45-758 Opole, Poland; g.krolczyk@po.opole.pl
5. Mechanical Engineering, Imperial College London, Exhibition Rd., London SW7 2AZ, UK
6. Mechanical Engineering, School of Engineering, University of Birmingham, Birmingham B15 2TT, UK
* Correspondence: c.pruncu@imperial.ac.uk

Received: 16 May 2019; Accepted: 11 June 2019; Published: 13 June 2019

Abstract: The aluminium matrix composites (AMCs) have become a tough competitor for various categories of metallic alloys, especially ferrous materials, owing to their tremendous servicing in the diversified application. In this work, additional efforts have been made to formulate a mathematical model, by using dimensionless analysis, able to predict the mechanical characteristics of the AMCs that have already been optimized and characterized by the authors. Here, the experimental and statistical data obtained from the Taguchi L18 orthogonal array and analysis of variance (ANOVA) have been used. They permit collection of the output responses and allow the identification of significant process parameters, respectively, which thereafter were used to design the mathematical model. Second order polynomial equations have been obtained from the specific output response and the relevant input parameter were incorporated with the highest level of contribution. The obtained quadratic equations indicate the regression values (R^2) equal to unity, hence, proving the performances of the fit. The results demonstrate that the developed mathematical models present very high accuracy for predicting the output responses.

Keywords: fused deposition modelling; investment casting; mathematical modelling; aluminium matrix composite

1. Introduction

In the last two decades, the rapid advancement of technology has contributed to large modification in the manufacturing sector. During this period, the demand for materials that can sustain the extreme level of service conditions increased globally. Specifically, in aerospace and automobile sectors, the requirement of materials having high strength, toughness, hardness, and prolonged service life was always a challenge. Apart from these properties, one of the major requirements is 'light weight'. Different studies have reported the needs of lighter material as one of the motivations behind the invention of reinforced materials, commonly, referred to as metal matrix composites [1–3].

Amongst various categories of metal matric composites, the one based on the aluminium (Al) matrix is in high demand, owing to its excellent thermal, mechanical, tribological, chemical, and structural characteristics [4–6]. Further, there exists a wide range of manufacturing processes, which can be used for the fabrication of the tailor-made composites with desirable properties [7]. Specifically, aluminium matrix composites (AMCs) are basically popular because of the low weight/density ratio, high wear resistance, cost effectiveness, high elastic modulus, and excellent strength [8–11]. Further, Sajjadi et al. reveal Al-Al_2O_3 as the most popular type of AMCs because it contains micro Al_2O_3 particles within the matrix of Al [12]. Such as, the Al-Al_2O_3 based composites have continually extended their applicability within industrial applications [13,14].

Traditionally, the reinforcements are introduced to the metallic matrix via an ex-situ method [15,16], wherein the matrix and reinforcements are mixed with each other outside the mould cavity or die. This method of reinforcement results in poor wettability between the reinforcement and the matrix due to the increased surface area and presence of surface contamination on the reinforcements [17]. In order to overcome the interface issue, recent trends have been shifted towards the use of reinforcements within the cavity or mould itself [18]. As defined in the literature, there are various routes, commercially available, for the preparation of AMC. The most widely used commercial routes are the stir casting and powder metallurgy [19–21]. There, the machinability of AMCs is very poor, in contrast to pure Al, due to the brittle reinforcements in the matrix. The investment casting (IC) process has shown its superiority, over the other solutions; in-terms of producing highly complex and near-net shaped parts with very fine surface finish [22]. Additionally, this process is simple, cost effective, and allows manufacturing a wide range of materials; however, the stretched production cycles represents one of the critical challenges for the IC [23,24]. As continuum improvement, the hybridization of IC and stir casting process has become the most popular methods to develop superior metal matrix composites (MMCs) [25,26]. The intrinsic weaknesses of IC process (such as: low strength of wax pattern, un-economical injection moulding cost, high die design cost, and longer production runs) can be eliminated by using fused deposition modelling (FDM) process for pattern making [27–30]. FDM works on the same principle as the Additive Manufacture (AD), wherein the thin plastic slices are deposited at a defined distance. The interface, therefore, can result in poor surface finish due to an integral stair-casing. Boschetto and Veniali suggested the barrel finishing of formed FDM parts as an efficient method to enhance their surface finish [31].

The collaboration of FDM and IC has been extensively researched in the literature, which duly cited the myriads of merits [32–36]. In the recent years, the authors have investigated a novel method for the production of the in-situ based AMCs, through the use of FDM assisted by the IC process. In this respect, the authors developed in-house composite polymeric composites that were used for the production of sacrificial patterns for the IC process. As observed in [1,2,37], the manufactured Al castings consist of Al_2O_3 distributions, which permit the validation of the authenticity of the adopted methodology. Further, the input process parameters; refer to Appendix A (i.e., Table A1), have been optimized by using Taguchi L18 orthogonal array-based design of experimentation techniques in response of dimensional accuracy [1], surface hardness [2], and surface roughness [3]. In this work, mathematical models, based on the obtained results of [1–3], for all the aforementioned output responses have been developed by using dimensionless modelling, Buckingham's π-approach. Further, regression equations have been implemented against the best features of the input process parameters, as per analysis of variance (ANOVA).

2. Materials and Methods

Figure 1 presents the methodology adopted in this research. By using a Fish bone diagram (see details on Figure 2), we highlight the main process parameters associated to the IC which can affect the quality features of the IC components. The number of IC slurry layer (N_{SL}) has been judicially selected as an input parameter due to its significance highlighted in the literature [38–40]. The following are the procedural steps followed to obtain AMCs, which refer to the original Taguchi L18 orthogonal array (Table A1), as given in the Appendix A:

- The alternative feedstock filaments (F_P) have been prepared using PA, Al_2O_3, and Al in different %wt. proportions with the help of single screw extrusion process.
- The formed filaments were used for the development of sacrificial patterns of cubical shape with three different volumes (V_P), such as 17,576 mm^3, 27,000 mm^3, and 39,304 mm^3. They were produced at low, high, and solid density of FDM process (D_P) by using uPrint-SE system of Stratasys Inc. (Edina, MN, USA). In the works, reported previously, it has been seen that the change in the in-fill density affects the mechanical and tribological performances of the developed AMCs [1–3]. The prime reason behind the selection of FDM technology is due to its affordability and suitability for hybridization within the IC process [23,24,41]. Further, the selection of the process parametric levels from previous studies has been judicially selected, based on the pilot studies.
- Prior to shell moulding, the barrel finishing (BF) process was performed on the samples, for the refurbishment of resulted surface finish [31]. Here, barrel finishing time (BF_T) and barrel finishing media weight (BF_W) have been selected as input process parameters.
- Then, the IC moulds were prepared by coating the trees (consisting of riser, pouring basin, gating, and also the FDM printed sacrificial pattern) with refractory layers of silica. The number of IC slurry layers (N_{SL}) has also varied in accordance to Table A1 in the Appendix A.
- Autoclaving and baking were performed in one step at 1150 °C (by maintaining the pouring sprue in a vertical up position so that the Al_2O_3 filler particles could be arrested within the cavity only). At this range of temperature, the matrix of the sacrificial patterns evaporates, immediately, without causing mould cracks.
- Finally, pouring of molten Al-6063 has been carried out.

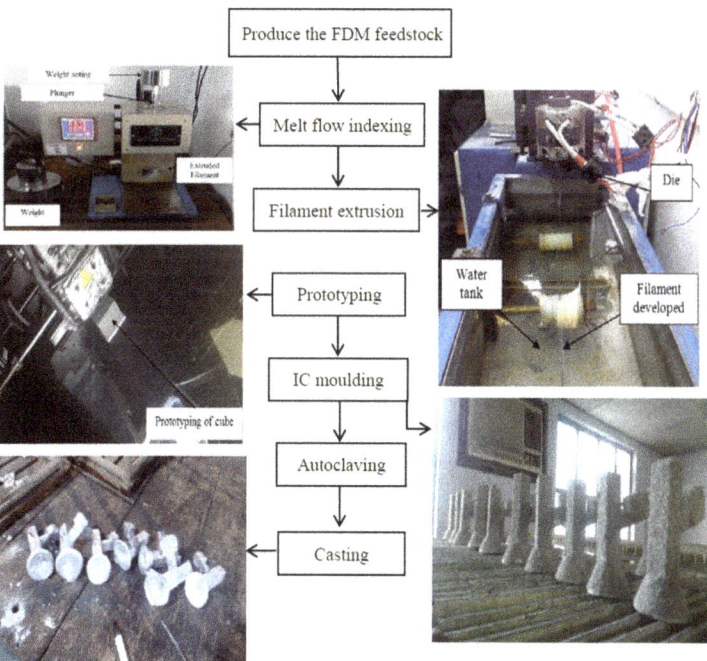

Figure 1. Step by step procedure of aluminium matrix composites (AMC) development.

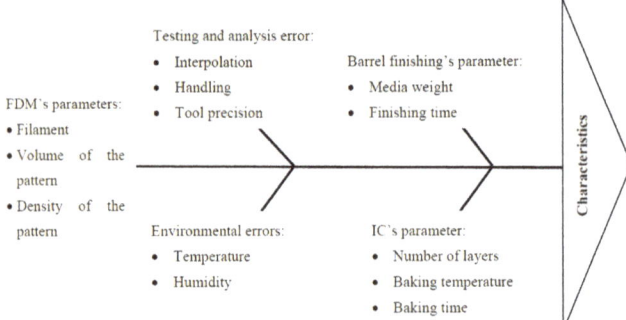

Figure 2. Fish bone diagram for prepared castings.

The castings manufactured were tested for surface hardness, dimensional accuracy and surface roughness by using HVS-1000BVM hardness tester (HV0.01 scale; ASTM-E384, Laizhou, China), Vernier Caliper (Mitutoyo: least count 0.01mm, Takatsu-ku, Kawasaki, Japan) and Mitutoyo SJ-210 (Japan, ISO: 1997) surface roughness tester, respectively. For microstructural evaluation, the Scanning Electron Microscopy (SEM, JEOL, Peabody, MA, USA) analysis has been performed on the casting manufactured in the experiment #16, #17 and #18 associated to Table A1. It has been seen that the Al_2O_3 particles presented in Al matrix allow to enhance the quality characteristics of the castings, especially the hardness on the surface. Figure 3 shows the SEM micrographs and their associated Energy Dispersive Spectroscopy (EDS) spectrums (JEOL, USA). The measurements indicate the presence of Al, O, Si, Fe, and C-peaks, which confirm the existence of alumina. These elements identified on the EDS measurements (i.e., Al, O, and C) are the common sign of alumina surface [42]. They were noted as well as the presence of elements Fe and Si, which denote some small impurity.

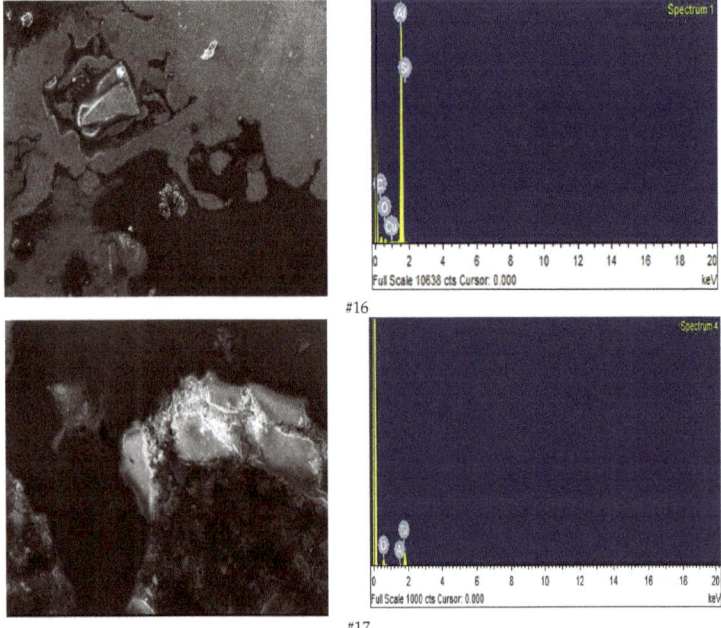

Figure 3. Cont.

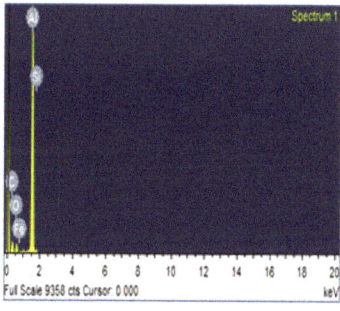
#18

Figure 3. SEM micrograph and EDS spectrum of experiment #16, #17, and #18.

3. Dimensionless modelling: Buckingham Pi Approach

Dimensionless modelling of the experimental data is considered an efficient method in order to formulate analytic mathematical functions that are out of a highly complex experimental system associated to numerous process parameters [43]. The concept of dimensionless analysis helps to reduce the influence of variables by means of physical equations [44–46]. To date, dimensionless modelling with the help of Buckingham Pi approach has been extensively investigated for a wide range of scientific and engineering applications including fluid dynamics [47], energy [48], electronics [49], heat transfer [50] and others. According to the Buckingham approach, any practical problem containing "n" factor sand further "m" dimensions, then the subtraction of n and m will result the counts of independent factors, which could be assumed. Presently, "n" and "m" are 7 and 3, respectively. Therefore, the problem will consist of $\pi 1$, $\pi 2$, $\pi 3$ and $\pi 4$ that are the dimensional magnitudes. Furthermore, the mathematical formulae derived for the assumed independent parameters help to develop the dimensional relationships by following a set of standard steps [51,52]. Standard quantities of the same physical nature (mass, length, and time) are used based on fundamental units. Consequently, it can be said that these systems belong to the same class. To generalize, a set of systems of units that differ only in the magnitude (but not in the physical nature) of the fundamental units are called a class of systems of units [53]. Unlike other statistical approaches, the mathematical modelling in the case of Buckingham' Pi approach could be very tedious if a proper set of producers is not considered. Based on [53], following are the step-by-step descriptions of the modelling process adopted in the present work:

i. First of all, the units of the input and the output process parameters have been unified and converted into physical quantities (such as M, L, and T). Further, it is of utmost importance to highlight that any kind of categorical parameter, either input or output, is not suitable for the modelling. Moreover, upon such conversions, it should be considered that the replacement could be represented in-terms of M, L, and T formats. Therefore, in present work, the original Table A1 in the Appendix A has been modified in order to balance the units, as well as to convert the qualitative parameters into quantitative. For instance, the parameter "filament proportion" has been quantified in-terms of its tensile strength; density of the FDM pattern has been considered in terms of mass and volume; mould wall thickness has been converted from a number of layers to thickness of the wall, etc. Table 1 is the final prepared modified version of Table A1.
The obtained dimensions of input and output parameters would be:
Hardness (H) as $ML^{-1}T^{-2}$, Dimensional accuracy as L,
Surface roughness as L,
Filament proportion (P) in-terms of tensile strength of filament as MLT^{-2},
Volume of FDM reinforced pattern (V) as L^3,
Density of FDM pattern (ϱ) as ML^{-3},
BF cycle time (t) as T,

ii. BF media weight (W) as M and the Number of IC slurry layers (l) resulting into mould wall thickness as L.

ii. Then, it is mandatory to find out the significance level of the input process parameters for the measured outcomes. In the present case, ANOVA has been implemented with the help of MINITAB-17 based statistical software in order to identify the significance and contribution of input parameters. Table 2 shows the contribution percentage of input process parameters for surface hardness, dimensional accuracy, and surface roughness.

iii. Before starting to formulate the π equations (let us say 'x'), it is necessary to identify the 'x − 1' top performing input parameters. For instance, in the case of surface hardness, when 'x' is equal to 4 that allows to develop 4 π-equations, three top performing input parameters have to be identified.

iv. Now, the top performing input parameters and the output parameters being analyzed represent the π equations.

v. After calculating the π equations, the $\pi1$ (related to the output parameter) is solved as a function of other πs ($\pi2$, $\pi3$, and $\pi4$, consisted of input parameters).

vi. Once the step-v is completed, a constant 'K' has been considered whose value has been driven from a second order quadratic equation of the fitness curve that connect the output response and the most contributing input parameter.

vii. Further, the fitness curve should be plotted between the measured output values and the corresponding values of the most significant input parameter, while keeping the rest of the parameters constant. Alternatively, in the present case, the plots have been drawn between the three levels of the input process parameters and the average of the corresponding output result. For instance, in case of Figure 4, the average of hardness for experiment #1, #4, #7, #10, #13, and #16 has been plotted against first level of F_D (5.12×10^{-6} N/mm^3) and the average of hardness for experiment #2, #5, #8, #11, #14, and #17 has been plotted against second level of F_D (7.63×10^{-6} N/mm^3). Similar procedure has been adopted for the third level of the F_D.

viii. Noticeably, the regression (R^2) ~ 1 indicates the best fitness of the data.

3.1. Hardness

In the present study, hardness is considered a function of all input process parameters that is expressed by Equation (1).

So,

$$H = f(P, V, \varrho, t, W, l) \qquad (1)$$

Based on the Table 2; the least significant parameters for this particular parameter are BF cycle time, BF media filament proportion, and weight that will directly go in "π" groups. The "π" eqns. for hardness can be written as:

$$\pi_1 = H \, (F)^{a1} \, (t)^{b1} \, (W)^{c1} \qquad (2)$$

$$\pi_2 = \varrho \, (F)^{a2} \, (t)^{b2} \, (W)^{c2} \qquad (3)$$

$$\pi_3 = l \, (F)^{a3} \, (t)^{b3} \, (W)^{c3} \qquad (4)$$

$$\pi_4 = V \, (F)^{a4} \, (t)^{b4} \, (W)^{c4} \qquad (5)$$

After substituting the decided dimensions in the "π" groups, Equations (6), (8), (10), and (12) are formed. Now, in order to solve these further, the resulted equations are equated to zero. For instance, the $\pi1$ will be solved as follows:

$$\pi_1 = ML^{-1}T^{-2} \, (ML^{-1}T^{-2})^{a1} \, (T)^{b1} \, (M)^{c1} \qquad (6)$$

Equating the basic dimensions to zero:
M: $1 + a_1 + c_1 = 0$
L: $-1 - a_1 = 0$
T: $-2 - 2a_1 + b_1 = 0$
We get,
$a_1 = -1$, $b_1 = 0$ and $c_1 = 0$
So, Equation (2) can be re-written as:

$$\pi_1 = H/F \tag{7}$$

Similarly, on solving π_2;

$$\pi_2 = ML^{-3} (ML^{-1}T^{-2})^{a_2} (T)^{b_2} (M)^{c_2} \tag{8}$$

Similarly, equating the basic dimensions to zero:
M: $1 + a_2 + c_2 = 0$
L: $-3 - a_2 = 0$
T: $-2a_2 + b_2 = 0$
We get,
$a_2 = -3$, $b_2 = -6$ and $c_2 = 2$
So, Equation (3) can be re-written as;

$$\pi_2 = \varrho/F^3 t^6 \tag{9}$$

On solving π_3;

$$\pi_3 = L (ML^{-1}T^{-2})^{a_3} (T)^{b_3} (M)^{c_3} \tag{10}$$

Equating the basic dimensions to zero:
M: $a_3 + b_3 = 0$
L: $1 - a_3 = 0$
T: $-2a_3 + b_3 = 0$
We get,
$a_3 = 1$, $b_3 = 2$ and $c_3 = -1$
The Equation (4) for π_3 can be re-written as;

$$\pi_3 = lFT^2/W \tag{11}$$

Solving π_4;

$$\pi_4 = L^3 (ML^{-1}T^{-2})^{a_4} (T)^{b_4} (M)^{c_4} \tag{12}$$

Equating the basic dimensions to zero:
M: $a_4 + c_4 = 0$
L: $3 - a_4 = 0$
T: $0 - 2a_4 + b_4 = 0$
We get,
$a_4 = 3$, $b_4 = 6$ and $c_4 = -3$
The Equation (5) for π_4 can be re-written as;

$$\pi_4 = VF^3 t^6/W^3 \tag{13}$$

The final relationship between all four Equations of "π" can be assumed as;
$\pi_1 = f(\pi_2, \pi_3$ and $\pi_4)$

Table 1. Modified Taguchi L18 orthogonal array.

Exp. No.	Tensile Strength, N/mm²	Volume of Fused Deposition Modelling (FDM) Reinforced Pattern (mm³)	Density of FDM Pattern, N/mm³	BF Cycle Time (sec)	BF Media Weight (N)	Mould Wall Thickness Obtained, mm	H, N/mm² (Converted from HV with a Multiplying Factor of 9.807)	Δd, mm	R_a, mm (Converted from μm with a Dividing Factor of 0.001)
1	21.65	17576	5.12×10^{-6}	1200	98	11.5	877.72	0.026	4762
2	21.65	17576	7.63×10^{-6}	2400	147	13	900.28	0.033	5151
3	21.65	17576	9.16×10^{-6}	3600	196	15	1127.80	0.02	4778
4	21.65	27000	5.12×10^{-6}	1200	147	13	787.50	0.056	4371
5	21.65	27000	7.63×10^{-6}	2400	196	15	848.30	0.063	5582
6	21.65	27000	9.16×10^{-6}	3600	98	11.5	1127.80	0.053	6094
7	21.65	39304	5.12×10^{-6}	2400	98	15	756.11	0.043	5368
8	21.65	39304	7.63×10^{-6}	3600	147	11.5	901.26	0.08	5658
9	21.65	39304	9.16×10^{-6}	1200	196	13	984.62	0.016	6404
10	21.53	17576	5.12×10^{-6}	3600	196	13	915.97	0.016	4709
11	21.53	17576	7.63×10^{-6}	1200	98	15	940.49	0.076	4573
12	21.53	17576	9.16×10^{-6}	2400	147	11.5	1317.08	0.056	4658
13	21.53	27000	5.12×10^{-6}	2400	196	11.5	934.60	0.033	5297
14	21.53	27000	7.63×10^{-6}	3600	98	13	919.89	0.05	5889
15	21.53	27000	9.16×10^{-6}	1200	147	15	1024.83	0.06	6845
16	21.53	39304	5.12×10^{-6}	3600	147	15	824.76	0.033	8564
17	21.53	39304	7.63×10^{-6}	1200	196	11.5	1004.23	0.043	5721
18	21.53	39304	9.16×10^{-6}	2400	98	13	1041.50	0.046	5894

Table 2. Percentage contribution of input process parameters.

Source	Surface Hardness (H)	Dimensional Accuracy (Δd)	Surface Roughness (Ra)
F_P	7.69%	0.76%	4.16%
V_P	8.85%	16.95%	43.84% *
D_P	65.75% *	19.83%	3.03%
BF_T	1.03%	3.30%	6.45%
BF_W	0.8 %	31.71% *	2.94%
N_{SL}	14.14%	8.97%	5.72%
Residual Error	1.74%	18%	33.86%
Total	100%	100%	100%

* Highly contributing factor.

Or

$H/F = \left(\frac{\rho}{F^3 t^6}, \frac{lFt^2}{W} \text{ and } VF^3 t^6 / W^3\right)$

The above expression can be written as:

$$H = K \cdot \rho \cdot F^2 \cdot l \cdot t^2 \cdot V / W^4 \tag{14}$$

Here, "K" is the proportionality constant.

Experimentally, it has been found that a correlation between the hardness and "ρ" exists (refer Table 2). Hence, it was taken as representative factors to develop the mathematical model. The average values of the hardness obtained at different levels of "ρ" (throughout the Table 1) has been plotted (see details in Figure 4). In this case, a regression equation ($R^2 = 1$) with a second order has been determined. Based on the obtained linear equation, the final mathematical model that includes the hardness is given:

$$H = [(2E + 13\rho^2 - 3E + 8\rho + 1607.5)]F^2 \cdot L \cdot t^2 \cdot V / w^4 \tag{15}$$

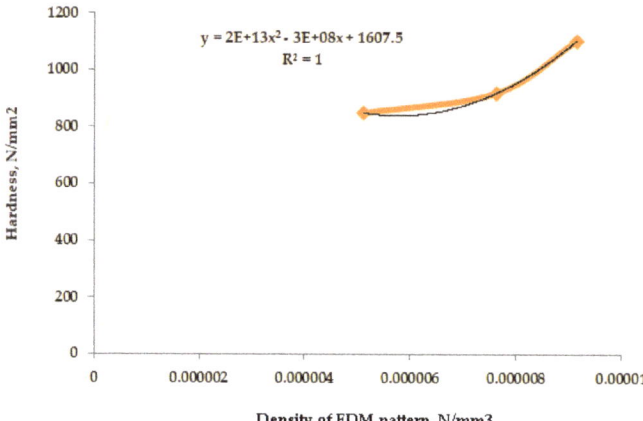

Figure 4. Hardness versus density of fused deposition modelling (FDM) pattern plot.

3.2. Dimensional Accuracy

In a similar way, dimensional accuracy is considered as a function of all input process parameters that is expressed by Equation (16).

$$\Delta d = f(F, V, \rho, t, W, l) \tag{16}$$

From Table 2, the least significant parameters are BF cycle time, number of IC slurry layers, and filament proportion, that will directly go in "π" groups. The "π" equation for dimensional accuracy can be written as:

$$\pi_1 = \Delta d \, (F)^{a1} \, (t)^{b1} \, (L)^{c1} \tag{17}$$

$$\pi_2 = W \, (F)^{a2} \, (t)^{b2} \, (L)^{c2} \tag{18}$$

$$\pi_3 = \varrho \, (F)^{a3} \, (t)^{b3} \, (L)^{c3} \tag{19}$$

$$\pi_4 = V \, (F)^{a4} \, (t)^{b4} \, (L)^{c4} \tag{20}$$

The same set of mathematical iterations has been repeated for dimensional accuracy and the relationship between the all four "π" equations is given in Equation (21) as below:

$$\Delta d/l = f\left(\frac{W}{Ft^2 l}, \frac{\rho l^2}{Ft^2} \text{ and } \frac{V}{l^3}\right) \tag{21}$$

On solving the above expression, we get:

$$\Delta d = K \cdot \varrho \cdot W \cdot V / F^2 \cdot l \cdot t^4 \tag{22}$$

BF media weight, which is the most significant parameter (refer to Table 2) with regards to dimensional accuracy, of the casted composites, has been taken as the representative parameter to develop the mathematical model. For this, the average values of the dimensional accuracy obtained at different levels of "BF_W" (throughout the Table 1) has been plotted; refer to Figure 5. Then, a regression equation ($R^2 = 1$) with a second order has been determined. Based on the obtained linear equation, the final mathematical model for dimensional accuracy is given as:

$$\Delta d = [(-5E - 06W^2 + 0.0014W - 0.0345] \, \varrho \cdot V / F^2 \cdot l \cdot t^4 \tag{23}$$

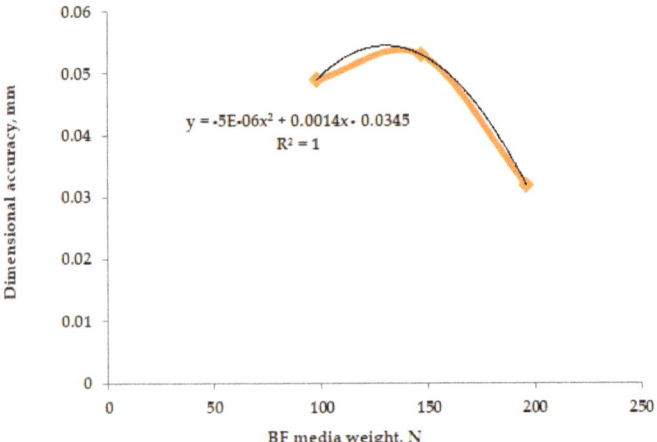

Figure 5. Dimensional accuracy versus barrel finishing (BF) media weight plot.

3.3. Surface Roughness

Further, Equation (24) represents the surface roughness, as a function of all input process variable:

$$Ra = f(F, V, \rho, t, W, l) \tag{24}$$

Based on the Table 2; density of FDM pattern, filament proportion, and BF media weight are the least significant parameters for surface roughness that will directly go in "π" groups. The "π" equation for dimensional accuracy can be written as:

$$\pi_1 = Ra\,(W)^{a1}\,(\rho)^{b1}\,(F)^{c1} \tag{25}$$

$$\pi_2 = V\,(W)^{a2}\,(\rho)^{b2}\,(F)^{c2} \tag{26}$$

$$\pi_3 = t\,(W)^{a3}\,(\rho)^{b3}\,(F)^{c3} \tag{27}$$

$$\pi_4 = l\,(W)^{a4}\,(\rho)^{b4}\,(F)^{c4} \tag{28}$$

Now, repeating the same set of mathematical operations the final expression that describes the relationship between all the four "π" is given as Equation (29):

$$R_a(\rho/W)^{1/3} = f\left(\frac{V\rho}{W},\, \frac{t}{(\rho)^{\frac{1}{6}}(FW)^{\frac{1}{3}}}\, \text{and}\, l\left(\frac{W}{\rho}\right)^{1/3}\right) \tag{29}$$

Equation (29) can be written as:

$$Ra = K \cdot (V \cdot t \cdot l \cdot \rho^{\wedge}(1/6))/F^{1/3} \cdot W^{2/3}, \tag{30}$$

Similar to the dimensional accuracy, volume of FDM reinforced pattern which is the most significant parameter (refer to Table 2) with regard to surface roughness, of the casted composites, has been taken as the representative parameter to develop the mathematical model. For this, the average values of the surface roughness obtained at different levels of "V_P" (throughout Table 1) has been plotted; refer to Figure 6. Then, a regression equation ($R^2 = 1$) with a second order has been determined. From the obtained linear equation, the final mathematical model for surface roughness is given as:

$$R_a = \left[\left(-2E - 6V^2 + 0.2101V + 1846.5\right)\right] \cdot \left(t \cdot l \cdot \rho^{\frac{1}{6}}\right)/F^{1/3} \cdot W^{2/3} \tag{31}$$

These results obtained in the present work are found to be in-line with the observations presented in the literature [42,45].

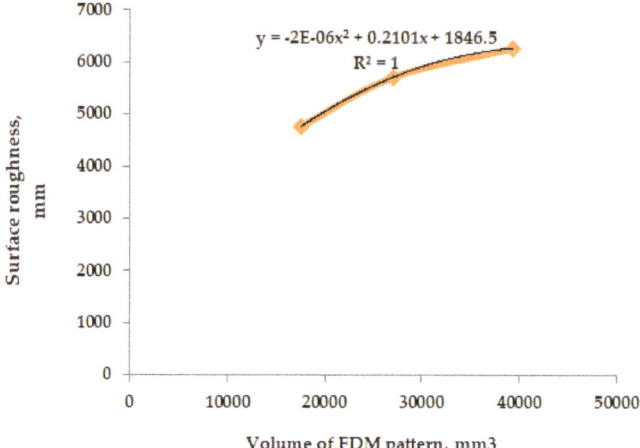

Figure 6. Surface roughness versus volume of FDM pattern plot.

4. Conclusions

In this work, Vashy-Buckingham's π-theorem was employed successfully for the development of the mathematical models related to the hardness, dimensional accuracy, and surface roughness of AMCs; material that was produced through FDM assisted by the IC process. The ANOVA simulation were embedded in the present methodology in order to generate a standard database and to recognize the significance process parameters, respectively. Further, all three mathematical models developed are of second order polynomial equations, with a regression value equal to 1, which prove the reliability of the models.

Author Contributions: Conceptualization, S.S. and R.S.; methodology, R.S., S.S; software, S.S., C.P; validation, S.S., P.A., C.P. and C.I.P; formal analysis, S.S., P.A; writing—original draft preparation, S.S., R.S., C.P., C.I.P; writing—review and editing, S.S., C.P., G.K,; supervision, C.P., R.S; project administration, S.S., R.S; funding acquisition, C.P., C.I.P.

Funding: This research received no external funding.

Conflicts of Interest: The authors declare no conflict of interest.

Appendix A

Table A1. Design of experimentation as per original Taguchi L18 orthogonal array.

Exp. No.	F_P	V_P (mm³)	D_P	BF_T	BF_W	NS_L	R_a (μm)	S/N ratio (dB)	Δd (mm)	S/N ratio (dB)	HV	S/N ratio (dB)
1	C1	26 × 26 × 26	Low density	20	10	7	4.762	−13.55	0.026	31.34	89.5	39.01
2	C1	26 × 26 × 26	High density	40	15	8	5.151	−14.27	0.03	29.45	91.8	39.18
3	C1	26 × 26 × 26	Solid	60	20	9	4.778	−13.58	0.02	33.30	115	41.18
4	C1	30 × 30 × 30	Low density	20	15	8	4.371	−12.82	0.06	23.37	80.3	38.06
5	C1	30 × 30 × 30	High density	40	20	9	5.582	−14.93	0.063	23.80	86.5	38.72
6	C1	30 × 30 × 30	Solid	60	10	7	6.094	−15.69	0.053	25.32	115	41.22
7	C1	34 × 34 × 34	Low density	40	10	9	5.368	−14.59	0.043	27.06	77.1	37.73
8	C1	34 × 34 × 34	High density	60	15	7	5.658	−15.05	0.08	21.89	91.9	39.25
9	C1	34 × 34 × 34	Solid	20	20	8	6.404	−16.13	0.016	35.22	100.4	39.92
10	C2	26 × 26 × 26	Low density	60	20	8	4.709	−13.45	0.016	35.22	93.4	39.38
11	C2	26 × 26 × 26	High density	20	10	9	4.573	−13.20	0.076	22.29	95.9	39.62
12	C2	26 × 26 × 26	Solid	40	15	7	4.658	−13.36	0.056	24.72	134.3	42.60
13	C2	30 × 30 × 30	Low density	40	20	7	5.297	−14.48	0.033	29.45	95.3	39.56
14	C2	30 × 30 × 30	High density	60	10	8	5.889	−15.40	0.050	25.90	93.8	39.41
15	C2	30 × 30 × 30	Solid	20	15	9	6.845	−16.70	0.060	24.35	104.5	40.37
16	C2	34 × 34 × 34	Low density	60	15	9	8.564	−18.65	0.033	29.20	84.1	38.29
17	C2	34 × 34 × 34	High density	20	20	7	5.721	−15.15	0.043	27.06	102.4	40.20
18	C2	34 × 34 × 34	Solid	40	10	8	5.894	−15.40	0.046	26.44	106.2	40.48

Where, F_P, V_P, D_P, BF_T, BF_W, NS_L, Ra, Δd, HV, and S/N represent the filament proportion, volume of the pattern, density of the pattern, barrel finishing time, barrel finishing media weight, number of IC slurry layers, surface roughness, dimensional accuracy/deviation, Vickers hardness, signal/noise, respectively. Further, C1 and C2 are the compositions of $PA_x/Al_2O_{3y}/Al_z$ (where x is 60% by wt.; y is 10% and 12% by wt., respectively; and z: 28% and 30% by wt., respectively).

References

1. Singh, S.; Singh, R. Investigations for dimensional accuracy of AMC prepared by using Nylon6-Al-Al$_2$O$_3$ reinforced FDM filament in investment casting. *Rapid Prototyp. J.* **2016**, *22*, 445–455. [CrossRef]
2. Singh, S.; Singh, R. Effect of process parameters on micro hardness of Al–Al$_2$O$_3$ composite prepared using an alternative reinforced pattern in fused deposition modelling assisted investment casting. *Robot. Comput. Integr. Manuf.* **2016**, *37*, 162–169. [CrossRef]
3. Singh, S.; Singh, R. Some investigations on surface roughness of aluminium metal composite primed by fused deposition modeling-assisted investment casting using reinforced filament. *J. Braz. Soc. Mech. Sci. Eng.* **2017**, *39*, 471–479. [CrossRef]
4. Koli, D.K.; Agnihotri, G.; Purohit, R. Advanced aluminium matrix composites: The critical need of automotive and aerospace engineering fields. *Mater. Today: Proc.* **2015**, *2*, 3032–3041. [CrossRef]

5. Iwai, Y.; Honda, T.; Miyajima, T.; Iwasaki, Y.; Surappa, M.K.; Xu, J.F. Dry sliding wear behavior of Al_2O_3 fibre reinforced aluminum composites. *Compos. Sci. Technol.* **2002**, *60*, 1781–1789. [CrossRef]
6. Barekar, N.; Tzamtzis, S.; Dhindaw, B.K.; Patel, J.; Babu, N.H.; Fan, Z. Processing of aluminum-graphite particulate metal matrix composites by advanced shear technology. *J. Mater. Eng. Perform.* **2009**, *18*, 1230–1240. [CrossRef]
7. Kathiresan, M.; Sornakumar, T. Friction and wear studies of die cast aluminum alloy-aluminum oxide-reinforced composites. *Ind. Lubr. Tribol.* **2010**, *62*, 361–371. [CrossRef]
8. Miyajima, T.; Iwai, Y. Effects of reinforcement on sliding wear behavior of aluminium matrix composite. *Wear* **2003**, *255*, 606–616. [CrossRef]
9. Prasad, S.V.; Asthana, R. Aluminum metal-matrix composites for automotive applications: Tribological considerations. *Tribol. Lett.* **2004**, *17*, 445–453. [CrossRef]
10. Surappa, M.K. Aluminium matrix composites challenges and opportunities. *Sadhana* **2003**, *28*, 319–325. [CrossRef]
11. Mohan, S.; Srivastava, S. Surface behaviour of as Cast Al-Fe intermetallic composites. *Tribol. Lett.* **2006**, *22*, 45–51. [CrossRef]
12. Sajjadi, S.A.; Ezatpour, H.R.; Parizi, M.T. Comparison of microstructure and mechanical properties of A356 aluminum alloy/Al_2O_3 composites fabricated by stir and compo-casting processes. *Mater. Des.* **2012**, *34*, 106–111. [CrossRef]
13. Ralph, B.; Yuen, H.C.; Lee, W.B. The processing of metal matrix composites—An overview. *J. Mater. Process. Technol.* **1997**, *63*, 339–353. [CrossRef]
14. Kurşun, A.; Bayraktar, E.; Robert, M.H. Low cost manufacturing of aluminium-alumina composites. *Adv. Mater. Process. Technol.* **2015**, *1*, 515–528. [CrossRef]
15. Woo, K.D.; Huo, H.W. Effect of high energy ball milling on displacement reaction and sintering of Al–Mg/SiO_2 composite powders. *Met. Mater. Int.* **2006**, *12*, 45. [CrossRef]
16. Tjong, S.C.; Ma, Z.Y. Microstructural and mechanical characteristics of in situ metal matrix composites. *Mater. Sci. Eng. R Rep.* **2000**, *29*, 49–113. [CrossRef]
17. Maleki, A.; Niroumand, B.; Meratian, M. Effects of processing temperature on in-situ reinforcement formation in Al (Zn)/Al_2O_3 (ZnO) nanocomposite. *Metall. Mater. Eng.* **2015**, *21*, 283–291. [CrossRef]
18. Afkham, Y.; Khosroshahi, R.A.; Rahimpour, S.; Aavani, C.; Brabazon, D.; Mousavian, R.T. Enhanced mechanical properties of in situ aluminium matrix composites reinforced by alumina nanoparticles. *Arch. Civ. Mech. Eng.* **2018**, *18*, 215–226. [CrossRef]
19. Harrigan, W.C. Commercial processing of metal matrix composites. *Mater. Sci. Eng. A* **1998**, *244*, 75–79. [CrossRef]
20. Degischer, H.P. Innovative light metals: Metal matrix composites and foamed aluminium. *Mater. Des.* **1997**, *18*, 221–226. [CrossRef]
21. Neussl, E.; Sahm, P.R. Selectively reinforced component produced by the modified investment casting process. *Compos. Part A* **2001**, *32*, 1177–1183. [CrossRef]
22. Mazahery, A.; Abdizadeh, H.; Baharvandi, R. Development of high-performance A356/nano-Al_2O_3 composites. *Mater. Sci. Eng. A* **2009**, *518*, 61–64. [CrossRef]
23. Singh, S.; Singh, R. Fused deposition modelling based rapid patterns for investment casting applications: A review. *Rapid Prototyp. J.* **2016**, *22*, 123–143. [CrossRef]
24. Kumar, P.; Ahuja, I.S.; Singh, R. Experimental investigations on hardness of the biomedical implants prepared by hybrid investment casting. *J. Manuf. Process.* **2016**, *21*, 160–171. [CrossRef]
25. Kisasoz, A.; Guler, K.A.; Karaaslan, A. Infiltration of A6063 aluminium alloy into SiC–B_4C hybrid preforms using vacuum assisted block mould investment casting technique. *Trans. Nonferrous Met. Soc. China* **2012**, *22*, 563–1567. [CrossRef]
26. Reddy, C.; Zitoun, E. Tensile behavior of 6063/Al_2O_3 particulate metal matrix composites fabricated by investment casting process. *Int. J. Appl. Eng. Res.* **2010**, *1*, 542–552.
27. Rooks, B. Rapid tooling for casting prototypes. *Assem. Autom.* **2012**, *22*, 40–45. [CrossRef]
28. Kakde, K.U.; Tumane, A.S. Development of customized innovative product using fused deposition modeling technique of rapid prototyping and investment casting. In Proceedings of the National Conference on Innovative Paradigms in Engineering and Technology, Nagpur, Maharashtra, India, 28 January 2012; pp. 27–30.

29. Singh, R.; Singh, S.; Mahajan, V. Investigations for dimensional accuracy of investment casting process after cycle time reduction by advancements in shell moulding. *Procedia Mater. Sci.* **2014**, *6*, 859–865. [CrossRef]
30. Singh, R.; Singh, S.; Singh, G. Dimensional accuracy comparison of investment castings prepared with wax and abs patterns for bio-medical application. *Procedia Mater. Sci.* **2014**, *6*, 851–858. [CrossRef]
31. Boschetto, V.G.; Veniali, F. Modelling micro geometrical profiles in fused deposition process. *Int. J. Adv. Manuf. Technol.* **2012**, *61*, 945–956. [CrossRef]
32. Blake, P.; Fodran, E.; Koch, M.; Menon, U.; Priedeman, B.; Sharp, S. FDM of ABS patterns for investment casting. In Proceedings of the 1997 International Solid Freeform Fabrication Symposium, Austin, TX, USA, 11–13 August 1997.
33. Hafsa, M.N.; Ibrahim, M.; Wahab, M.; Zahid, M.S. Evaluation of FDM pattern with ABS and PLA material. In *Applied Mechanics and Materials*; Trans Tech Publications: Switzerland, Switzerland, 2014; Volume 465, pp. 55–59.
34. Harun, W.S.; Sharif, S.; Idris, M.H.; Kadirgama, K. Characteristic studies of collapsibility of ABS patterns produced from FDM for investment casting. *Mater. Res. Innov.* **2009**, *13*, 340–343. [CrossRef]
35. Harun, W.S.; Safian, S.; Idris, M.H. Evaluation of ABS patterns produced from FDM for investment casting process. *Comput. Method Exp. Mater. Characterisation IV* **2009**, *1*, 319–328.
36. Idris, M.H.; Sharif, S.; Harun, W.S. Evaluation of ABS patterns produced from FDM for investment casting process. In Proceedings of the 9th Asia Pasific Industrial Engineering & Management Systems Conference, Kaohsiung, Taiwan, 3–5 December 2008; pp. 3–5.
37. Singh, S.; Singh, R. Study on tribological properties of Al–Al_2O_3 composites prepared through FDMAIC route using reinforced sacrificial patterns. *J. Manuf. Sci. Eng.* **2016**, *138*, 021009. [CrossRef]
38. Singh, R.; Singh, S. Effect of process parameters on surface hardness, dimensional accuracy and surface roughness of investment cast components. *J. Mech. Sci. Technol.* **2013**, *27*, 191–197. [CrossRef]
39. Sun, S.C.; Bo, Y.; Liu, M.P. Effects of moulding sands and wall thickness on microstructure and mechanical properties of Sr-modified A356 aluminum casting alloy. *Trans. Nonferrous Met. Soc. China* **2012**, *22*, 1884–1890. [CrossRef]
40. Jiang, W.; Fan, Z.; Chen, X.; Wang, B.; Wu, H. Combined effects of mechanical vibration and wall thickness on microstructure and mechanical properties of A356 aluminum alloy produced by expendable pattern shell casting. *Mater. Sci. Eng. A* **2014**, *619*, 228–237. [CrossRef]
41. Bikas, H.; Stavropoulos, P.; Chryssolouris, G. Additive manufacturing methods and modelling approaches: A critical review. *Int. J. Adv. Manuf. Technol.* **2015**, *83*, 389–405. [CrossRef]
42. Shanmughasundaram, P. Investigation on the Wear Behaviour of Eutectic Al-Si Alloy-Al_2O_3-Graphite Composites Fabricated Through Squeeze Casting. *Mater. Res.* **2014**, *17*, 940–946. [CrossRef]
43. Singh, S.; Singh, R. Wear modelling of Al-Al_2O_3 functionally graded material prepared by FDM assisted investment castings using dimensionless analysis. *J. Manuf. Process.* **2015**, *20*, 507–514. [CrossRef]
44. Wang, J. Predictive depth of jet penetration models for abrasive water jet cutting of alumina ceramics. *Int. J. Mech. Sci.* **2007**, *49*, 306–316. [CrossRef]
45. Anders, D.; Munker, T.; Artel, J.; Weinberg, K. A dimensional analysis of front-end bending in plate rolling applications. *J. Mater. Process. Technol.* **2012**, *212*, 1387–1398. [CrossRef]
46. Singh, R.; Khamba, J.S. Mathematical modeling of surface roughness in ultrasonic machining of titanium using Buckingham-$\prod$ approach: A Review. *Int. J. Abras. Technol.* **2009**, *2*, 3–24. [CrossRef]
47. Tinker, D.C.; Osborne, R.J.; Pitz, R.W. *Annular-Electrode Spark Discharges in Flowing Oxygen: Buckingham Pi Analysis*; American Inst. of Aeronautics and Astronautics: Reston, VA, USA, 2019.
48. Ekici, C.; Teke, I. Developing a new solar radiation estimation model based on Buckingham theorem. *Results Phys.* **2018**, *9*, 263–269. [CrossRef]
49. Tavakoli, S.; Sadeghi, J.; Griffin, I.; Fleming, P.J. PI controller tuning for load disturbance rejection using constrained optimization. *Int. J. Dyn. Control* **2018**, *6*, 188–199. [CrossRef]
50. Salmani, F.; Mahpeykar, M.R.; Rad, E.A. Estimating heat release due to a phase change of high-pressure condensing steam using the Buckingham Pi theorem. *Eur. Phys. J. Plus* **2019**, *134*, 48. [CrossRef]
51. Bakhtar, F.; White, A.J.; Mashmoushy, H. Theoretical treatments of two-dimensional two-phase flows of steam and comparison with cascade measurements. *Proc. Inst. Mech. Eng. Part C J. Mech. Eng. Sci.* **2005**, *219*, 1335–1355. [CrossRef]

52. Buckingham, E. On physically similar systems; illustrations of the use of dimensional equations. *Phys. Rev.* **1914**, *4*, 345. [CrossRef]
53. Zohuri, B. *Dimensional Analysis beyond the Pi Theorem*; Springer: Berlin, Germany, 2017.

© 2019 by the authors. Licensee MDPI, Basel, Switzerland. This article is an open access article distributed under the terms and conditions of the Creative Commons Attribution (CC BY) license (http://creativecommons.org/licenses/by/4.0/).

Article

Optimal Machining Strategy Selection in Ball-End Milling of Hardened Steels for Injection Molds

Irene Buj-Corral [1], Jose-Antonio Ortiz-Marzo [1], Lluís Costa-Herrero [1], Joan Vivancos-Calvet [1] and Carmelo Luis-Pérez [2,*]

[1] Universitat Politècnica de Catalunya (UPC)-Escola Tècnica Superior d'Enginyeria Industrial de Barcelona (ETSEIB), 08034 Barcelona, Spain; irene.buj@upc.edu (I.B.-C.); jose.antonio.ortiz@upc.edu (J.-A.O.-M.); lluis.costa@upc.edu (L.C.-H.); joan.vivancos@upc.edu (J.V.-C.)
[2] Universidad Pública de Navarra-Dpto. de Ingeniería, 31006 Navarra, Spain
* Correspondence: cluis.perez@unavarra.es

Received: 12 February 2019; Accepted: 8 March 2019; Published: 14 March 2019

Abstract: In the present study, the groups of cutting conditions that minimize surface roughness and its variability are determined, in ball-end milling operations. Design of experiments is used to define experimental tests performed. Semi-cylindrical specimens are employed in order to study surfaces with different slopes. Roughness was measured at different slopes, corresponding to inclination angles of 15°, 45°, 75°, 90°, 105°, 135° and 165° for both climb and conventional milling. By means of regression analysis, second order models are obtained for average roughness Ra and total height of profile Rt for both climb and conventional milling. Considered variables were axial depth of cut a_p, radial depth of cut a_e, feed per tooth f_z, cutting speed v_c, and inclination angle Ang. The parameter a_e was the most significant parameter for both Ra and Rt in regression models. Artificial neural networks (ANN) are used to obtain models for both Ra and Rt as a function of the same variables. ANN models provided high correlation values. Finally, the optimal machining strategy is selected from the experimental results of both average and standard deviation of roughness. As a general trend, climb milling is recommended in descendant trajectories and conventional milling is recommended in ascendant trajectories. This study will allow the selection of appropriate cutting conditions and machining strategies in the ball-end milling process.

Keywords: surface finish; high speed milling (HSM); roughness; modeling

---

1. Introduction

In order to increase productivity and reduce costs, it is important to choose appropriate cutting conditions in high speed milling (HSM) processes because they will influence surface roughness and the dimensional precision obtained. For example, the tool inclination angle significantly influences the surface roughness obtained. When the tool is perpendicular to the workpiece's surface, cutting speed is zero at the tool tip [1,2]. This implies that the tool tends to crush the material instead of cutting it.

In mathematical modeling of machining processes several methods can be used, such as statistical regression techniques, artificial neural network modeling techniques (ANN), and fuzzy set theory-based modeling [3]. Neural networks provide a relationship between input and output variables by means of mathematical functions, to which different weights are applied. A training algorithm is defined that consists of adjusting the weights of a network that minimize error between actual and desired outputs [4]. In recent times, neural networks have been used for modeling and predicting surface roughness in different machining operations. For example, Feng et al. modeled roughness parameters related to the Abbott–Firestone curve by means of ANN in honing operations [5] and in turning processes [6]. Özel et al. [7] and Sonar et al. [8] also employed ANN for modeling average roughness Ra, in turning processes. Moreover, simulations of machined surfaces have also been extensively investigated. Among many other studies, T. Gao et al. [9] developed a new method for the prediction of the machined surface topography in the milling process and Honeycutt and Schmitz [10] employed time domain simulation and experimental results for surface location error and surface roughness prediction. Vallejo and Morales-Menendez [11] used neural networks for modeling Ra in peripheral milling, with different input variables, such as feed per tooth, cutting tool diameter, radial depth of cut, and Brinell hardness. Zain et al. [12] modeled surface roughness with cutting speed, feed rate and radial rake angle as input variables in peripheral milling, and Quintana et al. [13] employed neural networks for studying average roughness in vertical milling. Regarding ball-end milling processes, Zhou et al. [14] used grey relational analysis (GRA) with neural network and particle swarm (PSO) algorithm to model 3D root mean square deviation of height value Sq, and compressive residual stresses, with tilt angle, cutting speed and feed as variables.

With regard to the modeling of milling processes by conventional regression models, several models have been developed, but most studies do not consider the variability which occurs as a consequence of the slope variations and which is developed in this study. Vivancos et al. [15] obtained mathematical models for arithmetic average roughness in ball-end milling operations by means of design of experiments, while Dhokia et al. [16] used design of experiments in ball-end milling to obtain models as a function of speed, feed and depth of cut. Oktem et al. [17] searched for minimum values in end milling taking into account cutting speed, feed rate, axial and radial depth of cut, and machining tolerance as input variables. In addition, they compared a response surface model with a neural network model [18]. It was observed that ANN lead to more accurate models than response surface methodology (RSM). Karkalos et al. [19] also compared regression models with ANN models in ball-end milling, with cutting speed, feed and depth of cut as variables and surface roughness as response. They found a higher correlation coefficient for ANN models than for RSM models. Vakondios et al. [20] obtained third order regression models for average maximum height of the profile Rz, as a function of axial depth of cut, radial depth of cut, feed rate and inclination angle, taking into account different manufacturing strategies. Wojciechowski and et al. [21] obtained a model for determining cutter displacements in ball-end milling. They took into account cutting conditions, surface inclination angle, run out, and the tool's deflection. They found that both the cutter's runout and surface inclination strongly influence cutter displacement. Wojciechowski and Mrozek optimized cutting forces and efficiency of the ball-end milling as a function of cutting speed and surface inclination angle [22]. Regarding Taguchi design of experiments, Pillai et al. [23] optimized machining time and surface roughness as a function of tool path strategic, spindle speed and feed rate in end milling with a single flute tool.

The main purpose of this study is to select an optimal machining strategy between climb and conventional milling in ball-end milling processes. For doing this, first mathematical models for roughness as a function of main process parameters were found. Unlike other works, in the present paper inclination angle of the surface to be machined is taken into account. Specifically, regression models and neural network models were obtained for parameters average roughness Ra, and total height of profile Rt. Finally, an optimal machining strategy was selected between climb and conventional milling for the different inclination angles considered. This will help molds and dies manufacturers to select appropriate strategies and cutting conditions in finish operations of surfaces with different inclination angles.

2. Experimental Procedure

2.1. Milling Tests

In the present study a factorial design of experiments was used for selecting experimental conditions in the ball-end milling process. The purpose of experimental tests is to analyze variability in the machining process of parts for injection molds, by means of several measurements performed on different areas of the machined workpieces with different inclinations. Two strategies were considered: climb milling and conventional milling.

The workpieces were manufactured in an HSM center with vertical-spindle Deckel Maho DMU 50 Evolution (DMG Mori Seiki Co, Nakamura-ku, Nagoya, Japan) with Heidenhain control TNC 430 (Dr. Johannes Heidenhain GmbH, Traunreut, Germany), as shown in Figure 1, and tool holder MST Ref. DN40AD-CTH20-75. Tool details are presented in Table 1.

Figure 1. Manufactured part and high-speed machining center employed.

Table 1. Tool details.

Tool Type	End Mill VC2SBR0300 KOBELCO Series MIRACLE (Kobe Steel, Chūō-ku, Kobe, Japan)
Tool material	(Al, Ti)N coated micro grain carbide
Number of flutes	2
Diameter (mm)	6

Tools employed were new or had little wear (average flank wear VB < 0.1 mm), in order to avoid influence of wear on surface roughness. Only three axes (X-Y-Z) were used.

Semi-cylindrical workpieces were machined in order to assess the effect of slope on surface roughness (Figure 2a). The material used for manufacturing the parts was a hot work tool steel W-Nr. 1.2344, hardened steel (50-54 HRC), with an approximate composition of 0.39% C, 1.10% Si, 0.40% Mn, 5.20% Cr, 1.40% Mo and 0.95% V.

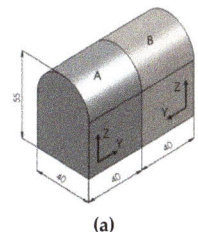 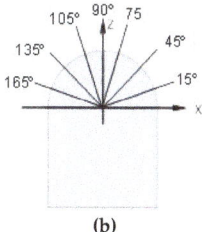

(a) (b)

Figure 2. Schematic drawing of (a) machined workpiece (units in mm), (b) measured position angles.

A central composite design was chosen for modeling the behavior of both Ra and Rt, consisting of a two level factorial design with 4 factors (2^4 = 16 experiments), with 4 central points. Since first-order models turned out to be inadequate for modeling both behavior of Ra and Rt, 8 star points were added, thus providing an orthogonal design with star points located at an axial distance of 1.60717. Selected factors were feed per tooth (f_z), axial depth (a_p), cutting speed (v_c), and radial depth (a_e). The study was developed for finish machining. Low and high levels for the different factors are shown in Table 2.

Table 2. Low and high levels for factors a_e, a_p, f_z and v_c.

Levels	a_p	a_e	f_z	v_c
Low	0.100	0.100	0.020	150.0
High	0.300	0.300	0.060	250.0

For each experiment, roughness was measured at different angular positions corresponding to different inclination angles of the workpiece's surface, as explained in Section 2.2 (Figure 2).

2.2. Roughness Measurement

Roughness was measured along different generatrices of the semi-cylindrical part in Figure 2a, corresponding to different angular positions (15°, 45°, 75°, 90°, 105°, 135° and 165°) in Figure 2b. Moreover, influence of milling strategy, either climb (down) milling (Figure 3a) or conventional (up) milling (Figure 3b), on surface roughness was also analyzed (Figure 3).

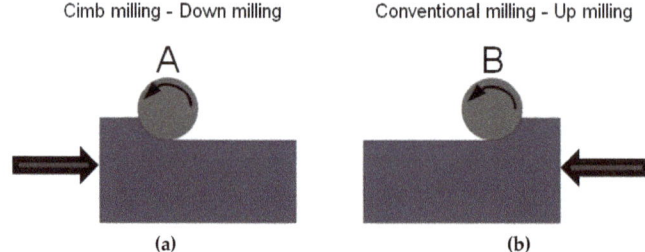

Figure 3. Schematic drawing of milling strategies. (**a**) climb (down) milling; (**b**) conventional (up) milling.

Roughness parameters *Ra* and *Rt* were measured using a Taylor-Hobson Form Taylsurf Series 2 profile roughness tester (as Figure 4b shows). An evaluation length of 4.8 mm (6 × 0.8 mm) was used, and a 2 µm radius stylus tip was used in conjunction with a 0.8 Gaussian cut-off filter and a bandwidth ratio of 320:1 to evaluate the *Ra* and *Rt* parameters. A stylus speed of 0.5 mm/s was used in conjunction with a 0.8 mN static stylus force and the stylus cone angle used was 90°.

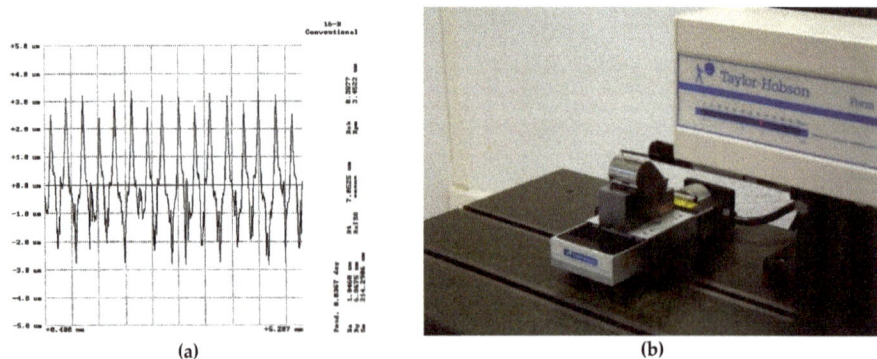

Figure 4. (**a**) Example of roughness profile; (**b**) Taylor-Hobson Form Taylsurf Series 2 profile roughness tester.

Figure 4a shows an example of a roughness profile. A quite regular profile with higher peaks than valleys was observed, which corresponds to ball-end milling. Although a high number of roughness parameters were measured, as shown in Figure 4a, parameters Ra and Rt were selected in order to obtain results related to a high averaging parameter (Ra) and a low averaging parameter (Rt). In addition, both roughness parameters are commonly used in roughness characterization [13,24].

2.3. Photographs

A Leica S8AP0 binocular magnifier (Leica Camera AG, Wetzlar, Germany) was used to obtain photographs of the workpiece's surface at 80× magnification.

3. Surface Roughness Results

In Table 3, as an example, roughness values of experiment 16 are compared, considering both A and B manufacturing strategies, respectively, at different angles which correspond to ascendant and descendant trajectories. Experiment 16 was chosen because it corresponds to high a_p, a_e, f_z, and v_c values (cutting conditions shown in Table 2), which lead to higher roughness values. In the images, changes of surface topography can be observed as a function of machining strategy (conventional or climb milling), position angle of the machined surface, and whether the tool displacement along f_z trajectory is ascendant or descendant. According to the methodology explained in Section 2, different slopes of the machined semi-cylindrical workpieces were considered. For 15°, 45° and 75° in climb milling (Figure 3a), corresponding to 165°, 135° and 105° in conventional milling (Figure 3b), the tool displacement is ascendant. For 105°, 135° and 165° in climb milling, corresponding to 75°, 45° and 15° in conventional milling, the tool displacement is descendant.

Table 3. Ra and Rt values using climb milling (Figure 3a) and conventional milling (Figure 3b) for experiment 16 in different angular positions.

Parameter	Climb Milling (Figure 3a)							Conventional Milling (Figure 3b)						
	15°	45°	75°	90°	105°	135°	165°	15°	45°	75°	90°	105°	135°	165°
Ra (μm)	1.56	1.68	0.88	1.23	0.80	0.73	0.75	1.22	1.06	0.70	1.05	0.83	0.83	1.15
Rt (μm)	7.04	7.05	3.82	6.54	4.17	3.08	3.65	5.41	4.72	4.52	7.05	6.00	5.18	5.25

In climb milling, roughness values remain almost constant between 15° and 45° and decrease significantly from 45° to 75° in the ascendant trajectory. Values increase at 90° because of a lack of cutting speed and decrease at 105°. In the descendant trajectory, values decrease slightly between 105° and 135° and remain almost constant between 105° and 165°. In conventional milling, similar results were obtained. As a general trend, lower roughness values were obtained for conventional milling than for climb milling in the ascendant trajectory, and higher roughness values were obtained for conventional milling than for climb milling in the descendant trajectory.

In Figure 5, machined surfaces of experiment 16 are presented.

In experiment 16, for each angle considered, surface topography obtained in climb milling is similar to that obtained in conventional milling. However, Table 3 shows that in general, when f_z trajectory is ascendant, roughness is lower for conventional milling (165° to 135°) than for climb milling (15° to 45°). On the other hand, when f_z trajectory is descendant, roughness is lower for climb milling (135° to 165°) than for conventional milling (45° to 15°). At 90°, instead of straight cutting marks, semicircular cutting marks are observed, suggesting that the tool does not cut properly because of zero cutting speed [1,2].

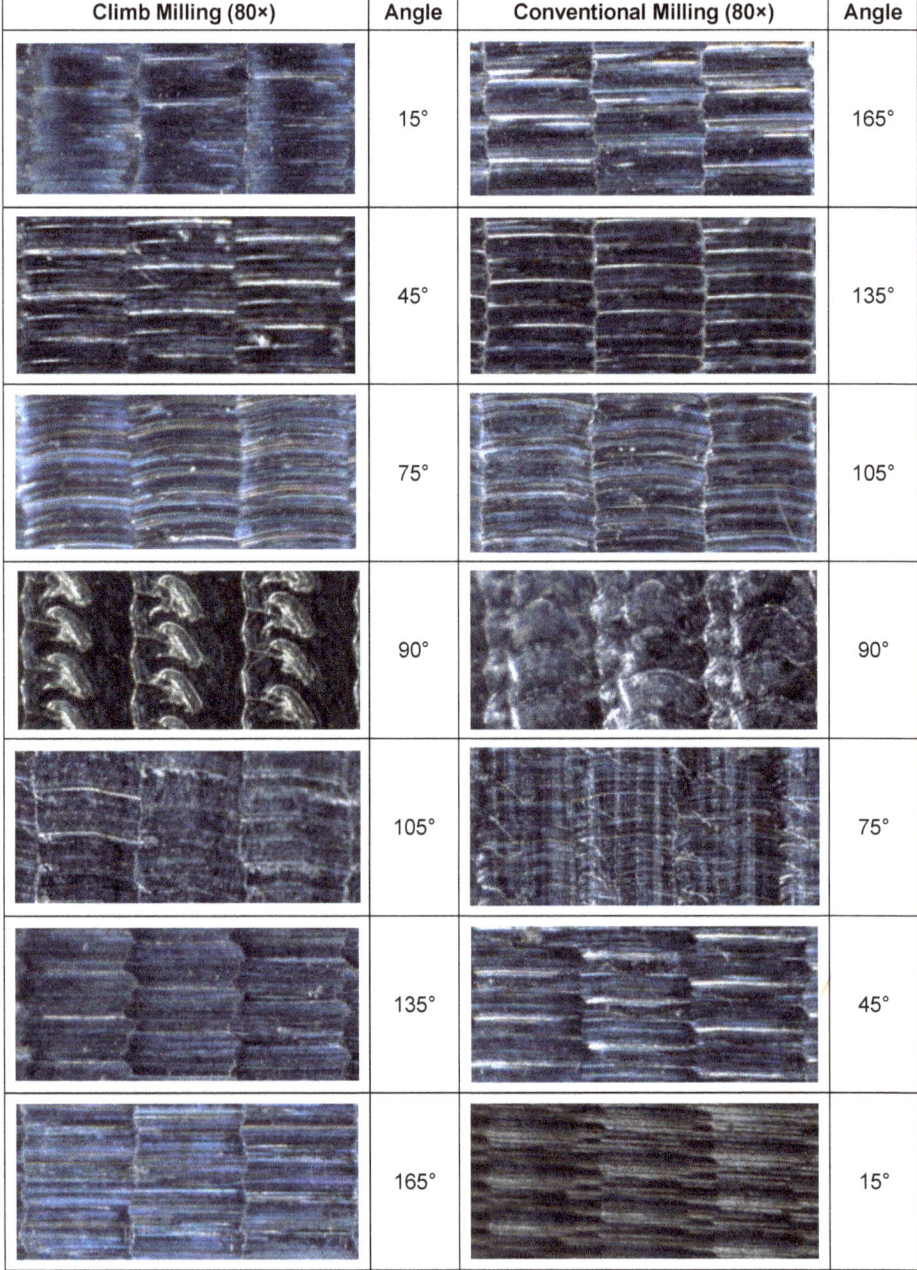

Figure 5. Machined surfaces corresponding to experiment 16.

4. Models for Surface Roughness

In this study, first the main cutting conditions that minimize Ra and Rt roughness parameters and their variability were selected. In addition to strategies and cutting conditions, inclination of the machined surface was considered, as there seems to be a lack of knowledge on the attained roughness in the manufacturing process of molds when different slopes have to be machined. Within the range of a_e and f_z values studied, surface topography is mainly determined by roughness in the transversal direction, which is perpendicular to tool marks in the feed f_z direction. Along tool marks the roughness level is remarkably low, since $f_z < a_e$ [24,25]. For this reason, 2D roughness was studied along the transversal direction (perpendicular to tool marks).

Vivancos et al. [15] previously analyzed this behavior by considering four factors (a_p, a_e, f_z and v_c) in regression models and by taking into account average roughness values in the whole workpiece without considering influence of each position angle separately. In order to obtain a more accurate analysis, it is necessary to consider the effect of each surface slope on obtained roughness, which is one of the core points of this work. Vakondios et al. [20] considered surface inclination in regression models for average maximum height of the profile, Rz. In the present study, regression analysis was carried out considering not only cutting conditions but also position angle of the surface on two different roughness parameters, Ra and Rt. Both regression and neural networks models were obtained. All regression analyses were carried out using Statgraphics®Centurion XVI. Regarding neural network models, the results found in this study were obtained by using the Neural Network Toolbox™ (São Paulo, Brazil) of Matlab™ (Mathworks, Natick, MA, USA). In addition, the optimal cutting strategy between climb and conventional milling was selected for different cutting conditions and inclination angles.

4.1. Regression Models and Analysis of Arithmetic Average Roughness, Ra

Ra was modeled by means of regression analysis, taking into account variability due to cylindrical geometry of the workpiece studied in this present work. In order to model the behavior of Ra for both manufacturing strategies (climb milling and conventional milling), second-order models were selected after analyzing p-values obtained from the lack-of-fit test performed with the first order modeling (3.0×10^{-12} and 3.04×10^{-4}, respectively). Since these p-values for the lack-of-fit are less than 0.05, there is a statistically significant lack-of-fit at the 95.0% confidence level, which means that first order models do not adequately represent the data. R^2 and adjusted-R^2 were 68.43% and 67.25% for climb milling, respectively, while R^2 and adjusted-R^2 were 69.40% and 68.26% for conventional milling, respectively.

Since there is lack of fit with the first order model, second order models were considered. For Ra in climb milling, the R^2 and adjusted-R^2 are 79.48% and 77.64%, respectively, and equations were obtained so that adjusted-R^2 is maximized. Four main effects (a_e, Ang, v_c and f_z) turned out to be relevant in the model in order to obtain the highest adjusted-R^2. Parameters a_e and $a_e{}^2$ turned out to be the most significant for a confidence level of 95% ($\alpha = 0.05$) (p-values ≤ 0.01). As can be observed in Figure 6, surface roughness remains almost constant with respect to a_p, v_c and f_z. Moreover, it can be shown that Ra has a quadratic tendency with regard to a_e, where a_e is the parameter that most influences Ra. Therefore, minimization of a_e will lead to a reduction in roughness values. This can be attributed to the fact that a_e determines width of machining marks, and in addition f_z values are low. In the study the rest of the factors are kept at their central values. Moreover, it can be shown that Ra has a quadratic tendency with regard to Ang.

Equations (1) and (2) show the proposed modeling for Ra using both climb and conventional milling. For Ra in conventional milling, R^2 and adjusted-R^2 are 76.52% and 73.84%, respectively. Four main effects (a_e, v_c, Ang and a_p) turned out to be relevant in the model in order to obtain the highest adjusted-R^2. Similar to the results obtained in climb milling, a_e and $a_e{}^2$ were the most significant factors at a confidence level of 95% ($\alpha = 0.05$) (p-values ≤ 0.01).

As can be observed in Figure 6, surface roughness has a quadratic tendency with regard to a_e, and a slight slope with respect to both a_p and v_c. In this case, factor f_z was not significant in the model that provides the highest adjusted-R^2. Moreover, a quadratic tendency with regard to the angle was observed. Conventional milling (Figure 6b) follows a similar tendency to climb milling (Figure 6a) regarding a_e, which is the most significant parameter. However, this influence is smaller than that obtained in climb milling.

$$\begin{aligned}
Ra\_Climb = &\ 0.379019 - 1.42452 \times a_p - 0.661785 \times a_e - 0.194863 \times f_z - 0.00189356 \times v_c \\
&+ 0.00239249 \times Ang - 0.269701 \times a_p^2 - 1.87232 \times a_p \times a_e \\
&+ 1.01875 \times a_p \times f_z + 0.00672571 \times a_p \times v_c + 0.00594601 \times a_p \times Ang \\
&+ 14.563 \times a_e^2 - 17.0402 \times a_e \times f_z + 0.00579143 \times a_e \times v_c - 0.01599 \\
&\times a_e \times Ang + 50.1135 \times f_z^2 + 0.0134214 \times f_z \times v_c - 0.030112 \times f_z \times Ang \\
&+ 0.00000135074 \times v_c^2 - 0.0000158617 \times v_c \times Ang + 0.0000152557 \times Ang^2 \\
& R^2 = 79.48\% \quad Adj - R^2 = 77.14\%
\end{aligned} \quad (1)$$

$$\begin{aligned}
Ra\_Convent = &\ 0.344666 + 0.376502 \times a_p + 0.060028 \times a_e - 9.41584 \times f_z + 0.00130472 \times v_c \\
&- 0.0054896 \times Ang + 1.19437 \times a_p^2 - 2.68455 \times a_p \times a_e - 0.828125 \times a_p \times f_z \\
&- 0.000939821 \times a_p \times v_c + 0.00201332 \times a_p \times Ang + 9.38685 \times a_e^2 \\
&- 22.5379 \times a_e \times f_z + 0.00357411 \times a_e \times v_c + 0.00083119 \times a_e \times Ang \\
&+ 80.4796 \times f_z^2 + 0.0369991 \times f_z \times v_c + 0.00481444 \times f_z \times Ang \\
&- 0.00000658429 \times v_c^2 + 6.90684 \times 10^{-7} \times v_c \times Ang + 0.0000217804 \times Ang^2 \\
& R^2 = 76.53\% \quad Adj - R^2 = 73.84\%
\end{aligned} \quad (2)$$

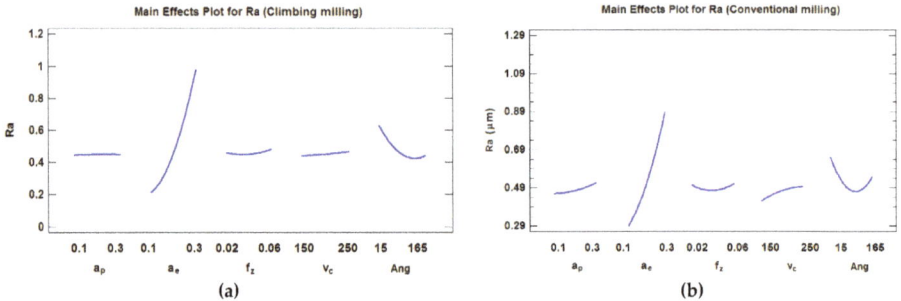

Figure 6. Main effects plot for Ra (considering the position angle) in (**a**) climb milling and (**b**) conventional milling.

4.2. Regression Models and Analysis of Maximum Peak-to-Valley Roughness Rt

Similar to the results obtained for Ra, the behavior of Rt was modeled, taking into account variability due to cylindrical geometry of the workpiece studied. In order to model the behavior of Rt in both manufacturing strategies (climb and conventional milling), second-order models were selected after analyzing p-values obtained from the lack-of-fit test performed with the first order modeling (5.52 × 10^{-5} and 1.14 × 10^{-26}, respectively). In all cases, obtained equations were simplified in order to obtain models with the highest adjusted-R^2.

For Rt, in climb milling the R^2 and adjusted-R^2 are 78.04% and 75.53%, respectively. Four main effects (a_e, Ang, f_z and a_p,) were present in the model in order to obtain the highest adjusted-R^2. The parameters a_e and a_e^2 were the most important parameters at a confidence level of 95% (α = 0.05) (p-values $\leq$ 0.01) (Figure 7a). For Rt, in conventional milling R^2 and adjusted-R^2 were 63.03% and 58.80%, respectively. Three main effects (a_e, a_p, and Ang) were present in the model in order to obtain the highest adjusted-R^2. Similar to the result obtained for climb milling, a_e and a_e^2 were the most important parameters at a confidence level of 95% (α = 0.05) (p-values $\leq$ 0.01) (Figure 7b).

Equations (3) and (4) show the regression analysis for *Rt*, taking the angle into account and considering both climb milling and conventional milling.

$$
\begin{aligned}
Rt\_Climb = \quad & 1.812 - 1.27897 \times a_p - 1.59259 \times a_e + 11.4864 \times f_z - 0.0153509 \times v_c \\
& +0.0158467 \times Ang - 4.28786 \times a_p^2 - 2.19768 \times a_p \times a_e + 23.8839 \times a_p \times f_z \\
& +0.00801286 \times a_p \times v_c + 0.0177314 \times a_p \times Ang + 64.6309 \times a_e^2 \\
& -114.621 \times a_e \times f_z + 0.0180725 \times a_e \times v_c - 0.0715339 \times a_e \times Ang \\
& +390.288 \times f_z^2 - 0.0134036 \times f_z \times v_c - 0.197239 \times f_z \times Ang \\
& +0.0000386847 \times v_c^2 - 0.0000518305 \times v_c \times Ang + 0.0000505297 \times Ang^2 \\
& R^2 = 78.04\% \quad Adj - R^2 = 75.53\%
\end{aligned} \quad (3)
$$

$$
\begin{aligned}
Rt\_Convent = \quad & 4.04698 + 4.45254 \times a_p + 0.0118769 \times a_e - 100.633 \times f_z - 0.00524991 \times v_c \\
& -0.021916 \times Ang + 2.41174 \times a_p^2 - 7.9592 \times a_p \times a_e - 38.4004 \times a_p \times f_z \\
& -0.00523589 \times a_p \times + 0.0158633 \times a_p \times Ang + 46.2882 \times a_e^2 - 155.4 \times a_e \times f_z \\
& +0.00680554 \times a_e \times v_c + 0.0256059 \times a_e \times Ang + 1002.59 \times f_z^2 \\
& +0.263647 \times f_z \times v_c + 0.0953535 \times f_z \times Ang - 0.00000371912 \times v_c^2 \\
& -0.0000202955 \times v_c \times Ang + 0.0000517796 \times Ang^2 \\
& R^2 = 63.02\% \quad Adj - R^2 = 58.8\%
\end{aligned} \quad (4)
$$

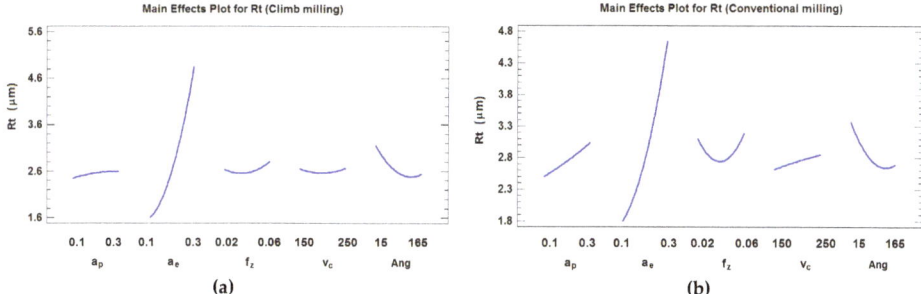

Figure 7. Main effects plot for *Rt* (considering the position angle) in (a) climb milling and (b) conventional milling.

As can be observed in Figure 6, a_e is the most influential parameter on *Rt* in both climb and conventional milling, which is similar to the results obtained for *Ra* in the present paper and for *Rz* parameter in other works [19,20]. Surface roughness has a quadratic behavior with respect to a_e in climb and conventional milling, and a slight slope with both a_p and v_c in conventional milling. In climb milling, surface roughness remains almost constant with respect to a_p, f_z and v_c. The fact that a_e has a greater influence on roughness than f_z in ball-end milling processes can be explained by the fact that at low radial depth of cut a_e, the influence of feed per tooth f_z is minimized by the tool performing very close successive passes in the a_e direction. Very close parallel grooves will be obtained. Thus, very similar roughness values will be achieved regardless of f_z employed for the same a_e value [24].

4.3. ANN Modeling for Ra and Rt

An artificial neural network (ANN) was also employed in this present study for modeling both *Ra* and *Rt*. This ANN was made up of an input layer, a hidden layer, and an output layer. The neural network considered in this work has a 5-1-4 configuration, which corresponds with five inputs (the four cutting conditions uses in regression analysis (a_e, a_p, f_z, and v_c) and the position angle of the surface (*Ang*), which is related to the slope of the surface to be machined. The network has one neuron in the hidden layer, and four outputs, one for each of the roughness parameters and machining strategies considered. Equation (5) shows the roughness parameters Ra and Rt for both machining strategies as a function of a_p, a_e, f_z, v_c, and Ang.

$$\begin{bmatrix} Ra_{Climb.} \\ Ra_{Conv.} \\ Rt_{Climb.} \\ Rt_{Conv.} \end{bmatrix} = \frac{1}{1+e^{-(-0.0449 \times a_p - 1.90361 \times a_e + 0.0855 \times f_z - 0.111 \times v_c + 0.266 \times Ang + 1.899)}} \begin{bmatrix} -2.3137 \\ -2.6528 \\ -1.9889 \\ -1.9144 \end{bmatrix} + \begin{bmatrix} 1.3164 \\ 1.8037 \\ 1.0583 \\ 1.1065 \end{bmatrix} \quad (5)$$

where Climb. corresponds to climb milling and Conv. Corresponds to conventional milling.

The design of experiments, previously shown in Table 2, was used to train the ANN. It was decided to choose one neural network with four outputs, since the results obtained were similar to those obtained for independent networks for each output. With this ANN a correlation value of 0.914 was obtained. This value is similar to that obtained by other authors with ANN models [19]. Hence, ANN 5-1-4 provides a relatively simple model with high precision, which in a compact way allows approximation of Ra and Rt roughness parameters in both machining strategies studied. This might be attributed to the fact that roughness parameters are related and they show similar variability.

4.4. Optimal Manufacturing Strategy Selection

In order to compare both machining strategies, a diagram of both average roughness values and standard deviations of roughness values obtained at different inclination angles for the 28 experiments considered is shown in Figures 7 and 8, for Ra and Rt, respectively. From these figures it is possible to determine which machining strategy is more appropriate for the cutting conditions selected in this present work.

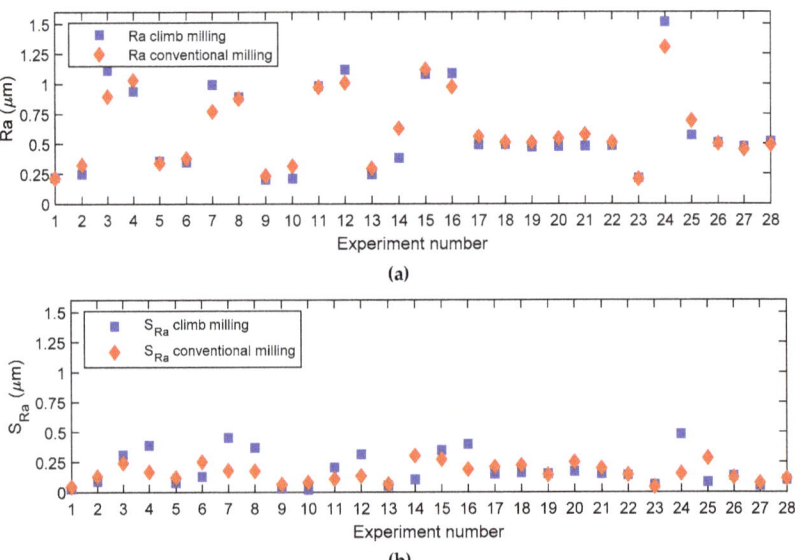

Figure 8. Experimental deviation plots for Ra considering both manufacturing strategies: (a) Mean, (b) standard deviation.

Figure 8a shows that average Ra values are very similar for both climb milling and conventional milling, if the same experiment is taken into consideration. However, in surfaces with variable inclinations, such as those found in injection molds, it is interesting not only to minimize roughness average values, but also its variability for different inclination angles. This will lead to a more uniform surface roughness. Then, in order to minimize variability (Figure 8b), the use of conventional milling is recommended in experiments 3, 4, 7, 8, 11, 12, 15, 16, and 24. Those experiments have a general tendency to exhibit high a_e values (a_e = 0.3 mm). Using climb milling is recommended in experiments 2, 5, 6, 9, 10, 14, 17, 18, 20, 21, 25, and 27, which in general correspond with low and medium a_e values (a_e = 0.1 mm and a_e = 0.2 mm, respectively).

For the remainder of experiments, similar values were obtained for both conventional and climb milling. Figure 9a also shows that average Rt values are similar for both machining strategies. However, variability (Figure 9b) determines that conventional milling is recommended in experiments 7, 11, 12, 13, 15, 16, and 24. As a general trend, those experiments correspond to high a_e values (a_e = 0.3 mm), with high v_c values (v_c = 250 m/min). Climb milling is recommended in experiments 2, 3, 4, 5, 6, 10, 14, 17, 18, 19, 20, 21, 22, 25, and 26, which correspond to maximum a_e with minimum v_c, minimum ae with maximum v_c, or medium a_e with medium v_c values. For the rest of experiments, the values obtained are similar.

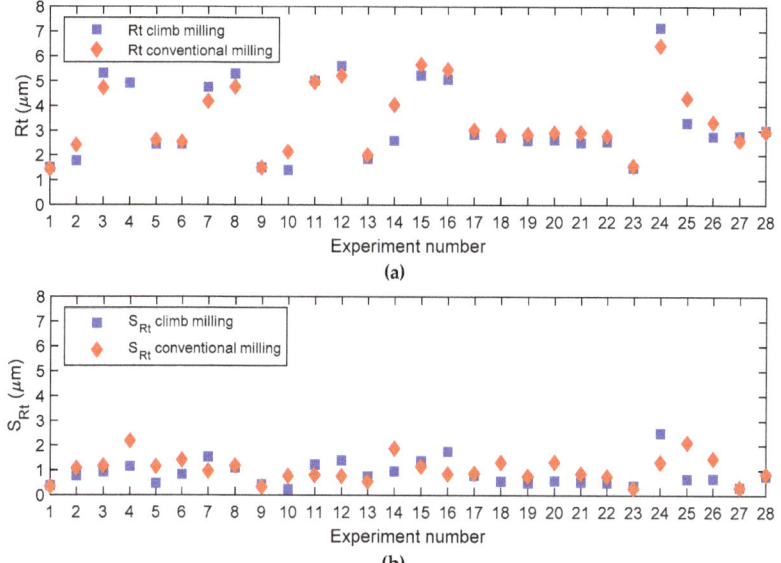

Figure 9. Experimental deviation plots for Rt considering both manufacturing strategies: (a) Mean, (b) standard deviation.

Ra and Rt average values do not vary significantly between climb and conventional milling. Given that mold manufacturers require roughness uniformity at different inclination angles of the machined surface, the most appropriate process will be chosen between conventional and climb milling, taking variability into account in the experiments studied (Table 4). Therefore, a manufacturing strategy will be selected that minimizes variability of roughness values in different angular positions. If Ra and Rt show opposite tendencies, a manufacturing strategy will be preferred that minimizes Rt, since Ra is a high-averaging parameter and, therefore, tends to mask errors on the machined surface. This does not happen with Rt. In the case where both strategies lead to the same Rt variability, then the strategy minimizing Ra variability will be chosen.

Table 4 summarizes the type of machining strategy that is recommended for each cutting condition and for each cutting strategy. The table shows that in 17 of 28 cutting conditions tested, climb milling is preferred. Conventional milling is only preferred in 8 cutting conditions, which in general corresponds with high a_e with high v_c. For the rest of the experiments, it makes no difference whether one or the other machining strategy is used. As was stated earlier, minimization of Rt has priority with respect to minimization of *Ra*. Std means standard deviation of roughness values for the different inclination angles. Conv. Milling stands for conventional milling.

Table 4. Optimal machining strategy selection.

a_p (mm)	a_e (mm)	f_z (mm)	v_c (m/min)	Minimum (Std Ra)	Minimum (Std Rt)
0.1	0.1	0.02	150	Conv./Climb	Conv./Climb
0.3	0.1	0.02	150	Climb Milling	Climb Milling
0.1	0.3	0.02	150	Conv. Milling	Climb Milling
0.3	0.3	0.02	150	Conv. Milling	Climb Milling
0.1	0.1	0.06	150	Climb Milling	Climb Milling
0.3	0.1	0.06	150	Climb Milling	Climb Milling
0.1	0.3	0.06	150	Conv. Milling	Conv. Milling
0.3	0.3	0.06	150	Conv. Milling	Conv. / Climb
0.1	0.1	0.02	250	Climb Milling	Conv./Climb
0.3	0.1	0.02	250	Climb Milling	Climb Milling
0.1	0.3	0.02	250	Conv. Milling	Conv. Milling
0.3	0.3	0.02	250	Conv. Milling	Conv. Milling
0.1	0.1	0.06	250	Conv./Climb	Conv. Milling
0.3	0.1	0.06	250	Climb Milling	Climb Milling
0.1	0.3	0.06	250	Conv. Milling	Conv. Milling
0.3	0.3	0.06	250	Conv. Milling	Conv. Milling
0.2	0.2	0.04	200	Climb Milling	Conv./Climb
0.2	0.2	0.04	200	Climb Milling	Climb Milling
0.2	0.2	0.04	200	Conv./Climb	Climb Milling
0.2	0.2	0.04	200	Climb Milling	Climb Milling
0.039	0.2	0.04	200	Climb Milling	Climb Milling
0.361	0.2	0.04	200	Conv./Climb	Climb Milling
0.2	0.039	0.04	200	Conv./Climb	Conv./Climb
0.2	0.361	0.04	200	Conv. Milling	Conv. Milling
0.2	0.2	0.008	200	Climb Milling	Climb Milling
0.2	0.2	0.072	200	Conv./Climb	Conv./Climb
0.2	0.2	0.04	119.641	Climb Milling	Conv./Climb
0.2	0.2	0.04	280.359	Conv./Climb	Conv./Climb

Regarding influence of angle, for both strategies (climb and conventional milling), when angle increases roughness decreases. However, it should be taken into account that high angles in climb milling (descendant trajectory) correspond to low angles in conventional milling (descendant trajectory), and low angles in climb milling (ascendant trajectory) correspond to high angles in conventional milling (ascendant trajectory). With all this, it is recommended to use climb milling in descendant trajectories and conventional milling in ascendant trajectories.

5. Conclusions

In the present study, as a general tendency climb milling is preferred to conventional milling. In general, conventional milling is only recommended at a high radial depth of cut with high cutting speed values. In order to reduce roughness values, in ascendant trajectories conventional milling is preferred and in descendant trajectories climb milling is recommended.

From the results obtained, it was determined that radial depth of cut was the most relevant factor on *Ra* and *Rt* for both climb and conventional milling. Axial depth of cut, cutting speed and feed per tooth have a slight influence on roughness within the range studied in this study. Regression models for average roughness showed high adjusted-R^2 values (above 73%) in all cases. Moreover, a correlation value of 0.914 was obtained with the neural network model employed.

Experimental roughness values obtained with both strategies (climb and conventional milling) were similar. However, in complex surfaces with variable inclination, such as those of injection molds, it is recommended not only to minimize roughness average values, but also its variability for different inclination angles. This will lead to more uniform surfaces. In the present study, it was found that the standard deviation of roughness parameters varies depending on the machining strategy chosen, for the different experiments carried out.

Author Contributions: All the authors of this present manuscript have approximately equally contributed to most of the research tasks.

Funding: This research was funded by the Spanish Ministry of Science and Technology, grant number DPI 2003-04727.

Acknowledgments: The authors thank Ramón Casado-López and Alejandro Domínguez-Fernández for their help with experimental tests and roughness measurements.

Conflicts of Interest: The authors declare no conflict of interest.

References

1. Denkena, B.; Böß, V.; Nespor, D.; Gilge, P.; Hohenstein, S.; Seume, J. Prediction of the 3D Surface Topography after Ball End Milling and its Influence on Aerodynamics. *Procedia CIRP* **2015**, *31*, 221–227. [CrossRef]
2. Toh, C.K. Surface topography analysis in high speed finish milling inclined hardened steel. *Precis. Eng.* **2014**, *28*, 386–398. [CrossRef]
3. Mukherjee, I.; Ray, P.K. A review of optimization techniques in metal cutting processes. *Comput. Ind. Eng.* **2006**, *50*, 15–34. [CrossRef]
4. Asiltürk, I.; Çunkas, M. Modeling and prediction of surface roughness in turning operations using artificial neural network and multiple regression method. *Expert Syst. Appl.* **2011**, *38*, 5826–5832. [CrossRef]
5. Feng, C.X.; Wang, X.; Yu, Z. Neural networks modeling of honing surface roughness parameters defined by ISO 13565. *J. Manuf. Syst.* **2003**, *21*, 395–408. [CrossRef]
6. Feng, C.X.; Wang, X.F. Development of empirical models for surface roughness prediction in finish turning. *Int. J. Adv. Manuf. Technol.* **2002**, *20*, 348–356.
7. Özel, T.; Esteves Correia, A.; Paulo Davim, J. Neural network process modeling for turning of steel parts using conventional and wiper inserts. *Int. J. Mater. Product Technol.* **2009**, *35*, 246–258. [CrossRef]
8. Sonar, D.K.; Dixit, U.S.; Ojha, D.K. The application of a radial basis function neural network for predicting the surface roughness in a turning process. *Int. J. Adv. Manuf. Technol.* **2006**, *27*, 661–666. [CrossRef]
9. Gao, T.; Zhang, W.; Qiu, K.; Wan, M. Numerical Simulation of Machined Surface Topography and Roughness in Milling Process. *J. Manuf. Sci. Eng.* **2015**, *128*, 96–103. [CrossRef]
10. Honeycutt, A.; Schmitz, T.L. Surface Location Error and Surface Roughness for Period-N Milling Bifurcations. *J. Manuf. Sci. Eng.* **2017**, *139*, 061010. [CrossRef]
11. Vallejo, A.J.; Morales-Menendez, R. Cost-effective supervisory control system in peripheral milling using HSM. *Annu. Rev. Control* **2010**, *34*, 155–162. [CrossRef]
12. Zain, A.M.; Haron, H.; Sharif, S. Prediction of surface roughness in the end milling machining using Artificial Neural Network. *Expert Syst. Appl.* **2010**, *37*, 1755–1768. [CrossRef]
13. Quintana, G.; Rudolf, T.; Ciurana, J.; Brecher, C. Using kernel data in machine tools for the indirect evaluation of surface roughness in vertical milling operations. *Robot. Comput.-Integr. Manuf.* **2011**, *27*, 1011–1018. [CrossRef]
14. Zhou, J.; Ren, J.; Yao, C. Multi-objective optimization of multi-axis ball-end milling Inconel 718 via grey relational analysis coupled with RBF neural network and PSO algorithm. *Measurement* **2017**, *102*, 271–285. [CrossRef]
15. Vivancos, J.; Luis, C.J.; Costa, L.; Ortiz, J.A. Optimal machining parameters selection in high speed milling of hardened steels for injection moulds. *J. Mater. Process. Technol.* **2004**, *155*, 1505–1512. [CrossRef]
16. Dhokia, V.G.; Kumar, S.; Vichare, P.; Newman, S.T. An intelligent approach for the prediction of surface roughness in ball-end machining of polypropylene. *Robot. Comput.-Integr. Manuf.* **2008**, *24*, 835–842. [CrossRef]
17. Oktem, H.; Erzurumlu, T.; Erzincanli, F. Prediction of minimum surface roughness in end milling mold parts using neural network and genetic algorithm. *Mater. Des.* **2006**, *27*, 735–744. [CrossRef]
18. Erzurumlu, T.; Oktem, H. Comparison of response surface model with neural network in determining the surface quality of moulded parts. *Mater. Des.* **2007**, *28*, 459–465. [CrossRef]

19. Karkalos, N.E.; Galanis, N.I.; Markopoulos, A.P. Surface roughness prediction for the milling of Ti-6Al-4V ELI alloy with the use of statistical and soft computing techniques. *Measurement* **2016**, *90*, 25–35. [CrossRef]
20. Vakondios, D.; Kyratsis, P.; Yaldiz, S.; Antoniadis, A. Influence of milling strategy on the Surface roughness in ball end milling of the aluminium alloy Al7075-T6. *Measurement* **2012**, *45*, 1480–1488. [CrossRef]
21. Wojciechowski, S.; Chwalczuk, T.; Twardowski, P.; Krolczyk, G.M. Modeling of cutter displacements during ball end milling of inclined surfaces. *Arch. Civil Mech. Eng.* **2015**, *15*, 798–805. [CrossRef]
22. Wojciechowski, S.; Mrozek, K. Mechanical and technological aspects of micro ball end milling with various tool inclinations. *Int. J. Mech. Sci.* **2017**, *134*, 424–435. [CrossRef]
23. Pillai, J.U.; Sanghrajka, I.; Shunmugavel, M.; Muthurmalingam, T.; Goldberg, M.; Littlefair, G. Optimisation of multiple response characteristics on end milling of aluminium alloy using Taguchi-Grey relational approach. *Measurement* **2018**, *124*, 291–298. [CrossRef]
24. Buj-Corral, I.; Vivancos-Calvet, J.; Domínguez-Fernández, A. Surface topography in ball-end milling processes as a function of feed per tooth and radial depth of cut. *Int. J. Mach. Tools Manuf.* **2012**, *53*, 151–159. [CrossRef]
25. Chen, J.S.B.; Huang, Y.K.; Chen, M.S. Feedrate optimization and tool profile modification for the high-efficiency ball-end milling process. *Int. J. Mach. Tools Manuf.* **2005**, *45*, 1070–1076. [CrossRef]

© 2019 by the authors. Licensee MDPI, Basel, Switzerland. This article is an open access article distributed under the terms and conditions of the Creative Commons Attribution (CC BY) license (http://creativecommons.org/licenses/by/4.0/).

Article

Intelligent Optimization of Hard-Turning Parameters Using Evolutionary Algorithms for Smart Manufacturing

Mozammel Mia [1], Grzegorz Królczyk [2,*], Radosław Maruda [3] and Szymon Wojciechowski [4]

1. Department of Mechanical and Production Engineering, Ahsanullah University of Science and Technology, Dhaka 1208, Bangladesh; mozammelmiaipe@gmail.com
2. Faculty of Mechanical Engineering, Opole University of Technology, St. Mikołajczyka 5, 45-001 Opole, Poland
3. Faculty of Mechanical Engineering, University of Zielona Gora, 4 Prof. Z. Szafrana Street, 65-516 Zielona Gora, Poland; r.maruda@ibem.uz.zgora.pl
4. Faculty of Mechanical Engineering and Management, Poznan University of Technology, 3 Piotrowo St., 60-965 Poznan, Poland; sjwojciechowski@o2.pl
* Correspondence: g.krolczyk@po.opole.pl; Tel.: +48-77-449-8429; Fax: +48-77-449-8461

Received: 2 February 2019; Accepted: 11 March 2019; Published: 15 March 2019

Abstract: Recently, the concept of smart manufacturing systems urges for intelligent optimization of process parameters to eliminate wastage of resources, especially materials and energy. In this context, the current study deals with optimization of hard-turning parameters using evolutionary algorithms. Though the complex programming, parameters selection, and ability to obtain the global optimal solution are major concerns of evolutionary based algorithms, in the present paper, the optimization was performed by using efficient algorithms i.e., teaching–learning-based optimization and bacterial foraging optimization. Furthermore, the weighted sum method was used to transform the diverse responses into a single response, and then multi-objective optimization was performed using the teaching–learning-based optimization method and the standard bacterial foraging optimization method. Finally, the optimum results reported by these methods are compared to choose the best method. In fact, owing to better convergence within shortest time, the teaching–learning-based optimization approach is recommended. It is expected that the outcome of this research would help to efficiently and intelligently perform the hard-turning process under automatic and optimized environment.

Keywords: intelligent optimization; hard turning; surface roughness; cutting temperature; evolutionary algorithm

1. Introduction

Smart manufacturing (SM) is regarded as the next generation manufacturing revolution—Industry 4.0 [1,2]. In this technology, the manufacturing system is optimized to the highest level to extract the highest benefits in terms of production economics, quality, and time. Use of advanced technologies such as sensors, smart materials, production and process planning, cloud systems etc., and their interaction with humans determines the success of a smart manufacturing unit [3]. In a nutshell, SM is a technology that allows the process improvement via optimization and exploitation of advanced technologies that establish it as a next generation manufacturing model.

Intelligent optimization of real-life manufacturing systems is key factor to smart manufacturing. The implementation of an intelligent optimization technique in the soft part of manufacturing units facilitates an effective production control. Among prevailing optimization techniques (statistical, neural, evolutionary, machine learning, etc.) the evolutionary methods have recently been employed successfully in various engineering sectors. Traditionally, the effectiveness and efficiency of selecting parameters of machining processes are determined by the trial-and-run method or by the experience of the machine operators [4,5]. As per the requirements of smart manufacturing as in 'turning' systems, the cutting parameters i.e., cutting speed, feed rate, and depth-of-cut need to be optimized. This information can be synchronized with the database of the manufacturing unit. In fact, as an upgrade based on instantaneous requirements, the control factors can be optimized in real-time. In this particular segment, the learning capability of modeling methods (i.e., artificial intelligence, machine learning) possesses the potential to enhance requirement-based manufacturing to reduce human intervention [6].

Successful implementation of intelligent/evolutionary methods in the manufacturing realm can be found in literature. For instance, Rao et al. [7] applied the novel teaching–learning-based optimization (TLBO) method to optimize multiple mechanical design problems. In studying the welding of Cr–Mo–V steel, Rao and Kalyankar [8] applied TLBO alongside the Taguchi-based optimization. In another study, Pawar and Rao [9] applied TLBO in abrasive water jet machining, milling, and grinding to optimize the machining parameters. They compared the results of TLBO with other methods of optimization and found better results of TLBO. In another paper, Rao and Kalyankar [10] optimized the parameters of modern machining processes namely the Ultra Sound Machining (USM), Abrasive Jet Machining (AJM), and Wire Electrical Discharge Machining (WEDM) using TLBO. Gupta et al. [11] employed one statistical method (i.e., Response Surface Methodology (RSM)) and one evolutionary method (i.e., Particle Swarm Optimization (PSO)) to optimize the machining parameters. Mukhopadhyay et al. [12] employed an artificial neural network and genetic algorithm for the modeling and optimization of the wire electrical discharge machining process. In addition, they have implemented the hybrid modeling to extract better machining optimization results. Kim and Lee [13] optimized the induction-assisted milling process using finite element analysis, signal-to-noise ratio, and analysis of variance. The optimized responses were surface roughness, tool wear, and surface roughness in selected machining environment.

In high-performance precision engineering application, the quality of products produced by turning process is evaluated by the roughness parameters of machined surface [14]. The increased pressure from the industries to produce parts with very low surface-roughness values forces researchers to find ways of reducing surface roughness. Among many alternatives, optimization of process parameters can refine the surface roughness value. This means the appropriate parameter settings of control factors can generate surfaces with a surface roughness value that is lower than the roughness found in conventional processes and/or hard machining. Hybridization of PSO–bacteria foraging optimization (BFO) was reported in the optimization of additive manufacturing parameters for the fused deposition modeling Raju et al. [15]. As can be seen, the advanced computational methods for optimization have been reported in multifarious sectors such as welding, additive manufacturing, machining, modern machining processes, etc. However, very few articles reported the adoption of evolutionary algorithms for hard turning.

Besides the surface roughness parameter, the overall machining outcomes are largely influenced by the cutting temperature. The intensive friction induced by the plastic deformation during the chip generation cause the mechanical energy to be transformed into heat energy. As such, the chip–tool interface and work–tool interface temperature rises. An increase in temperature results in the expedited wear rate of tool; hence, the machining economy is compromised. Also, the premature failure or sudden breakage of the tool can result in the adherence of tool broken debris onside the machine surface—rejection of the product. In that perspective, controlling of cutting zone temperature is inevitable.

Use of cooling/lubricating agents during cutting is widely accepted as a temperature-controlling system and surface-quality improver. However, the practice of coolant/lubricant is costly, and it causes serious harm to the environment and to human health. In that respect, the clean and sustainable manufacturing system that must be a smart manufacturing system too can be attained by the intelligent optimization of process parameters so that the benefits of using coolant/lubricant are alternatively achieved by the adoption of optimum control factor settings.

In the current paper, the control factors of hard-turning operation are optimized using intelligent evolutionary algorithms. The optimization is performed using two methods: (i) teaching–learning-based optimization and (ii) bacterial foraging optimization. Finally, these two methods are compared between themselves to select the best method for the further implementation. The key expected outcomes are optimum cutting speed, feed rate, and depth-of-cut for the lowest surface roughness parameters and cutting temperature.

2. Materials and Method

The hard-turning operation was conducted by machining of hardened high-carbon steel i.e., AISI 1060 in a center lathe (Origin: SJR Machinery Co., Ltd., Nantong, China, Max. W/P length: 1 m). The material stock had dimensions of length 300 mm and diameter 100 mm. The cutting tool used was coated tungsten carbide (WC) insert with Chemical Vapor Deposition (CVD) coating with TiCN/Al_2O_3/TiN. The insert's ISO designation was SNMM 120408. When the insert was used with the tool holder PSBNR 2525M12 (Sandvik Coromant, Sandviken, Sweden), the tool cutting edge was 75°, the clearance angle was 6°, and orthogonal rake angle was −6°. The hardening of work material was performed by austenizing followed by oil quenching and, lastly, tempering. The temperatures for the respective stages were 900 °C, 30 °C, and 370 °C. After the heat treatment, the hardness of the workpiece was 40 ± 2 Rockwell C.

The machining runs were varied according to the variations made in machining parameters. Investigated machining parameters were cutting speed, feed rate, and depth-of-cut. The Taguchi L8 orthogonal array was used for the design of experiment (DOE) to reveal 8 experiments (Table 1) based on 4 levels of cutting speed, and 2 levels of feed and cutting depth. This DOE reduced the number of experiments by 50% compared to full factorial DOE—thus, a step towards conservation of resources. The selection of machining parameters values are based on the knowledge of literature and current industrial practice.

The representative indices of surface quality, two surface roughness parameters, were recorded after each machining run. Those parameters were (i) arithmetic mean deviation of surface roughness, R_a, and (ii) maximum height of profile of surface roughness, R_z. Their measurement was conducted by using SRG 4500 roughness tester (cut-off length 0.8 mm, Thread Check Inc., New York, NY, USA). The cutting temperature was measured by using a tool–work thermocouple. Initially, the mili-volt reading of the thermocouple was recorded, and after that the mili-volt value was converted into temperature in Celsius scale. Proper calibration was done before using the thermocouple. For details on temperature measurement by tool–work thermocouple, refer to authors' other work [16]. The measured responses are listed in Table 1.

Table 1. Taguchi L8 orthogonal array and values of responses found from machining runs.

Experiment Number	Cutting Speed, v_c (m/min)	Feed Rate, f (mm/rev)	Depth of Cut, a_p (mm)	Surface Roughness, R_a (μm)	Surface Roughness, R_z (μm)	Cutting Temperature, θ (°C)
1	45	0.1	1.0	2.60	14.36	404
2	45	0.2	1.5	4.21	21.75	543
3	60	0.1	1.0	3.87	22.20	488
4	60	0.2	1.5	2.78	12.35	622
5	75	0.1	1.5	3.51	16.48	585
6	75	0.2	1.0	2.41	11.85	638
7	90	0.1	1.5	1.70	10.26	674
8	90	0.2	1.0	2.73	15.45	699

3. Intelligent Optimization Algorithms

3.1. Teaching–Learning-Based Optimization (TLBO)

The TLBO, reported by Rao et al. [7], is a population-based optimization method that mimics the behavior of teachers and learners by which process the teachers teach and the learners learn. This algorithm considers the teachers and learners as two fundamental components. Herein, the learning is accomplished in two forms: (i) learning from teachers and (ii) learning from other learners. The first one is called the teacher phase while the second one is named the learner phase. Evidently, the performance of the algorithm was evaluated by the grades of the learners which in turn is dependent on the quality of the teachers. Note that the learners are regarded as population and the subjects offered by the teachers are design parameters and the achieved grades are the 'fitness' value. Both phases are discussed below.

Teacher phase: In this section, the learners learn from the teachers only. Here, the quality of teachers influences the learning outcomes. As to be noted, the best quality learners are assigned as the teachers. The objective of teachers is to increase the mean results of the class. For instance, consider that in iteration "i" the numbers of subjects taught are "m", size of learners is "n" numbers, for specific subject "j" the mean of the result is $M_{j,i}$. Now the result of best learner k_{best} can be considered as the overall best result $X_{total\text{-}kbest\text{-}i}$. In deciding this best result, the whole subject spectrum is accounted. The learner with this best result is usually considered as teacher. The difference of the mean of the learners of a subject and the best learner (i.e., teacher) can be presented by Equation (1) [7].

$$Diff\_Mean_{j,k,i} = r_i \left(X_{j,kbest,i} - T_F M_{j,i} \right) \quad (1)$$

where $0 \leq r_i \leq 1$ is random number and the T_F indicates the teaching factor that to be determined by Equation (2) [7].

$$T_F = round[1 + rand(0,1)2 - 1] \quad (2)$$

where the distribution T_F follows equal probability distribution and it is not considered as parameter of TLBO.

The current solution is updated by the Equation (3) [7].

$$X'_{j,k,i} = X_{j,k,i} + Diff\_Mean_{j,k,i} \quad (3)$$

In this manner, the updated result is accepted if it is better. At the last stage of teacher phase, the updated results are saved and used as the input to the learner phase.

Learner phase: In this stage, the learners enhance the learning by the effective interaction among the other learners. In a random manner, one learner learns from another learner only if that learner has better knowledge than him/her as defined in Equations (4) and (5) [7]. As mentioned, the two random learners are P and Q, who had results $X_{total-P,i}$ and $X_{total-Q,i}$, have updated results $X'_{total-P,i}$ and $X'_{total-Q,i}$ at the end of the teacher phase. However, the $X'_{total-P,i} \neq X'_{total-Q,i}$.

$$X''_{j,P,i} = X'_{j,P,i} + r_i \left(X'_{j,P,i} - X'_{j,Q,i} \right) \text{ If } X'_{total-P,i} < X'_{total-Q,i} \quad (4)$$

$$X''_{j,P,i} = X'_{j,P,i} + r_i \left(X'_{j,Q,i} - X'_{j,P,i} \right) \text{ If } X'_{total-Q,i} < X'_{total-P,i} \quad (5)$$

where the $X''_{total-P,i}$ is accepted when it has better result. After the learner stage, all the better results are saved and afterward used as inputs to the next iteration of teacher phase. The flow chart of TLBO is demonstrated in Figure 1.

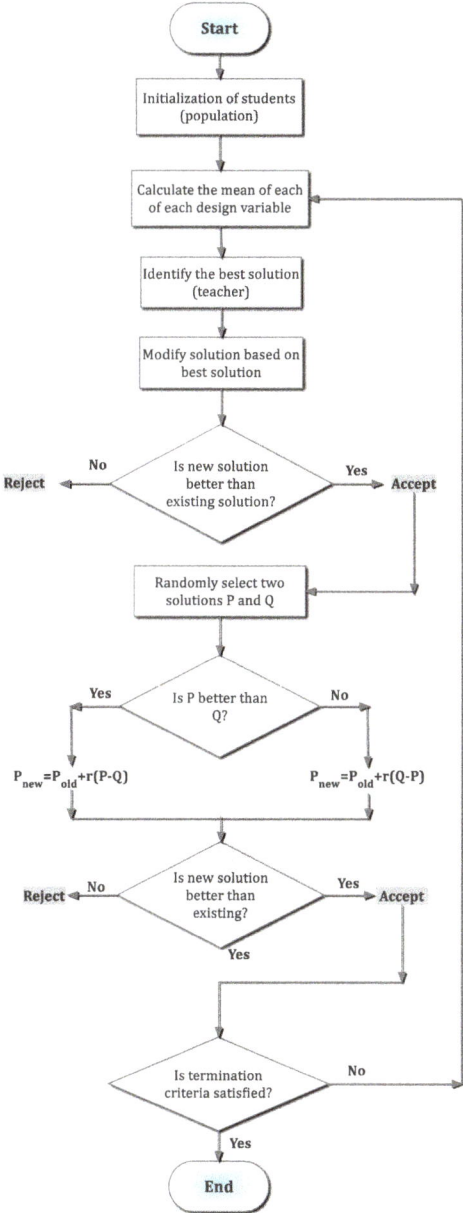

Figure 1. Flowchart for teaching—learning-based optimization (TLBO).

3.2. Bacteria Foraging Optimization (BFO)

In recent times, the adaptation of the bacteria foraging optimization method [17], primarily inspired from the attitudes of E. Coli's, is notably reported with significant success. This method is considered as an intelligent method that is meta-heuristic in nature. The nature of response of the bacteria with respect to changes in the surrounding environment, especially in the chemical gradient, stands out as the backbone of this method. In concept, the movement of the bacteria cells i.e., agents is driven by the awareness for food in the environment. In this manner, the cells are moved towards the optimal condition. For ease of understanding, the procedure of BFO is listed here.

Chemotaxis: The representative step for the behaviors like swimming as well as tumbling. In this step, the basic mechanism is to search the nutrients in any random directions. At any point, when the nutrient gradient is run into, the behavior of bacteria is dominated as the swim rather than tumble. Henceforth, the chemotaxis can be expressed as:

Function $\theta^i(j,k,l)$ represents the position of ith bacterium that possesses the jth chemotaxis having k^{th} reproduction and l^{th} elimination and dispersal. Also, mathematically the directional adjustment is modified according to Equation (6) [17].

$$\theta(j+1,k,l) = \theta(j,k,l) + C(i) \cdot \frac{\phi(i)}{\sqrt{\phi^T(i) \cdot \phi(i)}} \tag{6}$$

Here, the success of the optimization is largely dependent on the traits of foraging process of bacteria.

Swarming: In this stage, the randomly moved bacteria are organized in sophisticated uniformity as colonies. This altogether swarming is caused by the signals given by the cell, and it is mathematically presented by Equation (7) [17].

$$\begin{aligned}
J_{CC}(\theta, P(j,k,l)) &= \sum_{i=1}^{S} J_{CC}^i(\theta, \theta(j,k,l)) \\
&= \sum_{i=1}^{S} \left[-d_{attract} exp \left(-\omega_{attract} \sum_{n=1}^{P} (\theta_n - \theta_n^i)^2 \right) \right] \\
&+ \sum_{i=1}^{S} \left[-d_{repellent} exp \left(-\omega_{repellent} \sum_{n=1}^{P} (\theta_m - \theta_n^i)^2 \right) \right]
\end{aligned} \tag{7}$$

In Equation (2), the value of cost function is denoted by $J_{cc}(\theta, P(j,k,l))$. Note that the addition of varying cost function to the value of cost function results in actual cost function—that needs to be minimized. Furthermore, the S and P represent the numbers of total bacteria and parameters for optimization respectively. The parameters such as $d_{attract}$, $w_{attract}$, $d_{repellent}$, $w_{repellent}$ need to be selected correctly for each bacterium.

Reproduction: The health condition of the bacteria is defined by Equation (8) [17]. This phase is characterized by the reproduction of comparatively better-fitted bacteria into two bacteria. Note that the least-fitted bacteria die; as such, the overall population remains constant.

$$J_{health}^i = \sum_{j=1}^{N_c+1} J(j,k,l) \tag{8}$$

Elimination and dispersion: In case of scarcity of bacteria in any place, the bacteria of other places may face dispersal. This is due to the changing nature of habitable environment of the bacteria. In fact, such dispersal may cause destruction of chemotactic process. It is also possible that the dispersal causes the chemotaxis process to be assisted by adequate nutrient sources in proximity.

In the above fashion, the bacteria is never satisfied with the nutrients they get, and thereby they keep searching—it indicates the continuous nature of the chemotaxis, swarming, reproduction, and elimination and dispersal steps. The flow chart of BFO is shown in Figure 2.

4. Results and Discussion

Initially, the effects of the control factors i.e., cutting speed, feed rate, and depth-of-cut on the responses are portrayed graphically and discussed to understand the role of factors on the responses. Then, the surface-roughness parameters and cutting temperature found in hard turning were optimized by using the teaching–learning-based optimization and bacteria foraging optimization separately. Later, the optimum results are compared with respect to common parameters of interest. Eventually, the best optimization method is selected.

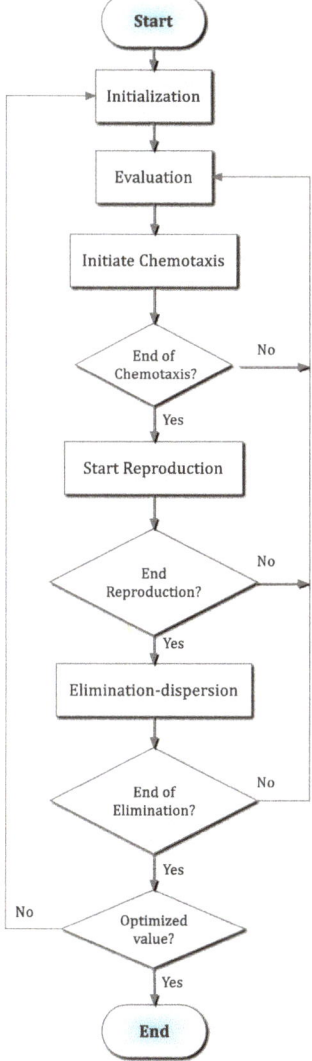

Figure 2. Flowchart for bacteria foraging optimization.

4.1. Results

The hard-turning operation is controlled by three factors—namely the cutting speed, feed rate, and depth-of-cut. Hence, these parameters mostly influence the eventual outcomes of machining. For that reason, the mean behaviors of the surface-roughness parameters as well as the cutting temperature are plotted in Figure 3. It is evident from Figure 3a that the increase in cutting speed is reflected by a decrease in the surface roughness, R_a. This is due to the fact the lower cutting speed is associated with increased chatter of machine tool. Moreover, the higher cutting speed is associated with lower coefficient of friction. In addition, the increased cutting speed causes the temperature to rise (Figure 3c), which, in turn, softens the material. As such, an ease of cutting is experienced [18]. A similar movement of surface roughness R_z with respect to cutting speed is also noticeable in Figure 3b. Note that the increase in cutting temperature with the increase in cutting speed is due to the conversion of mechanical energy (rotation of job in spindle) into heat energy.

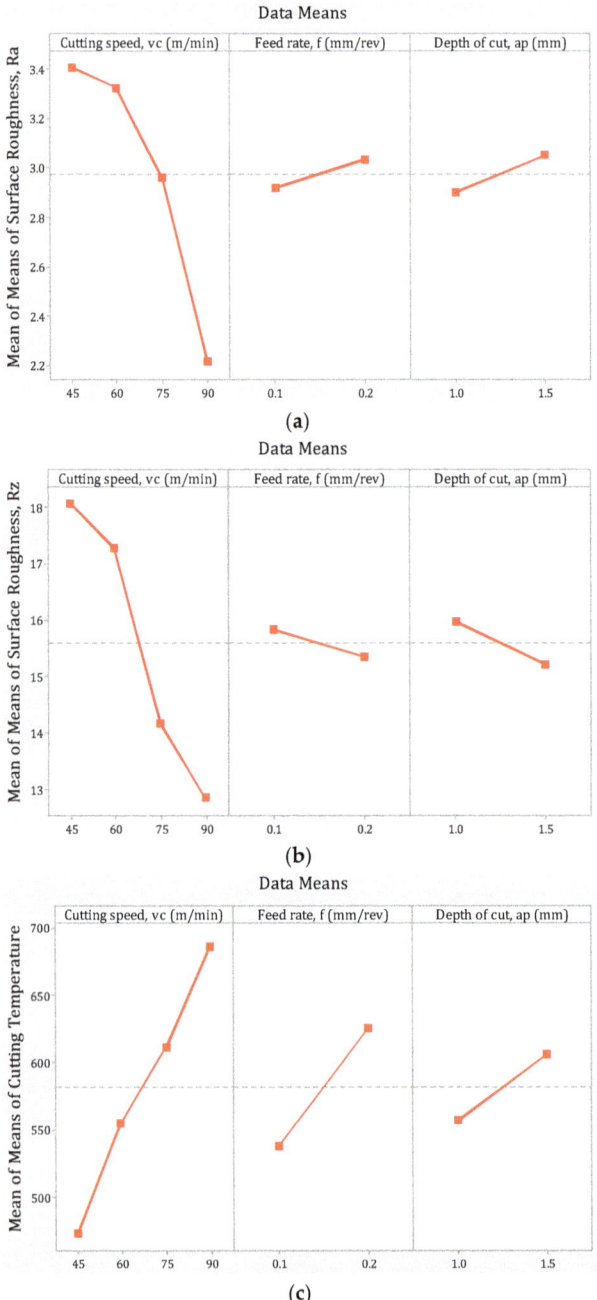

Figure 3. Effects of control factors on (**a**) mean of surface roughness, R_a, (**b**) mean of surface roughness, R_z, and (**c**) mean of cutting temperature.

The effects of feed rate on the surface-roughness parameters were opposing in nature (Figure 3a,b). For instance, an increase in feed rate caused a slight increase in R_a (this is according to the theoretical relation of f and R_a) while causing a slight decrease in R_z. Similarly, when the depth of cut was increased, the R_a increased by a slight amount but the R_z deceased by a small amount. This opposing nature can be interpreted by the fact that R_a is the average parameter of roughness while the R_z is the maximum height of roughness. As such, it is possible to get lower R_z when the R_a is increasing. Lastly, the increase in all three factors caused an increase in the cutting temperature. This is expected by the fact that the material removal rate (MRR) = $v_c \cdot f \cdot a_p$. This means an increase in speed, feed, and depth causes an increase in the amount of materials removed per unit of time. When the MRR increases, more energy is required to deform such increased material. Consequently, the thermal state of cutting zone experiences higher temperature.

It is also appreciable that the cutting speed has the highest contribution on both the surface-roughness parameters and cutting temperature. Compared to cutting speed, the feed rate and depth-of-cut have minor roles in defining the value of roughness and temperature. Hence, during optimization, the change in cutting speed is most significant to favorably align the value of the responses.

4.2. Optimization by TLBO and BFO

Before optimization, the two roughness parameters and the cutting temperature were converted into a single function (normalized) as shown in Equation (9).

$$Min\ Z = W_1(R_a/R_{amin}) + W_2(R_z/R_{zmin}) + W_3(\theta/\theta_{min}) \tag{9}$$

where the W_1, W_2, and W_3 are weight factors for average surface roughness parameter, maximum height surface-roughness parameter, and cutting temperature, respectively. However, their summation should be 1.0. Also, the R_{amin} and R_{zmin} are the minimum values of the average surface roughness parameter and maximum height surface roughness parameter respectively found in hard turning. For this current study, all three responses were considered to possess equal weight ($W_1 = W_2 = W_3 = 1/3$) as all of them are valuable to the manufacturers as they largely influence the machining outcomes.

The software used for performing the optimization was Matlab 2018b (The MathWorks, Natick, MA, USA) with intel i5 Processesor and 4 GB RAM. By nature, TLBO depends on the population size and the generation numbers which makes this algorithm parameter-less. To make the solution, the trials runs are conducted that eventually caused the population size to be 50 and generation to be 100. On the other hand, in the BFO method, the optimization was started with the initial parameters listed in Table 2. Initially, the 50 bacterial elements were taken into account to run the algorithm.

Table 2. Input parameters in the bacteria foraging optimization (BFO) algorithm.

Parameters	Values
Number of bacterial elements considered, S	50
Max defined chemotactic steps, N_c	50
Max defined reproduction steps, N_{re}	4
Total elimination–dispersal event, N_{ed}	2
Max allowed swim steps, N_s	4
Elimination–dispersal probability, P_{ed}	0.1

The optimum results by the TLBO and BFO methods are listed in Table 3. It is visible that the cutting speed of 80 m/min was fixed as the optimum cutting speed. Interestingly, from Figure 3, it is observable that the highest cutting speed (90 m/min) was responsible for the lowest surface roughness; however, at the same time, it caused the temperature to be highest. At this point, a trade-off is required to make the system congenial for both the surface roughness and the cutting temperature. And, because of this reason, the TLBO reported 80 m/min as the optimum cutting speed. For the same opposing nature (discussed earlier in Figure 3), the feed rate of 0.13 mm/rev was found as the optimum feed rate. However, the highest value of depth of cut (1.5 mm) was the optimum depth-of-cut. On other side, the BFO approach, within 16 s, revealed the optimum run. The optimum parameters by BFO are a cutting speed of 75 m/min, a feed rate of 0.10 mm/rev, and a depth-of-cut of 1.3 mm.

Table 3. Comparison of TLBO and BFO.

Parameters	TLBO	BFO
Cutting speed (m/min)	80	75
Feed rate (mm/rev)	0.13	0.10
Depth of cut (mm)	1.5	1.3
Best solution (minimum of Z)	0.54326	0.55262
Worst solution	0.56592	0.57854
Average time (s)	4 s	16 s

A comparison of TLBO and BFO revealed that the best solution was given by the TLBO method (Z is minimum = 0.54326). Further, this solution was found within the lowest time—for TLBO the time was 4 s while that for BFO was 16 s—and four times faster than the BFO. The convergence of the TLBO and BFO is illustrated in Figure 4. With the number of iterations, the overall fitness function of the TLBO aligns better to the optimum fitness function value. Hence, between the TLBO and BFO, the TLBO is recommended.

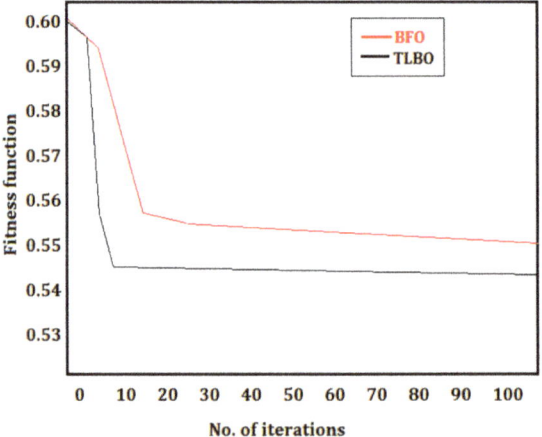

Figure 4. Convergence of bacteria foraging optimization (BFO) and teaching–learning-based optimization (TLBO).

5. Conclusions

In this study, two intelligent optimization algorithms were employed for the optimization of hard-turning parameters. Adoption of evolutionary optimization methods, with the assistance of high-level computing, can convert the conventional machining processes to be more effective, efficient, and cost-economic. From this study, the following conclusions can be drawn:

- Intelligent optimization is an important ingredient of smart manufacturing in which the learning capability of the method is required—which is present in both teaching–learning-based optimization and bacteria foraging optimization. Lack of implementation of these methods in hard turning motivated the current study, and eventually their successful implementation is shown here.

- The influences of cutting speed, feed rate, and cutting depth on the arithmetic mean deviation of surface roughness R_a, the maximum height of the profile of surface roughness R_z, and cutting temperature are investigated by portraying the main effects plot. It was found that the cutting speed played the most dominant role in defining the roughness parameter as well as the temperature. Moreover, an increase in cutting speed resulted in a decrease in the roughness values but an increase in the cutting temperature. This outcome necessitated a trade-off of factor values.
- Trade-off of the responses/factors was accomplished by employing the intelligent optimization method i.e., TLBO and BFO. Optimum results by the TLBO approach were a cutting speed of 80 m/min, feed rate of 0.13 mm/rev, and depth-of-cut of 1.5 mm; optimum parameter settings by BFO were a cutting speed of 70 m/min, feed rate of 0.10 mm/rev and depth-of-cut of 1.3 mm.
- The TLBO was found to be superior to the BFO in terms of better convergence and shorter time of computation—hence, the TLBO is recommended.
- Future research direction can be the adoption of evolutionary methods in the parametric optimization of additive manufacturing processes. Also, further research attention can be given to the integration of optimization methods with the real-time parameter optimization.

Author Contributions: Conceptualization, M.M., G.K.; data curation, M.M., S.W.; formal analysis, M.M., R.M.; investigation, M.M., R.M., S.W.; methodology, M.M., G.K.; project administration, M.M., G.K.; software, S.W.; writing—original draft, M.M.; writing—review and editing, M.M. and G.K.

Funding: This work was supported by Regional Operational Programme for the Opole Voivodeship financed by the Structural Funds of the European Union and the State budget of Poland: RPOP.01.01.00-16-0017/16. The financial support from the program of the Polish Minister of Science and Higher Education under the name "Regional Initiative of Excellence" in 2019–2022, project no. 003/RID/2018/19.

Acknowledgments: The authors are grateful to Munish Kumar Gupta, University Center for Research & Development, Chandigarh University, Gharuan, Punjab, India, for his help in carrying out the optimization.

Conflicts of Interest: The authors declare no conflict of interest.

References

1. Patalas-Maliszewska, J.; Kłos, S. An intelligent system for core-competence identification for industry 4.0 based on research results from german and polish manufacturing companies. In *International Conference on Intelligent Systems in Production Engineering and Maintenance*; Springer: Cham, Switzerland, 2017; pp. 131–139.
2. Zhong, R.Y.; Xu, X.; Klotz, E.; Newman, S.T. Intelligent manufacturing in the context of industry 4.0: A review. *Engineering* **2017**, *3*, 616–630. [CrossRef]
3. Davis, J.; Edgar, T.; Porter, J.; Bernaden, J.; Sarli, M. Smart manufacturing, manufacturing intelligence and demand-dynamic performance. *Comput. Chem. Eng.* **2012**, *47*, 145–156. [CrossRef]
4. Kujawińska, A.; Diering, M.; Rogalewicz, M.; Żywicki, K.; Hetman, Ł. Soft modelling-based methodology of raw material waste estimation. In *International Conference on Intelligent Systems in Production Engineering and Maintenance*; Springer: Cham, Switzerland, 2017; pp. 407–417.
5. Araújo, A.F.; Varela, M.L.; Gomes, M.S.; Barreto, R.C.; Trojanowska, J. Development of an Intelligent and Automated System for Lean Industrial Production, Adding Maximum Productivity and Efficiency in the Production Process. In *Advances in Manufacturing*; Springer: Cham, Switzerland, 2018; pp. 131–140.
6. Mia, M.; Dhar, N.R. Prediction and optimization by using SVR, RSM and GA in hard turning of tempered AISI 1060 steel under effective cooling condition. In *Neural Computing and Applications*; Springer: London, UK, 2017.
7. Rao, R.V.; Savsani, V.J.; Vakharia, D. Teaching–learning-based optimization: A novel method for constrained mechanical design optimization problems. *Comput.-Aided Des.* **2011**, *43*, 303–315. [CrossRef]
8. Rao, R.V.; Kalyankar, V. Experimental investigation on submerged arc welding of Cr–Mo–V steel. *Int. J. Adv. Manuf. Technol.* **2013**, *69*, 93–106. [CrossRef]
9. Pawar, P.; Rao, R.V. Parameter optimization of machining processes using teaching–learning-based optimization algorithm. *Int. J. Adv. Manuf. Technol.* **2013**, *67*, 995–1006. [CrossRef]

10. Rao, R.V.; Kalyankar, V. Parameter optimization of modern machining processes using teaching–learning-based optimization algorithm. *Eng. Appl. Artif. Intell.* **2013**, *26*, 524–531.
11. Gupta, M.K.; Sood, P.K.; Sharma, V.S. Machining Parameters Optimization of Titanium Alloy using Response Surface Methodology and Particle Swarm Optimization under Minimum-Quantity Lubrication Environment. *Mater. Manuf. Process.* **2016**, *31*, 1671–1682. [CrossRef]
12. Mukhopadhyay, A.; Barman, K.T.; Sahoo, P.; Davim, P.J. Modeling and Optimization of Fractal Dimension in Wire Electrical Discharge Machining of EN 31 Steel Using the ANN-GA Approach. *Materials* **2019**, *12*, 454. [CrossRef] [PubMed]
13. Kim, J.E.; Lee, M.C. A Study on the Optimal Machining Parameters of the Induction Assisted Milling with Inconel 718. *Materials* **2019**, *12*, 233. [CrossRef] [PubMed]
14. Matras, A.; Zębala, W.; Machno, M. Research and Method of Roughness Prediction of a Curvilinear Surface after Titanium Alloy Turning. *Materials* **2019**, *12*, 502. [CrossRef] [PubMed]
15. Raju, M.; Gupta, M.K.; Bhanot, N.; Sharma, V.S. A hybrid PSO–BFO evolutionary algorithm for optimization of fused deposition modelling process parameters. *J. Intell. Manuf.* **2018**, 1–16. [CrossRef]
16. Mia, M.; Dhar, N.R. Optimization of surface roughness and cutting temperature in high-pressure coolant-assisted hard turning using Taguchi method. *Int. J. Adv. Manuf. Technol.* **2017**, *88*, 739–753. [CrossRef]
17. Passino, K.M. Bacterial foraging optimization. *Int. J. Swarm Intell. Res.* **2010**, *1*, 1–16. [CrossRef]
18. Astakhov, V.P. Machining of hard materials: Definitions and industrial applications. In *Machining of Hard Materials*; Davim, J.P., Ed.; Springer Science & Business Media: Berlin, Germany, 2011; pp. 1–32.

© 2019 by the authors. Licensee MDPI, Basel, Switzerland. This article is an open access article distributed under the terms and conditions of the Creative Commons Attribution (CC BY) license (http://creativecommons.org/licenses/by/4.0/).

Article

Prediction of Tool Wear Using Artificial Neural Networks during Turning of Hardened Steel

Paweł Twardowski and Martyna Wiciak-Pikuła *

Faculty of Mechanical Engineering and Management, Poznan University of Technology, 3 Piotrowo St., 60-965 Poznan, Poland
* Correspondence: martyna.r.wiciak@doctorate.put.poznan.pl; Tel.: +48-790-412-919

Received: 11 June 2019; Accepted: 16 September 2019; Published: 22 September 2019

Abstract: The ability to effectively predict tool wear during machining is an extremely important part of diagnostics that results in changing the tool at the relevant time. Effective assessment of the rate of tool wear increases the efficiency of the process and makes it possible to replace the tool before catastrophic wear occurs. In this context, the value of the effectiveness of predicting tool wear during turning of hardened steel using artificial neural networks, multilayer perceptron (MLP), was checked. Cutting forces and acceleration of mechanical vibrations were used to monitor the tool wear process. As a result of the analysis using artificial neural networks, the suitability of individual physical phenomena to the monitoring process was assessed.

Keywords: artificial neural network; prediction; tool wear

1. Introduction

Currently, many methods are used to evaluate tool wear in real time. The cutting process monitoring system is a tool used to eliminate catastrophic tool failure (CTF). Assessment of the condition of the tool wear based on physical quantities that are associated with the cutting process is possible on the basis of many different methods. Research has been conducted over the years, which compares the effectiveness of assessing tool wear condition based on diagnostic inference methods. Regression models and pattern recognition are among the basic methods. Monitoring of the manufacturing processes is an issue that still requires improvement, despite the use of many modern systems in industry. One of the newer methods used to monitor the condition of the cutting edge is the empirical method, empirical mode decomposition (EMD), which is based on the decomposition of signals in the time domain. As reported in Olufayo et al. [1] paper, it was used to detect the cracking of the tool based on the measurement of cutting forces. The researchers presented an online industrial monitoring system that reliably obtained precise information on tool wear. On the basis of the coefficient of friction and average power, two forms of wear, i.e., tool edge chipping and tool edge wear, were detected in real time using the CUSUM algorithm (cumulative sum control chart), which gave satisfactory results as compared to an offline method. Wide usage in machining also has indirect monitoring of cutting wear based on cutting forces, evaluation of chip morphology, mechanical vibrations, and acoustic emission. An internet system for measuring and monitoring tool wear based on machine vision was designed and developed in line with the characteristics of a ball-end cutter. The validation of the experiment showed an error of only 2.5% in relation to the actual tool wear. In Wang et al. [2] investigations, acoustic emission signals were used to diagnose ceramic inserts during milling at high cutting speeds, where a multisensor system for classifying the used cutting wedge was additionally employed. The research used spectral analysis observation and wavelet feature extraction to evaluate tool wear. On the basis of the data obtained, the researchers developed a feed-forward backpropagation neural network (BPNN) model to predict tool wear. Learning the neural network gave a low error value of 0.00523 to classify the state of the tool. This is a very low average square error relative to actual consumption values, which confirms the effectiveness of the prediction based on acoustic emissions. Very often, cutting forces are used to diagnose the condition of the tool, and neural networks are used for diagnostic inference [3,4]. In addition to neural networks, the wavelet transformation and spectral grouping algorithms are also used for diagnostic inference such as in Aghazadeh et al. [5] paper. This experiment presented a tool conditioning monitoring (TCM) system using deep convolutional neural networks (CNNs) as an effective method of deep learning. Force and vibration signals from the experimental ETS (Emissions Trading System) dataset were used, which were independently selected to develop a monitoring system. In contrast to other learning models, these data-driven models were able to learn discriminative nonlinear feature representations. In this way, they could provide an efficient prediction model for error detection by learning feature representations directly from the input signals. Different methods of machine learning algorithms with force and vibration signals were compared and the smallest root mean square error (RMSE) was obtained for the CNN (0.0709 for force signals and 0.086 for vibration). Another example is the research of Kong et al. [6] paper that presented an effective model of wear width prediction using the Gaussian model with the radial basis function kernel principal component analysis (KPCA_IRBF). In this technique, the Gaussian noises can be modeled quantitatively in the GPR (Gaussian Process Regression) model. Many studies confirm that cutting forces are the most sensitive to changes in tool wear. However, their industrial application involves interference in the construction of machine tools or causes restrictions in the working space. Vibration sensors do not have such limitations, and therefore they are easy to assemble and do not interfere with the machine's construction. Therefore, diagnostic methods based on vibration measurements are constantly developed. One of the solutions is the use of multisensors because different sensors correlate better with subsequent stages of tool wear. This solution gives a full view of potential wear. After receiving the raw signal, signal processing and feature extraction methods are used, i.e., time domain analysis using autoregressive models (AR), moving average models (MA) or autoregressive moving average (ARMA) mixed models, and methods based on frequency domain analysis, wavelet transformation or empirical mode decomposition (EMD) method. Methods based on multiple monitoring models such as in Zhou et al. [7] paper are created based on multisensory systems. In addition, the rapid development of artificial intelligence (AI) and advanced methods of inference enable more and more effective application of these methods to predict tool condition [8,9].

Hassan et al. [10] proposed a monitoring system for online prediction and prevention of tool chipping during intermittent turning. A correlation between the chipping size and cutting parameters was designed to protect the machined surfaces. The work presented an integrated system based on acoustic emission (AE) signal processing in order to detect the tool pre-failure before tool chipping, and focused on cracks due to mechanical loads during an intermittent turning operation. The TKEO-HHT (Teager Kaiser Energy Operator-Hilbert–Huang Transform) technique was used which has the ability to deal with the nonstationary and nonlinearly AE_{RMS} signal in the pre-failure phase. This method successfully predicted tool chipping before failure with a processing time of 2 ms. The determined parameter, Ψ_{BW}, showed an exponential relationship with chipping, which made it possible to determine the threshold depending on the allowable chipping. The algorithm was optimized to provide sufficient time to stop the machine from damaging the workpiece. A new method of tool wear modeling is the application of a dedicated tribometer, which is able to simulate tribological conditions between the tool and workpiece. Rech et al. [11] investigated a contact pressure and sliding velocities (s_n, V_s) during turning. Tribological conditions were used to identify a wear model with a new tool geometry. The modeling method was based on an orthogonal cutting simulation (ALE) developed with Abaqus Implicit. In this work, the researchers reported that using the contact temperature as a parameter in the wear model was not a good idea. Instead, they decided to identify a wear model based on the contact pressure, s_n, and sliding velocity, V_s. Currently, this is in accordance with a trend in the field of tribology. They found that this model was very good to predict crater wear. This wear model has been implemented in numerical cutting model which is able to simulate cutting operations.

Currently, there is a lot of work that deals with monitoring wear during hard machining. One such work is Scheffer et al. [12] paper, where the researchers developed an accurate and flexible system for monitoring tool wear during hard turning. They designed an artificial intelligence (AI) model for monitoring crater and flank wear during hard turning. The purpose of developing the model was to obtain an intelligent and dynamic method. This modern approach to monitoring was based on parameters correlated directly with tool wear such as cutting force, vibration, and AE signals. In connection with this assumption, eight experiments were carried out with simultaneous measurement of cutting force, vibrations, AE, and temperature. An additional advantage of the chosen method was the ability to identify and isolate disturbances generated during the process, which was important because it was difficult to determine if the change in the sensor signal was due to wear or interference from the process. Additionally, they analyzed a self-organizing map (SOM) to identify interference that occurred during the process and they applied the method to achieve more efficient prediction of wear during hard machining. Another example of monitoring tool wear during hard turning is Ozel et al. [13], in which a neural network model was created for predicting tool wear and surface roughness. It demonstrates that the trend in process monitoring focuses not only on tool wear, but also on the evaluation of the machined surface to allow the best machining efficiency. This study utilized neural network modeling as compared with regression models. A neural network was obtained with the following seven inputs: workpiece hardness in Rockwell-C, cutting speed (m/min), feed rate (mm/rev), axial cutting length (mm), and mean values of three force components Fx, Fy, Fz (N). The small flank wear and surface roughness root mean square (RMS) errors on the test data showed the reliability of the method. The validation using neural networks gave better results than the use of regression models. The developed forecasting system was able to accurately predict surface wear and roughness. The wide range of use of aviation alloys contributed to the development of work in which tool wear is tested during machining of Inconel. One of the works is Capassoa et al. [14] investigation, in which the characteristics of tool wear during Inconel DA 718 turning with inserts with different coatings were examined. Tool wear was developed using three-dimensional (3D) volumetric wear progression. A predictive model was created based on both 3D and flank wear patterns. The model of tool wear with TiAlCrN/TiCrAl$_{52}$Si$_8$N PVD coating and AlTiN at different cutting speeds reached a value of the fitting factor R of over 93%, which meant that the method produced very good results. In addition to the new predictive models, they found that the tool with a PVD nanocomposite coating exhibited a substantial reduction in chipping, which confirmed superior wear resistance. In summary, the applied methodology proved that the volumetric wear prediction method was reliable.

There is a lack of studies comparing the use of different measured quantities as input data during the cutting process. If they already appear, it does not interfere with the network structure such as by changing the activation function or the number of neurons in the layer. In particular little information relates to the processing of hard materials, where the most advantageous information is about predicting tool wear. Therefore, studies have focused on comparing the effectiveness of predicting neural network models with different structures and with different input data.

In this paper, artificial neural networks are presented to predict tool wear based on various input data such as cutting forces and mechanical vibrations. Measurements of selected physical quantities were carried out during turning of hardened steel with constant cutting parameters.

2. Materials and Method

The turning of hardened bearing steel 100Cr$_6$ with a hardness of 61 ± 1 HRC was conducted. The tool material was oxide ceramics (Al$_2$O$_3$ + TiN). Mechanically fixed inserts SNGN120408 MC2 (Kennametal, Latrobe, PA, USA) were used for testing. The research was carried out on a universal lathe TUR560E (FAT, Wroclaw, Poland) with constant cutting parameters:

- cutting speed $v_c = 180$ m/min;
- rotational speed $n = 1400$ rev/min;
- feed $f = 0.08$ mm/rev;
- depth of cut $a_p = 0.1$ mm.

After each pass (length of the shaft $L = 150$ mm and cutting time of a single pass $t_s = 1.34$ min) the value of flank wear, VB_c, was measured (VB_c, flank wear of the tool corner), by means of a workshop microscope with a resolution of 0.01 mm.

During the turning operation, the following cutting force components were measured:

- F_x, F_f for feed direction;
- F_y, F_p for radial direction;
- F_z, F_c for main direction,

In addition, the acceleration of vibration was measured in the following different directions:

- A_x, A_f for feed direction;
- A_y, A_p for radial direction;
- A_z, A_c for main direction.

Figure 1 presents a simplified diagram of the measurement setup, which considers the location of sensors and additional components necessary for signal processing and analysis. Piezoelectric sensors were used to measure cutting forces and mechanical vibrations.

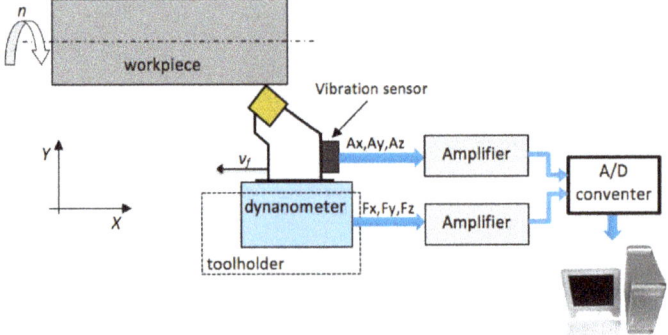

Figure 1. Scheme of measurement line used during the tests.

The maximum, minimum, and mean square values of cutting forces and vibration accelerations were selected as diagnostic measures. The mechanical vibrations were measured by a piezoelectric three-axis acceleration sensor fixed to the toolholder using a thread, while the cutting forces were measured using a piezoelectric measuring platform.

On the basis of digital signals sent to the computer, the mean square RMS values (Equation (1)) were evaluated:

$$M_{RMS} = \sqrt{\frac{1}{T_2 - T_1} \int_{T_1}^{T_2} [x(t)]^2 dt} \qquad (1)$$

where M_{RMS} is the mean square value for arbitrary diagnostic measure.

The time interval for determining the maximum, minimum, and RMS value was 4 s and the obtained measures were correlated with the corresponding tool wear values.

Under the same conditions, the wear process was carried out for 15 tool tips (15 tests). For each corner, the test was continued until the wear value $VB_c \approx 0.4$ mm was reached. The conventional tool life criterium that was adopted was $VB_c = 0.3$ mm.

3. Results

3.1. Tool Wear Analysis

Figure 2 shows the relationship between the flank wear, VB_c, and the cutting time, t_s, for all 15 tool corners. To determine the relation, $VB_c = at_s^3 + bt_s^2 + ct_s$, a third degree polynomial function was selected as the most representative for the tool wear process. This function reflects the results obtained in the best way, and the coefficient $R^2 = 0.98$, which indicates a high adjustment to the selected mathematical function.

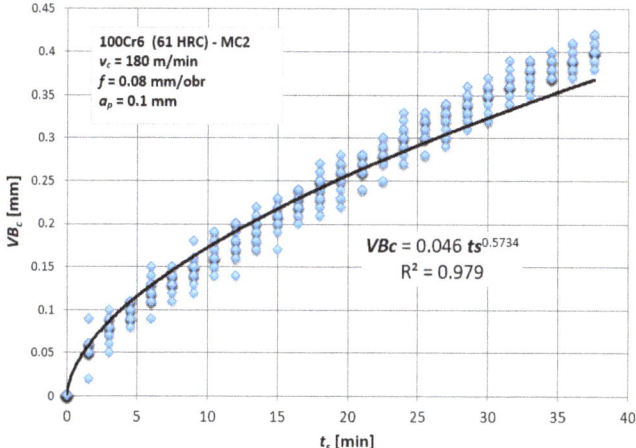

Figure 2. Tool wear, VB_c, as a function of cutting time, t_s, including all tests carried out.

The graph shows that the assumed tool life criterion, $VB_c = 0.3$ mm, is reached for $t_s = 25$ min. This criterion was selected based on previous experience related to the machining of hardened steels. Above this value, the probability of chipping of ceramic tool increased significantly, due to an increase in the level of mechanical vibration amplitudes.

The next step was to recognize the relationship between the tool wear and designated measures of diagnostic signals. Figure 3 depicts an example of the relation between the maximum value of the feed force, F_{f_max}, and tool wear, VB_c. This relationship was described by the linear function $F_{f\_max} = a \cdot VB_c + b$ and the coefficient $R^2 = 0.78$. For all other diagnostic measures, based on force measurements (i.e., F_{i_max}, F_{i_min}, and F_{i_RMS}), the best results were also obtained for the linear function.

However, there are different dependencies of the type, $F_i = a \cdot VB_c + b$, when the data will be divided into individual tool corners. The R^2 coefficient indicates correct matching of the assumed mathematical function to the experimental results. Figure 4a,b shows the individual R^2 coefficient for all analyzed tool tips and for two exemplary measures, F_{f_max} and F_{p_max}.

For example, for the F_{p_max} measure, the extreme waveforms were selected, i.e., for the extreme values of R^2, in order to illustrate changes in the amplitudes of the diagnostic measure as a function of tool wear (Figure 5). The analyzed changes of the F_{p_max} measures are described by a linear function with similar coefficients but with different dispersion of results. The larger the spread of results (the smaller the R^2), the more difficult it is to build a correct diagnostic model, although in the case of cutting forces the best results were obtained (the highest R^2 coefficient).

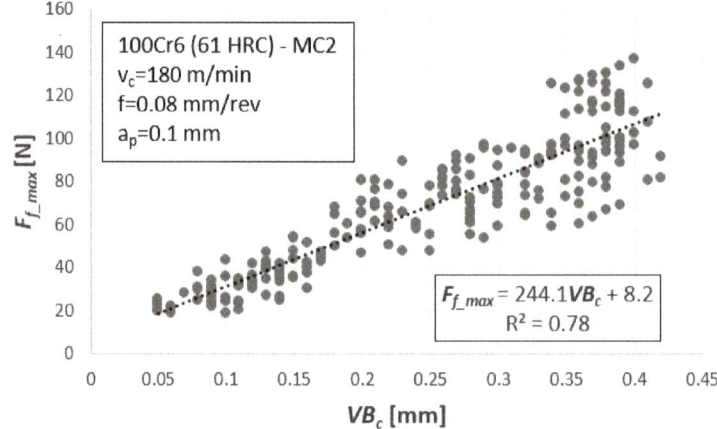

Figure 3. Feed force, Ff_max, as a function of flank wear, VBc, for all 15 cutting wedges.

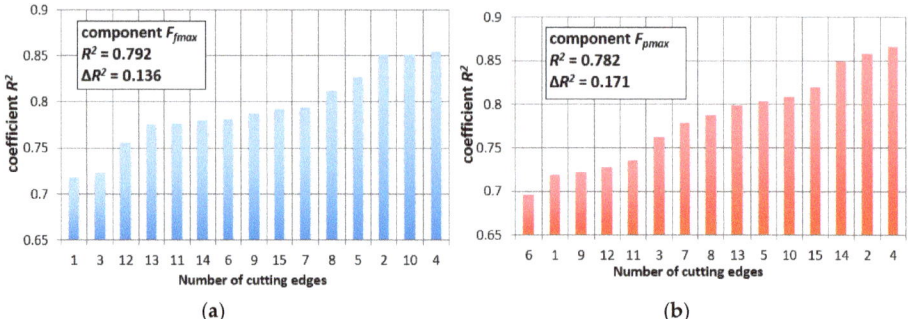

Figure 4. Matching coefficient, R^2, for the: (**a**) F_{f_max} measure and (**b**) F_{p_max}.

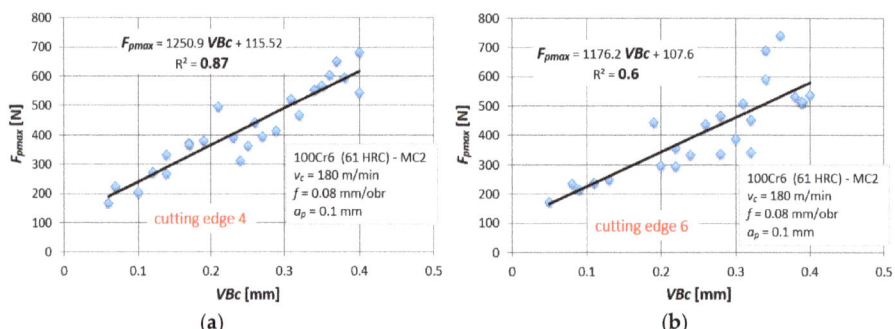

Figure 5. Cutting force, Fp_max, as a function of the tool wear, VBc, for tool tip (**a**) No. 4 and (**b**) No. 6.

The dependencies based on accelerations of vibrations look different and the dispersions of results are much larger. Figure 6 shows R^2 values for all tool tips and for exemplary measures, A_f and A_p. In comparison to cutting forces, extreme values are much higher and equal to $\Delta R^2 = 0.545$ for A_f and $\Delta R^2 = 0.49$ for A_p, respectively. These differences are illustrated in Figure 7. In the worst case, for $A_p - R^2 = 0.34$. The matching of the mathematical function to the actual results is unsatisfactory. The same analysis was carried out for all measures (in the X, Y, and Z directions), obtaining similar relationships.

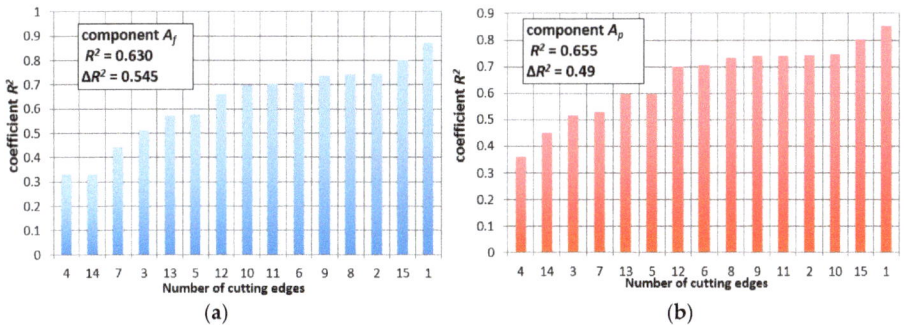

Figure 6. Matching coefficient R^2 for (**a**) the Af measure and (**b**) the Ap measure

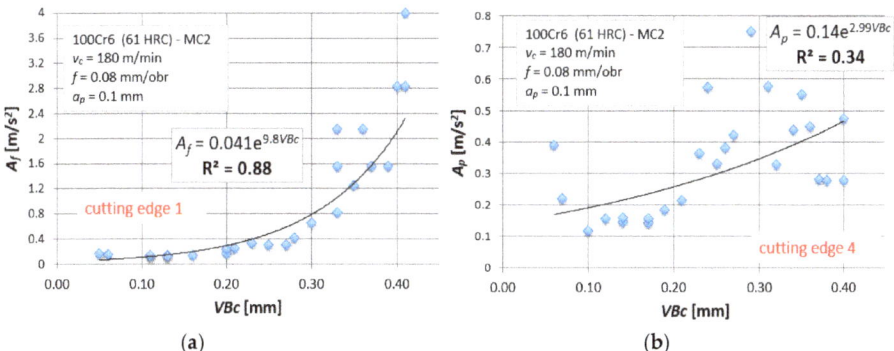

Figure 7. Acceleration of vibrations in feed direction, Af, as a function of the tool wear, VBc, for tool tips (**a**) No. 1 and (**b**) No. 4.

3.2. Diagnostic Model in the Form of a Regression Model

Section 3.1 describes the changes in diagnostic measures (for cutting forces and acceleration of vibrations) as a function of tool wear. The main purpose of the diagnosis of the cutting tool's condition is to recognize its degree of wear precisely based on the measured values of forces and vibrations. Nevertheless, a different approach to this issue can be applied. In industrial practice, very often two states are recognized, acceptable and unacceptable, i.e., when the tool tip should be replaced with a new one. For this purpose, the permissible tool wear value must be defined, i.e., tool life criterion. This work assumes: $VB_c < 0.3$ mm (a tool tip capable of machining) and $VB_c \geq 0.3$ mm (a blunt tool tip).

Figure 8 shows the two tool conditions (for all 15 tool corners) for two measures of cutting forces, F_{p_max} and F_{f_max}. Separation of the two areas, a tool tip capable of machining and a blunt tool tip, is the task of the monitoring system, which works based on various mathematical algorithms.

Such an analysis can be carried out because of several diagnostic measures that were selected in this work. The task is not complicated and in the diagnostic systems it is called the classification. However, a two-step evaluation of the tool condition is not always enough. Usually, recognition of the tool condition in the next cycle is the relevant information to withdraw the tool before exceeding the allowable wear. In this situation, we are dealing with prediction, and therefore a valid mathematical model for prediction that can assess the tool condition at any time. The simplest model is the one-variable regression equation shown in Figure 9.

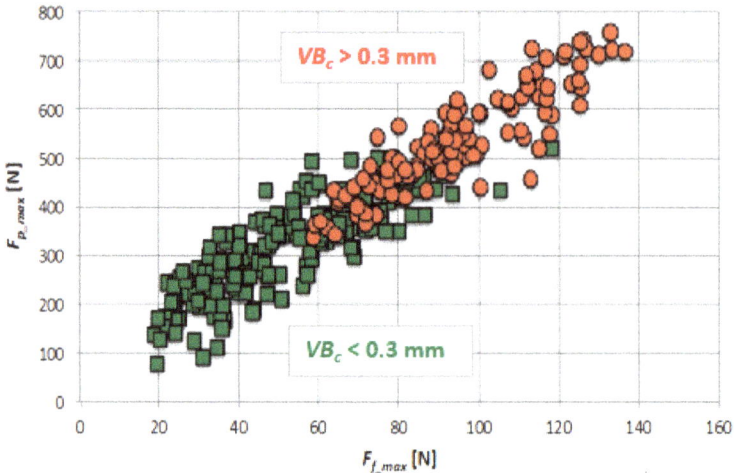

Figure 8. Two tool conditions for Fp_max and Ff_max measures (for 15 tool corners).

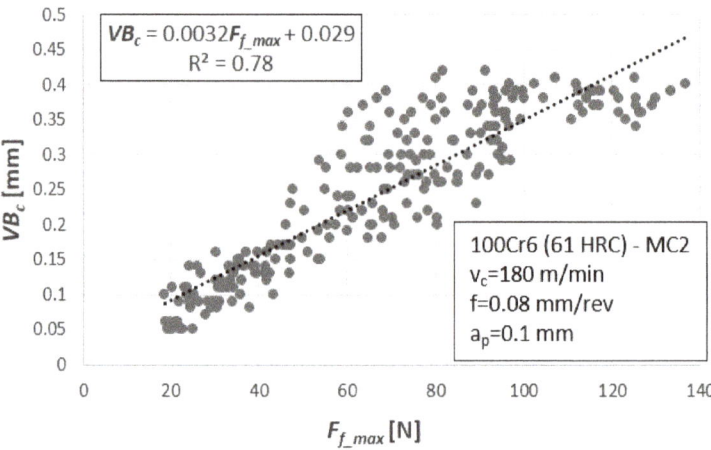

Figure 9. Diagnostic model in the form of a regression model for 15 tool tips.

For cutting forces components, the linear relationship of the type $y = a \cdot x + b$ (i.e., $VB_c = a \cdot F_i + b$) is best suited for assessing tool wear. In this context, it is enough to substitute the appropriate value of the cutting force component and read the value of the tool wear indicator. A similar procedure is applied when using diagnostic measures based on mechanical vibration signals. The only difference is that for vibrations the best-suited dependence is the logarithmic function of the type, $y = a \cdot \ln(x) + b$ (i.e., $VB_c = a \cdot \ln(A_i) + b$, Figure 10). The basic disadvantage of the one-variable regression model is low precision. Figure 10b shows an example, where the coefficient of determination, $R^2 = 0.34$, and dispersions of test results are very large. Therefore, it is difficult to carry out the correct verification process. The natural dispersion of experimental results means that the predicted values are not precise and are sometimes burdened with error. Hence it is better to use multivariable models or artificial intelligence algorithms, such as artificial neural networks.

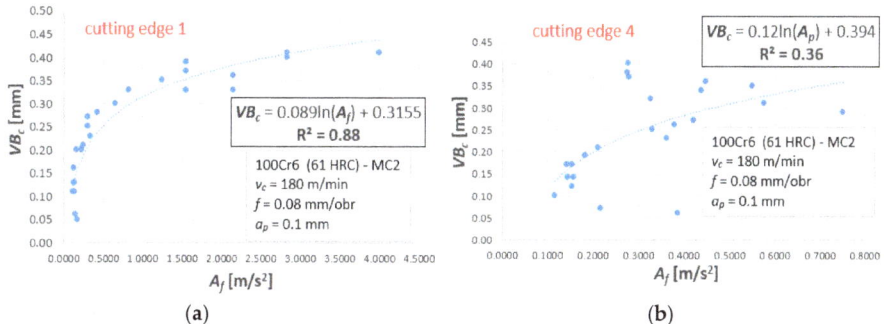

Figure 10. Exemplary one-variable regression model for Af: (**a**) tool corner No. 1 and (**b**) tool corner No. 4.

3.3. Diagnostic Verification Based on Artificial Neural Networks

After the analysis of individual diagnostic measures, neural network models were developed to recognize tool wear using multilayer perceptron (MLP) feedforward networks. The input data were diagnostic measures based on the analysis of vibration accelerations and the components of the cutting force. The structure of the used networks, on the entry of two or four neurons, in the hidden layer the number of neurons, varied from four to 20, while in the initial layer it was one neuron. The first step was focused on learning the neural networks using the Broyden—Fletcher—Goldfarb—Shanno algorithm (BFGS), which is considered one of the most effective. Input data were results for 13 tool tips working with identical cutting parameters. However, the results from the next two cutting edges were used to validate the developed neural networks. In order to select the best network, the activation functions were changed in the hidden and the initial layer: linear, logistic, hyperbolic, and exponential. The functions used are listed in Table 1.

Table 1. Activation functions applied at the network learning stage.

Function	Equation
Linear	x
Logistics	$\frac{1}{1-e^{-1}}$
Hyperbolic (Tanh)	$\frac{e^x - e^{-x}}{e^x + e^{-x}}$
Exponential	e^{-x}

3.3.1. Models of Neural Networks Used to Recognize the VBc Tool Wear Based on Accelerations of Vibrations

In the first place, the results of vibration accelerations were analyzed starting from the network learning stage. The networks created with different configuration of the activation functions in the hidden layer and the output layer. Among the 130 models developed, three models with the best learning and testing efficiency were selected. Table 2 shows the MLP neural networks together with a list of all parameters generated in the Statistica program. After entering data into the program, the number of random samples was assumed at the level of 70% for the training set, 15% for the test set, and 15% for the validation set.

Table 2. Selected neural networks based on vibration acceleration analysis (generated in Statistica).

No.	Network Name	Quality (Learning)	Quality (Testing)	Quality (Validation)	Activation Function (Hidden Layer)	Activation Function (Output Layer)
23	MLP 2-12-1	0.8826	0.8336	0.9119	Logistic	Exponential
72	MLP 2-15-1	0.8818	0.8449	0.9249	Exponential	Exponential
126	MLP 2-20-1	0.8797	0.8518	0.9188	Tanh	Exponential
No.	Network name	Learning algorithm	Error (learning)	Error (testing)	Error (validation)	Error function E_{sos}
23	MLP 2-12-1	BFGS 55	0.0012	0.0010	0.0007	SOS
72	MLP 2-15-1	BFGS 84	0.0008	0.0007	0.0007	SOS
126	MLP 2-20-1	BFGS 68	0.0007	0.0007	0.0007	SOS

The network number 126 has the best quality of testing around 85%, while the network number 23 has the best learning efficiency of approximately 88%. The number of neurons in the hidden layer, the number of cycles of the BFGS algorithm (Broyden–Fletcher–Goldfarb–Shanno algorithm, MLP network learning algorithm), and the function of the hidden layer and the initial layer affect the effectiveness.

The SOS error function, E_{SOS} (sum of squares), was used to determine the error during learning, testing, and validation. The error function determines the correspondence between calculated values and actual values. In Statistica, the error function is calculated as the sum of squared differences based on the Equation (2):

$$E_{SOS} = \sum_{i=2}^{n} (y_i - t_i)^2 \qquad (2)$$

where n is the number of examples (input and output pairs) used for learning, y_i is the network prediction (network output), and t_i is the "real" value (output according to data) for ith value.

The effectiveness of selected neural networks was evaluated based on the root mean square error (RMSE). After learning the network, new data was introduced and the effectiveness of the prediction of the tool wear, VB_c, was checked. A comparison of the effectiveness of individual networks based on new data is presented in Table 3.

Table 3. Root mean square error (RMSE) values for three MLP neural network (based on acceleration of vibrations).

Network Name	RMSE
MLP 2-12-1	0.051
MLP 2-15-1	0.050
MLP 2-20-1	**0.049**

The smallest error was obtained for a network with 20 neurons in the hidden layer (MLP 2-20-1). The mean square error using this network was 0.049. Figure 11 graphically illustrates the correlation between experimental data and the predicted data for the MLP 2-20-1 networks.

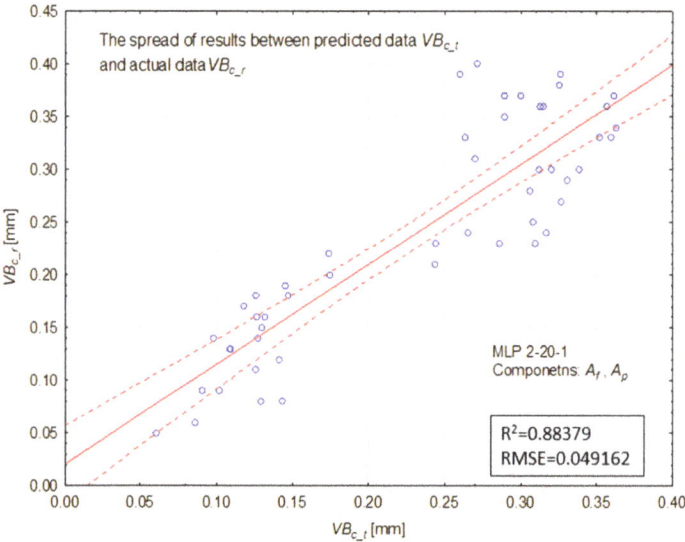

Figure 11. Comparison of predicted values, VBc_t, with experimental data, VBc_r, at the validation stage (MLP 2-20-1).

3.3.2. Models of Neural Networks Used to Recognize the VBc Tool Wear Based on the Cutting Force Components

The next step in checking the effectiveness of tool wear prediction value was the development of an ANN model with input values of the components of cutting forces. Two diagnostic measures were selected for the creation of the model, F_{p_max} and F_{f_max}. The 95 different models of neural networks were created corresponding to models based on vibration accelerations. Table 4 presents selected models for analyzing the effectiveness of tool wear prediction.

Table 4. Selected neural networks based on the cutting force components (generated in Statistica).

No.	Network Name	Quality (Learning)	Quality (Testing)	Learning Algorithm	Activation Function (Hidden Layer)	Activation Function (Output Layer)
4	MLP 2-10-1	0.8787	0.9245	BFGS 4	Linear	Linear
7	MLP 2-10-1	0.9228	0.9502	BFGS 149	Tanh	Linear
94	MLP 2-20-1	0.9259	0.9510	BFGS 140	Tanh	Logistic

The best learning quality was obtained for a network with twenty neurons in the hidden layer (No. 94), approximately 93%. Similarly, the same network achieved the best testing efficiency, approximately 95%. Noticeably, by increasing the number of neurons in the hidden layer, the performance of the model was improved. It is obvious that for the networks with the best training and testing quality, the smallest RMS error was obtained. Table 5 presents the comparison of the obtained RSM errors for individual models.

Table 5. The RMSE values for three MLP neural network (based on the cutting force components).

Network Name	RMSE
MLP 2-10-1 (4)	0.048
MLP 2-10-1 (7)	0.046
MLP 2-20-1	0.045

However, Figure 12 shows the dispersion of predicted and experimental values for the best performing model.

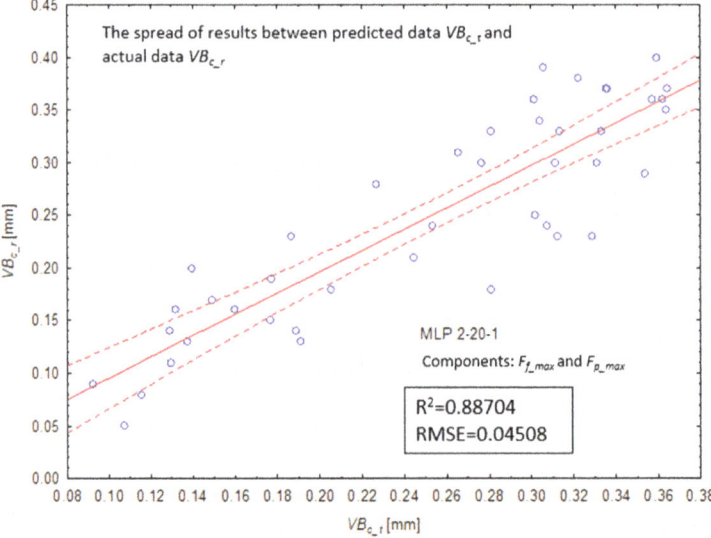

Figure 12. Comparison of predicted values, VBc_t, with experimental data, VBc_r, at validation stage (MLP 2-20-1).

3.3.3. Models of Neural Networks Used to Recognize the VBc Tool Wear Based on Four Variables

The final stage of the analysis was the creation of a model of neural networks based on four different variables. Diagnostic measures with the best correlation coefficient R^2 were selected as input data. Two measures of the cutting forces component were selected, F_{p_max} and F_{f_max}, and two measures of vibration acceleration, A_f and A_p. Thirty different models of neural networks were created. Table 6 presents three models selected for analyzing the effectiveness of tool wear prediction.

Table 6. Selected neural networks based on four different variables (generated in Statistica).

No.	Network Name	Quality (Learning)	Quality (Testing)	Learning Algorithm	Activation Function (Hidden Layer)	Activation Function (Output Layer)
5	**MLP 4-4-1**	0.9711	0.9766	BFGS 90	Tanh	Exponential
21	MLP 4-6-1	0.9697	0.9817	BFGS 60	Tanh	Linear
26	**MLP 4-6-1**	0.9706	0.9807	BFGS 62	Tanh	Tanh

Table 7 lists the root mean square error RMSE for three selected networks. However, Figure 13 presents the spread of results between the predicted and experimental values of tool wear based on the MLP 4-6-1 model (No.21).

Table 7. The RMSE values for three MLP neural network (based on four variables).

Network Name	RMSE
MLP 4-4-1	0.043
MLP 4-6-1 (21)	**0.040**
MLP 4-6-1 (26)	0.050

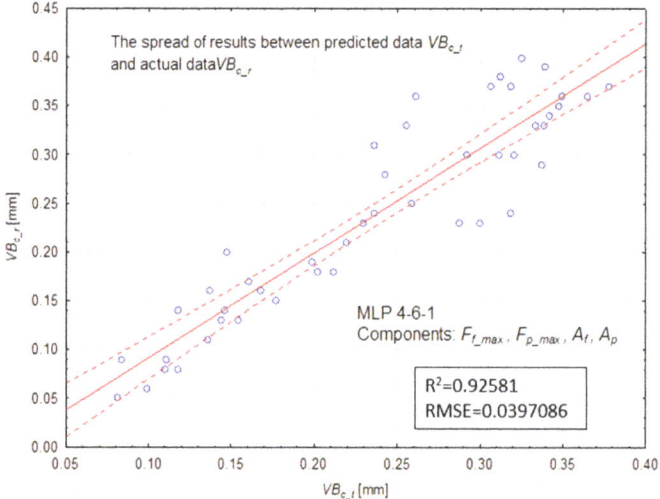

Figure 13. Comparison of predicted values, VBc_t, with experimental data, VBc_r, at the validation stage MLP 4-6-1 (based on Fp_max, Ff_max, Af, and Ap variables).

The best learning quality, approximately 97%, was obtained for the network No. 5, while the best test efficiency was a network with six neurons in the hidden layer of about 98% (No. 21). Similar to the previous analysis, after selecting the best quality models, new data were introduced to determine the effectiveness of the neural networks at the validation stage. The smallest RMS error was obtained for a network with six neurons in the hidden layer and a linear function of activation in the hidden layer of 0.040 (No.21). In this analysis, a significant impact of the activation function used on the network performance was noticed. The error values from all analyzed models are summarized in Figure 14.

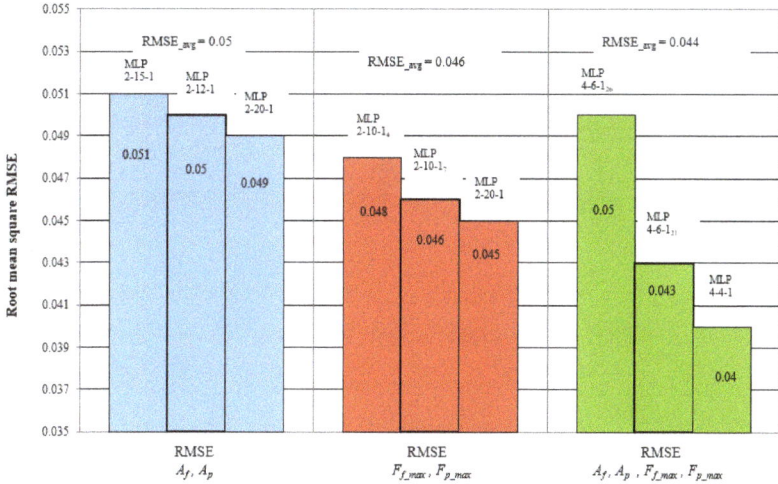

Figure 14. Comparison of RMS errors of the neural networks used in the validation stage.

On the basis of all the developed models of neural networks, the best network performance for four different variables was determined (for diagnostic measures based on vibration and cutting force signals). This solution has a practical drawback because it requires measurement of two physical quantities, vibration and forces. In practice, systems based on measurements of several physical quantities are problematic and cost intensive. Therefore, sometimes a better compromise is the use of simple measuring methods (as in the case of vibrations) in exchange for greater effectiveness in recognizing the condition of the cutting tool. There are some limits of compromise, but this should be checked in practical applications.

4. Conclusions

This paper presents a neural network model to predict tool wear based on cutting forces and mechanical vibrations. From this study, the following conclusions can be drawn:

1. The correlation coefficient between tool wear, VB_c, and the diagnostic measure assumes different values depending on the direction of vibration or force measurement. The best correlation coefficient was obtained for the radial cutting forces F_{p_max}. The coefficient value $R^2 = 0.87$ indicates a good correlation with the power mathematical function. On the basis of the coefficient, the input data for creating the network was also selected.
2. Proper selection of the number of neurons in the hidden layer and activation function in the hidden and initial layers significantly affect the effectiveness of predicting the tool wear value. Changes in the structure of the model at the beginning of its creation by the user help to achieve prediction at a satisfactory level. The artificial neural network, MLP, is an effective model for predicting tool condition during machining difficult-to-cut materials. The use of various diagnostic measures increases the efficiency of prediction.
3. The correlation coefficient obtained in the analysis of vibration accelerations was definitely lower than in the analysis of cutting forces. Nevertheless, the tool wear model, ANN, based on the measures of acceleration of vibrations A_p and A_f obtained the ability to forecast tool wear with the efficiency loaded by the mean square error RMSE = 0.049.
4. Wear prediction based on measurements of cutting force components F_{p_max} and F_{f_max} obtained slightly better results than at vibration accelerations. The error accomplished was RMSE = 0.045 mm. This means that both cutting forces and vibration acceleration are equally good for assessing tool wear during machining difficult-to-cut materials.
5. By creating different structures of the ANN model, the most effective prediction possibility for the model with the four input measures was obtained: F_{p_max}, F_{f_max}, A_p and A_f. The use of various diagnostic measures produced the best prediction results, RMSE error = 0.040 mm. In this case, the tangent activation function in the hidden layer and the linear activation function in the output value accomplished the best effects.

Author Contributions: P.T. and M.W.-P. conceived of the presented idea. P.T. carried out the experiment, processed the experimental data and performed the tool wear analysis. M.W.-P. verified the Artificial Neural Networks methods. P.T. and M.W.-P. contributed to the interpretation of the results. All authors discussed the results and contributed to the final manuscript.

Funding: This research was funded by Poznan University of Technology 02/22/SBAD/1501.

Conflicts of Interest: The authors declare no conflicts of interest.

References

1. Shi, X.; Wang, X.; Jiao, L.; Wang, Z.; Yan, P.; Gao, S. A real-time tool failure monitoring system based on cutting force analysis. *Int. J. Adv. Manuf. Technol.* **2018**, *95*, 2567–2583. [CrossRef]
2. Olufayo, O.; Abou-El-Hossein, K. Tool life estimation based on acoustic emission monitoring in end-milling of H13 mould-steel. *Int. J. Adv. Manuf. Technol.* **2015**, *81*, 39–51. [CrossRef]
3. Wang, G.; Guo, Z.; Yang, Y. Force sensor based online tool wear monitoring using distributed Gaussian ARTMAP network. *Sens. Actuators A Phys.* **2013**, *192*, 111–118. [CrossRef]
4. Liu, T.-I.; Jolley, B. Tool condition monitoring (TCM) using neural networks. *Int. J. Adv. Manuf. Technol.* **2015**, *78*, 1999–2007. [CrossRef]

5. Aghazadeh, F.; Tahan, A.; Thomas, M. Tool Condition Monitoring Using Spectral Subtraction Algorithm and Artificial Intelligence Methods in Milling Process. *Int. J. Mech. Eng. Robot. Res.* **2018**, *7*, 30–34. [CrossRef]
6. Kong, D.; Chen, Y.; Li, N. Gaussian process regression for tool wear prediction. *Mech. Syst. Signal Process.* **2018**, *104*, 556–574. [CrossRef]
7. Zhou, Y.; Xue, W. Review of tool condition monitoring methods in milling processes. *Int. J. Adv. Manuf. Technol.* **2018**, *96*, 2509–2523. [CrossRef]
8. Khorasani, A.; Yazdi, M.R.S. Development of a dynamic surface roughness monitoring system based on artificial neural networks (ANN) in milling operation. *Int. J. Adv. Manuf. Technol.* **2017**, *93*, 141–151. [CrossRef]
9. Kong, D.; Chen, Y.; Li, N. Hidden semi-Markov model-based method for tool wear estimation in milling process. *Int. J. Adv. Manuf. Technol.* **2017**, *92*, 3647–3657. [CrossRef]
10. Hassan, M.; Sadek, A.; Damir, A.; Attia, M.H.; Thomson, V. A novel approach for real-time prediction and prevention of tool chipping in intermittent turning machining. *Manuf. Technol.* **2018**, *67*, 41–44. [CrossRef]
11. Rech, J.; Giovenco, A.; Courbon, C.; Cabanettes, F. Toward a new tribological approach to predict cutting tool wear. *CIRP Ann. Manuf. Technol.* **2018**, *67*, 65–68. [CrossRef]
12. Scheffer, C.; Kratz, H.; Heyns, P.S.; Klocke, F. Development of a tool wear-monitoring system for hard turning. *Int. J. Mach. Tools Manuf.* **2003**, *43*, 973–985. [CrossRef]
13. Ozel, T.; Karpat, Y. Predictive modeling of surface roughness and tool wear in hard turning using regression and neural networks. *Int. J. Mach. Tools Manuf.* **2005**, *45*, 467–479. [CrossRef]
14. Capassoa, S.; Paivab, J.M.; Juniorb, E.L.; Settineric, L.; Yamamotod, K.; Amorime, F.L.; Torrese, R.D.; Covellif, D.; Fox-Rabinovichb, G.; Veldhuisb, S.C. A novel method of assessing and predicting coated cutting tool wear during Inconel DA 718 turning. *Wear* **2019**, *202949*, 432–433. [CrossRef]

© 2019 by the authors. Licensee MDPI, Basel, Switzerland. This article is an open access article distributed under the terms and conditions of the Creative Commons Attribution (CC BY) license (http://creativecommons.org/licenses/by/4.0/).

Article

Evaluating Hole Quality in Drilling of Al 6061 Alloys

Mohammad Uddin [1,2], Animesh Basak [3], Alokesh Pramanik [4], Sunpreet Singh [5], Grzegorz M. Krolczyk [6] and Chander Prakash [5,*]

1. School of Engineering, University of South Australia, Mawson Lakes 5095, SA, Australia; Mohammad.Uddin@unisa.edu.au
2. Future Industries Institute, University of South Australia, Mawson Lakes 5095, SA, Australia
3. Adelaide Microscopy Unit, University of Adelaide, Adelaide 5005, SA, Australia; animesh.basak@adelaide.edu.au
4. Department of Mechanical Engineering, Curtin University, Perth 6845, WA, Australia; alokesh.pramanik@curtin.edu.au
5. School of Mechanical Engineering, Lovely Professional University, Phagwara, Punjab 144411, India; snprt.singh@gmail.com
6. Department of Manufacturing Engineering and Automotive Products, Opole University of Technology, 76 Proszkowska St., 45-758 Opole, Poland; g.krolczyk@po.opole.pl
* Correspondence: chander.mechengg@gmail.com; Tel.: +91-987-880-5672

Received: 14 November 2018; Accepted: 29 November 2018; Published: 2 December 2018

Abstract: Hole quality in drilling is considered a precursor for reliable and secure component assembly, ensuring product integrity and functioning service life. This paper aims to evaluate the influence of the key process parameters on drilling performance. A series of drilling tests with new TiN-coated high speed steel (HSS) bits are performed, while thrust force and torque are measured with the aid of an in-house built force dynamometer. The effect of process mechanics on hole quality, e.g., dimensional accuracy, burr formation, surface finish, is evaluated in relation to drill-bit wear and chip formation mechanism. Experimental results indicate that the feedrate which dictates the uncut chip thickness and material removal rate is the most dominant factor, significantly impacting force and hole quality. For a given spindle speed range, maximum increase of axial force and torque is 44.94% and 47.65%, respectively, when feedrate increases from 0.04 mm/rev to 0.08 mm/rev. Stable, jerk-free cutting at feedrate of as low as 0.04 mm/rev is shown to result in hole dimensional error of less than 2%. A low feedrate along with high spindle speed may be preferred. The underlying tool wear mechanism and progression needs to be taken into account when drilling a large number of holes. The findings of the paper clearly signify the importance and choice of drilling parameters and provide guidelines for manufacturing industries to enhance a part's dimensional integrity and productivity.

Keywords: drilling; dynamometer; hole quality; forces; roundness; roughness; wear; chips; burr

1. Introduction

In manufacturing industries, hole drilling has been a signature process employed to create various geometric features as well as to ensure secure assembly with other components for enhanced product integrity, reliability, and life cycle [1,2]. In particular, the process has been one of the major fabrication processes in automotive and aerospace industries, when machining of lightweight metals and composites are concerned [3,4]. Tool-based and laser-assisted drilling are adopted to fabricate the holes with the desired hole quality [5]. While laser drilling is shown to create holes with high geometric precision, often high process temperature may potentially deteriorate the structural integrity of the part. Such a phenomenon has unanimously been touted as a major issue in drilling of the composites. For instance, high temperature causes melting and swelling around the hole area, thus leading to damage to the drilled part [6].

In particular, burr formation and poor surface quality in drilling negatively affect the dimensional accuracy, and cause additional difficulty, reworking, cost, and even damage, e.g., fatigue, in the assembly. Therefore, the drilled holes are often deburred to retain a component's functional reliability. It is reported that the deburring simply accounts for about 30% of total fabrication cost in an aircraft's fuselage assembly [7]. As such, the importance of minimization of burr formation and comprehensive techniques to achieve this have been stressed out, with an aim of developing a more robust process modelling and database.

Regardless of materials and techniques, the important process parameters, such as spindle speed and feedrate, significantly affect the drilling performance, in terms of material removal rate, thrust force, and torque. The effect of these parameters on the process mechanics and their optimization in the drilling of different types of materials has been studied [8]. It is shown that the thrust force and torque dictates the final outcome of the drilling. A singled-out consensus, though, is that high thrust force and torque result in poor hole quality and deterioration of tool life. It is, therefore, very important to assess and understand further the process mechanics in a drilling process.

Commercial piezoelectric force sensors, such as Kistler's dynamometer, are used to measure the cutting dynamics in terms of the thrust force and the torque. While they are highly accurate and reliable, they are very expensive for small–medium-sized manufacturing shop floors to afford. Also, as the sensor's dynamic response is affected by workpiece mass and geometry, the sensors measure static forces with potential drifting, causing erroneous force measurements. However, as an inexpensive option, strain-gauge-based mechanical force sensors are becoming a potential candidate, which can still offer reasonably accurate and reliable force measurements. In this case, elastic deformation of a mechanical element is sensed by a series of strain gauges, which are interfaced with an electrical instrumentation, and under loading, the force is measured and estimated as an equivalent electrical voltage output. The gauge sensors are highly sensitive to strain and can be easily attached to the mechanical structure. Recently, the current authors designed and developed an innovative octagonal-elliptical strain gauge-based sensor for measuring milling force data, and demonstrated its working functions [9]. As a simple and robust tool, the designed sensor is found to have a potential to adapt in drilling and evaluate its underlying performance.

With the aid of appropriate sensing and assessment tools, in the past, numerous analytical, numerical, and experimental approaches have been employed to characterize the drilling process, i.e., estimating the thrust force, torque, assessing hole quality, of aluminum alloy and composite materials. In drilling of the fiber-reinforced plastic, Wei et al. [10] reported that the thrust force and hole quality are strongly influenced by the feedrate while the effect of the cutting speed is relatively less. In drilling of Ti6Al4V, Glaa et al. [11] proposed and studied a numerical model in estimating force and torque by taking regenerative chatter and process damping into account, and evaluated the effect of process parameters.

Ko et al. [12,13] studied the effect of drill-bit geometry, suggesting a larger point angle and step drill to enhance the hole quality, i.e., reduced burr size. Similar observations are reported elsewhere in the work by Lauderbaugh [14]. Nauri et al. [15] has investigated tool wear in dry drilling via experimental analysis and optimization, stressing that abrasive and adhesion wear causes tool bluntness and consequently, breakage, thus resulting in the hole's dimensional inaccuracy. Kurt et al. [16] recommended a low cutting speed and feedrate for enhanced hole quality in the drilling of aluminum alloys.

The effect of the coolant, such as MQL (minimum quantity lubrication) liquid nitrogen, was studied, and it is found that while the coolant reduces the thrust force and tool wear, and improves the hole quality, the use of coolant may cause environmental hazards [17,18]. Along with by a sustainable manufacturing manifesto, the machining process is expected to be less hazardous for the operators, users, and environment. As such, drilling in dry conditions can often be preferred. In an extensive work, Ramulu et al. [19] observed that, in drilling with HSS drill bits, the temperature at the cutting zone increases with the increase of spindle speed and the decrease of feedrate. Increasing the spindle speed leads to increased tool wear, larger entrance and exit burrs, while an increased feedrate leads to an increased thrust force and torque, but smaller entrance and exit burrs. Such observation is somewhat contradictory to earlier findings by Chen and Elhman [20]. It is therefore apparent that the relationships between the drilling parameters and the thrust force, tool wear, and burr formation may vary with the underlying workpiece and drill-bit material. In other words, while the process mechanics, i.e., material removal and chip generation, seems to be generic, the process outcome can still change with the effective parameters and conditions employed [21]. This warrants further investigation to explore and validate such perspective in drilling.

Given the observed discrepancies, with the aid of an in-house designed and built affordable and accurate force dynamometer, the objective of the current study is to recap the drilling mechanics with an aim of comprehensively investigating the effect of the key process parameters—the spindle speed and the feedrate—on the thrust force and torque. As a final outcome, the drilled hole quality, in terms of hole diameter, roundness, surface roughness, and burr formation, is assessed and discussed in relation to the tool wear mechanism and chip formation characteristics. To observe a sustained evolution of tool wear, a series of holes are drilled out of an aluminum 6061 alloy workpiece.

2. Materials and Methods

Axial force (i.e., thrust force) and torque are two major indicators for the assessment of drilling dynamics. It has been demonstrated that compared to a circular ring, an octagonal structure with an internal elliptical hole generates high strain under loading, thus improving the sensitivity of strain gauge-based load cells [9]. In this study, we have designed and fabricated in-house an octagonal-ellipse shape force dynamometer to measure the thrust force and the torque in drilling. Figure 1 depicts a schematic diagram of the dynamometer structure along with strain gauge arrangement on it. The tangential force data is used to estimate the drilling torque. Top and bottom plates shown are attached to hold the workpiece to be drilled out. The details of the electronics including the bridge circuits and signal processing unit are not shown here for simplicity. The force dynamometer was statically calibrated on an Instron machine (Model: 5567, Norwood MA, USA). A linear relationship between the applied force and the output voltage is found with a fitting accuracy of 98%. To capture the dynamic behavior in drilling, the dynamometer is calibrated in a real drilling test. Figure 2 shows a representative dynamic axial force (F_a) and torque (T) while drilling a hole. Torque is estimated using the relationship of $T = F_t * r$ [22], where r = the radius of the drill bit and F_t = tangential force measured by the dynamometer. It is seen that the force and torque vary with time. In particular, torque increases with time as the depth of drilling increases, i.e., when the full contact between the drill bit and the hole surface reaches. This indicates that the force dynamometer used in this study can detect and measuring the dynamic and transient cutting force information.

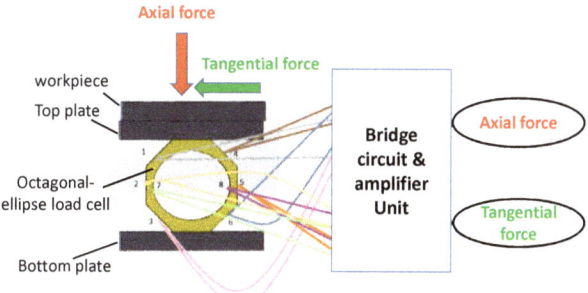

Figure 1. Arrangement of an octagonal-ellipse load cell and the connections of strain gauges to measure axial and tangential forces.

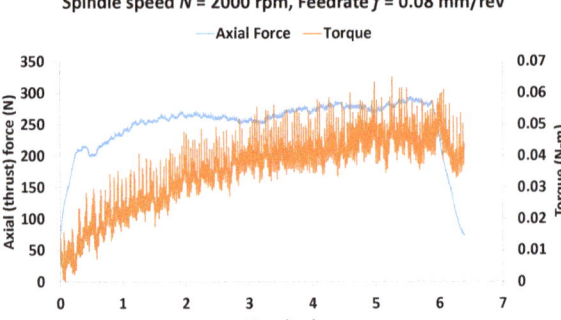

Figure 2. Representative dynamic axial force and torque measurement in drilling of a hole at spindle speed $N = 2000\times$ rpm and feedrate $f = 0.08$ mm/rev.

To assess hole quality, a series of drilling experiments was conducted on a 3-axis mill-drill machine (MetalMaster's MB-52VE, HAFCO, New South Wales, Australia). Drilling tests were performed in a dry condition. Parameters considered are shown in Table 1. These levels of parameters are often found to be used in conventional cutting of aluminum alloys in various manufacturing industries [23]. Figure 3 shows an experimental setup including an in-house built force dynamometer along with a data acquisition system. For a given set of process parameters, ten (10) holes are drilled out to investigate the effect of tool wear on hole quality. Therefore, there are six (6) sets of parameter combination made between the spindle speed and the feedrate (see Table 1). For each set of 10 holes, a new and sharp edge drill bit is used. As seen in Figure 3 (see inset image) a dedicated workpiece of 125 mm × 125 mm × 8 mm with pre-drilled holes of 10 mm is made and mounted onto the top plate so that, during drilling of each hole, the axial force is always pointed towards the central axis of the dynamometer. Workpiece is made of aluminum 6061 alloys, whose mechanical properties are shown in Table 2. and the drill bits used are two fluted, TiN-coated A002 high speed steel of 8 mm in diameter along with a cutting angle/drill point = 118°, cutting direction = right-hand. The depth of the hole drilled is 16 mm.

Table 1. Drilling parameters.

Parameters	Values
Spindle speed N (rpm)	1000, 1500, 2000
Feedrate f (mm/rev)	0.04, 0.08

For each hole, the transient forces are measured and recorded using the designed force dynamometer. The transient data with the drilling time are averaged out to determine the final force and torque. Drill-bit cutting edges and chips are observed and analyzed by an optical microscope (Leica's DVM500) and scanning electron microscope (SEM) (Merlin, Carl Zeiss, Oberkochen, Germany) to investigate tool wear and cutting mechanism as the number of drilled holes increases.

Table 2. Mechanical properties of workpiece material (Al 6061 alloys) used.

Parameters	Values
Young's modulus (GPa)	68.9
Poisson's ratio	0.33
Tensile strength (MPa)	124–290
Density (g/cm^3)	2.7
Thermal conductivity (W/m·K)	151–202
Specific heat capacity (J/Kg·K)	897

Hole diameter and roundness are measured by a coordinate measuring machine (CMM) (Brown & Sharpe's MicroXcel 7.6.5) manufactured by Hexagon Metrology (Melbourne, Australia). The machine is connected to measurement software PC-DMIS (Version 3.7), which is used to collect the measurement data for further processing. For each hole, 8 horizontal planes perpendicular to the depth direction from the top of the hole at an interval of 2 mm are chosen, where the CMM probe touches at least 10 points on the inner surface at approximately an equal angle of interval at each depth and measures the diameter on the plane by using the least square circle (LSC) method. CMM Measurement includes entry and exit sides of the hole. Average of the diameter measured at 8 planes is the final diameter. Hole roundness is defined as the radial distance between the minimum circumscribing circle and the maximum inscribing circle, which possesses the profile of the inner surface at a section perpendicular to the axis of rotation. Using the same CMM data obtained for diameter measurement, the final roundness is recorded as the average of roundness measured at 8 depth sections. Drilled surface roughness is measured using a Mitutoyo's Surftester (Model: SJ 211), where a cut-off length of 2 mm is considered. Roughness (R_a) measurement is taken on at least five locations along the hole depth direction and their average is recorded as the final value. Burr formation appears to be common in drilling, which affects the hole quality in terms of dimensional accuracy and performance of drilling. In drilling, burrs are generated at the entry and exit side of the hole. In this study, exit burr thickness and height has been measured by the optical microscope (Leica's DVM 500). Measurements are conducted at four locations equally distant to each other on the hole, and their average is considered the final recorded value.

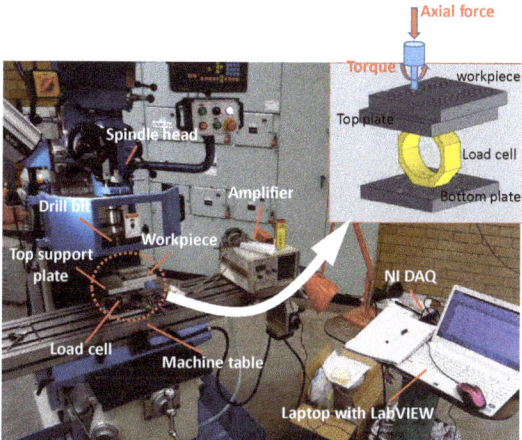

Figure 3. Experimental setup for drilling tests.

3. Results and Discussion

3.1. Axial Force and Torque

Figure 4 shows the variation of average axial (thrust) force and torque with respect to spindle speed and feedrate. For each hole, the average value estimated as the average of transient force data measured from the moment when the drill bit enters full into workpiece until the bit exits the hole completely. Clearly, both axial force and torque increase significantly with the increase of feedrate. For instance, when feedrate is increased from 0.04 mm/rev to 0.08 mm/rev, the increase of axial force is 29.23%, 44.94% and 34.02% at spindle speed of 1000× rpm, 1500× rpm and 2000× rpm, respectively. For the same change of feedrate, torque increases by 29.95%, 41.55% and 47.65%, at spindle speed of 1000× rpm, 1500× rpm and 2000× rpm, respectively. Larger feedrate means the drill bit experiences faster penetration axially, thus resulting in larger chip thickness and material removal rate. As a result, thrust force and torque increase. On the other hand, at a given feedrate, spindle speed appears to have insignificant influence on thrust force. For example, at a feedrate of 0.08 mm/rev, axial force increases by 4.74% and 9.18% when the spindle speed changes from 1000× rpm to 1500× rpm, and to 2000× rpm, respectively. In drilling of homogenous titanium alloy stacks, Wei et al. [10] observed that the change of thrust force with respect to spindle speed is minimum or negligible. Hence, this supports our results for drilling of aluminum alloys. In other words, drilling of titanium and aluminum alloys follow qualitatively the similar trend, and so is expected in terms of hole quality.

However, some moderate effect on torque is noticed. For instance, at a feedrate of 0.08 mm/rev, torque increases by 59.19% when spindle speed increases from 1000× rpm to 1500× rpm, and then remains nearly stable with a moderate increase of 10.95% as the spindle speed reaches to 2000× rpm. As the spindle speed increases further, temperature generated at the cutting zone softens the material, and hence, the drill bit requires less force for plastic deformation and shearing of material. As can be seen in Figure 4a, a slight increase of axial force (and torque) with increase of spindle speed can be due to the variation of degree of thermal softening and temperature rise because of actual spindle speed variation (i.e., commanded spindle speed may not be constant during drilling). The results clearly suggest that higher spindle speed and lower feedrate may be preferred; but tool wear effect, which will be discussed in the following section, must be taken into consideration simultaneously.

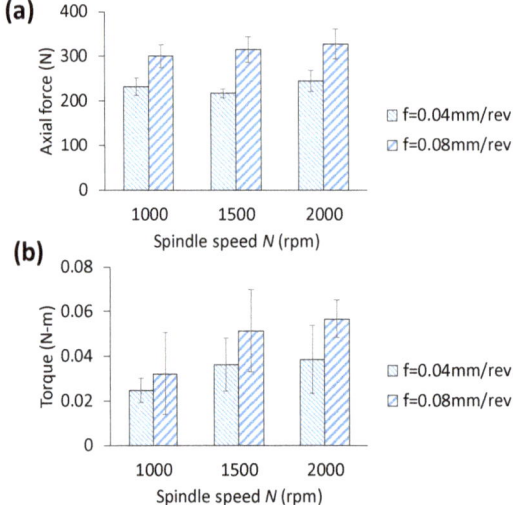

Figure 4. Effect of feedrate and spindle speed on (**a**) average axial force and (**b**) average torque. Error bar indicates standard deviation of force data for drilling of 10 holes.

3.2. Tool Wear Mechanism

Figure 5 illustrates SEM photos of the drill bit's cutting edges after 10th hole for each combination of spindle speed and feedrate. As compared to the chisel edge, the cutting-edge wear is the dominant factor impacting the drilling performance. Noticeable wear on the flank face includes adhesion due to built-up edge, abrasive and chipping or fracture. These types of wear are very common for cutting of soft material, such as, aluminum alloys. It can be seen from Figure 5 that, adhesion wear has been the obvious wear mechanism, regardless of the spindle speed and the feedrate studied. As the feedrate increases from 0.04 to 0.08 mm/rev, the abrasion wear takes place, causing the true flank wear and weakens the strength of the cutting edge. Large feedrate means higher material removal rate, causing larger thrust force onto the cutting edge. Consequently, the edge chipping along with plastic deformation starts to occur, which may lead to the breakage of the drill bit. In particular, TiN coating on the drill bit would be more vulnerable. This observation is consistent with the findings of drilling force and torque. It can, therefore, be imperative to say that the moderate feedrate can be recommended to avoid early initiation of the cutting-edge wear and failure. Though no significant measurable wear on the flank face is observed even after the 10th hole, it is expected that the severity of the cutting-edge wear will accelerate as the drilling time for producing more holes will increase. In addition to shear straining in primary shear zone, the complex interaction and temperature rise at the interface between the tool and chip rule the dominant adhesion wear evolution in drilling. Nuoari et al. [15] reported adhesion occurs in two stages as built-up edge (BUE) and build-up layer (BUL) as the drilling of more holes continues. Initial unstable BUE transforms into BUL due to pressure and temperature in contact zone, leading to potential diffusion of aluminum towards the tool, and micro-welding forms on tool surface. When BUL formation reaches tool edge and breaks due to dynamic non-continuous cutting, the tool edge becomes irregular and weakens, which potential may result in catastrophic fracture failure. SEM images shown in Figure 5 show clearly a change of BUE to BUL along with rough tool edges. As such, our results on tool wear are consistent with literature. Therefore, along with appropriate choice of spindle speed and feedrate, the use of highly wear resistant and low friction coated drill bit (e.g., (Ti-Al)N or diamond coating via CVD/PVD on tungsten carbide (WC) tool [24]) along with an effective cooling mechanism can be considered to minimize the severity of tool wear, and hence, improve drilling performance in terms of hole quality (which is discussed in the following sections), tool life and manufacturing productivity [15].

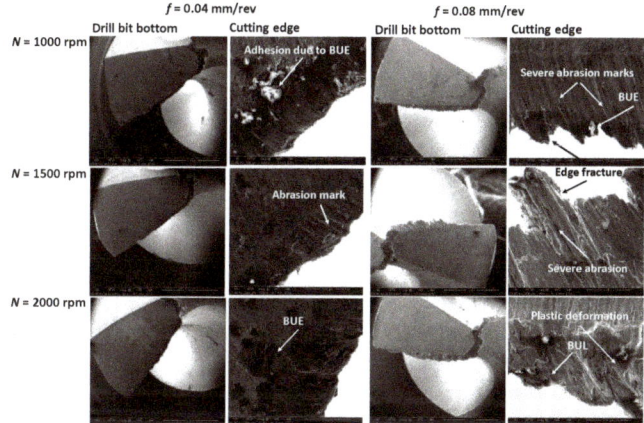

Figure 5. Scanning electron microscope (SEM) images of the drill-bit bottom and their magnified view of the cutting edge after 10th hole for different spindle speed and feedrate.

3.3. Hole Diameter

Figure 6 shows the variation of average hole diameter and % of difference from its nominal size with respect to spindle speed (N) and feedrate (f). As is obvious, for the range of speed and feedrate studied, hole size is always larger than the nominal and the maximum % of difference in diameter is less than 2%, i.e., the hole is less than 150 μm large from its nominal dimension (of 8 mm in diameter). Despite the diameter increase is relatively small, it appears that smaller feedrate is shown to reduce the dimensional difference while the spindle speed has no noticeable effect, except for the condition of $N = 1000\times$ rpm and $f = 0.04$ mm/rev, which indicates that low speed and low feedrate would be preferred. Lower feedrate means slower penetration rate and the cutting edge removes material with smaller chip thickness, allowing a stable and jerk-free drilling, and as a result, the hole diameter with less dimensional error is achieved. It is reported that faster spindle speed causes temperature rise at the cutting zone, and softens the material, thus facilitating a smoother drill surface with a good surface quality with an improved dimensional accuracy. Though the difference is not statistically significant, our results on hole diameter shown in Figure 6, indicate an improvement of dimensional accuracy as the spindle speed increases from 1000× rpm to 2000× rpm. The results are consistent to force and torque data. At a low feedrate, shear cutting is the dominant mechanism, resulting in a continuous chip generation and lower force, and as a result, hole deviation is minimum. Similar conclusions on hole size in drilling of Al alloys are observed elsewhere in literature [2,23]. Therefore, it is safe to say that, given a spindle speed, slower feedrate can suitably be selected to minimize the dimensional error. Although a slower feedrate compromises productivity, the decision must be made by establishing a fair balance between the productivity and the dimensional accuracy required.

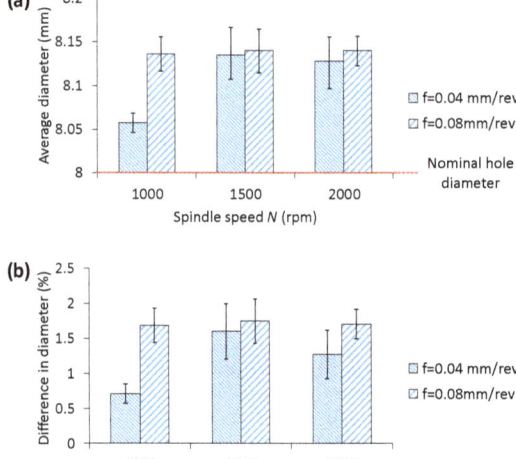

Figure 6. Effect of federate and spindle speed on (**a**) hole diameter and (**b**) % of difference with respect to its nominal size (=8 mm). Error bar indicates standard deviation of hole diameter for drilling of 10 holes.

3.4. Burr Formation

Figure 7 shows an example drilled hole with burrs at the exit side, and the geometric definition. Burr size at the entry side is found to be smaller than the exit side. Therefore, burr at the exit side is measured for drilling of ten holes at each combination of process parameters and presented for analysis. Figure 8 shows the change of the exit burr thickness and height with respect to spindle speed and feedrate. It can be seen that higher spindle speed and feedrate increase the burr size. In particular, the increase of burr size with the feedrate is higher when the spindle speed is larger. For instance, the change of burr thickness between feedrate of 0.04 mm/rev and 0.08 mm/rev increases from 18% to 50% when the spindle speed increases from 1000× rpm to 2000× rpm, respectively. On the other hand, for the same condition, the burr height jumps from 19.56% to 28%. As explained in earlier section, higher feedrate rate introduces higher thrust force, which causes larger and faster chip generation, and, as a result, the burr geometry becomes larger. Overall, spindle speed influences the burr thickness the most than the burr height. Figure 9 shows a representative topography of the exit side burr with respect to spindle speed and feedrate. It is to be noted that burr formation and its increase can be carefully observed when drilling with larger diameter. In other words, larger diameter tool increases cutting speed and dynamic rake angle, which cause plastic deformation in machining hardening layer and residual stress depth on the hole wall, hence resulting in increase of burr thickness and height [25]. The above suggests that lower spindle speed and feedrate must be chosen to minimize burr generation, thus saving cost for further rework in removing burrs. Past computational modelling and experimental investigation on drilling of aerospace aluminum and composite materials have reiterated the severity of burr formation, and made similar recommendations to ensure superior hole quality [26]. For instance, Sorrentino et al. [22] reported a reduction of push-out delamination factor (i.e., exit burr geometry) by 37% for drilling of CFRP (carbon fiber-reinforced polymer) when feedrate is changed from 0.3 mm/rev to 0.1 mm/rev.

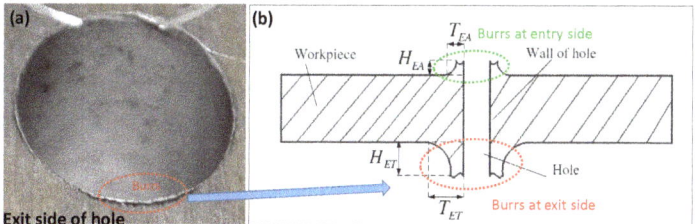

Figure 7. (a) Representative hole with burrs at the exit side of a hole (b) definition of burr geometry.

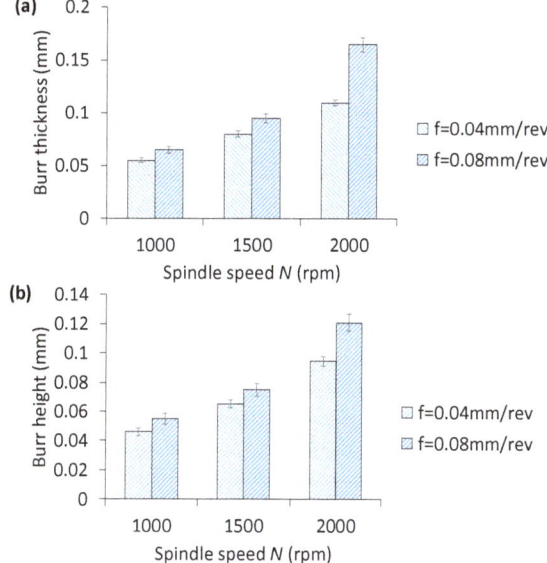

Figure 8. Effect of spindle speed and feedrate on (a) thickness and (b) height of exit burr. Error bar indicates standard deviation of burr size for drilling of 10 holes

3.5. Roundness and Roughness

Figure 10a show the roundness error with respect to spindle speed and feedrate. The roundness error increases significantly with the feedrate. For instance, when the feedrate is increased from 0.04 mm/rev to 0.08 mm/rev, the roundness increases by 78.78% at a spindle speed of 1000× rpm. The roundness error could primarily be because of burrs generated at the entry and the exit sides of the hole, the abrupt thrust force, and the dynamic instability of the drill bit. Higher thrust force due to a larger feedrate would be the dominant reason for an increased roundness error. For all ten holes drilled out, the roundness error is less than 60 μm, which is reasonably acceptable for small to medium size holes (of 8 mm diameter). On the other hand, the roundness error is less impacted by the spindle speed, but reduces at a large feedrate of 0.08 mm/rev. This is surprisingly interesting observation though, and conflicts with the trend of burr geometry with higher feedrate and spindle speed. Such variation could be due to the errors in roundness measurements by CMM. In other words, as the CMM probe touches the inner surface of the hole, the cutting debris potentially adhered to the surface affects the measurement, and hence, the overall roundness error. Even though hole's inner surface is cleaned by high speed air spray by an air gun, very minute debris may be stick to the surface. Effect of spindle speed and feedrate on hole roughness R_a is shown in Figure 10b. It is seen that roughness varies between 8.5 µm and 11.15 µm. Feedrate has no or minimum influence on roughness, while higher spindle speed is shown to give lower roughness. These results imply that low to moderate spindle speed and feedrate can safely be selected to achieve smooth hole surface finish. Clean and smooth hole surface is expected to pull-out strength and mechanical integrity of the underlying assembly structure. Observation of chip morphology which is shown in the next section further explains the mechanism for improved surface finish.

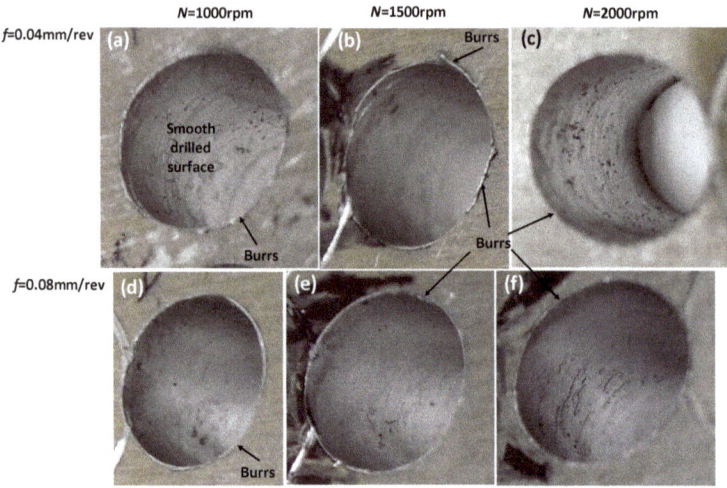

Figure 9. Representative topography of exit side of holes at parameter combination of (**a**) N = 1000 rpm and f = 0.04 mm/rev, (**b**) N = 1500 rpm and f = 0.04 mm/rev, (**c**) N = 2000 rpm, f = 0.04 mm/rev, (**d**) N = 1000 rpm, f = 0.08 mm/rev, (**e**) N = 1500 rpm, f = 0.08 mm/rev (**d**) N = 2000 rpm, f = 0.08 mm/rev.

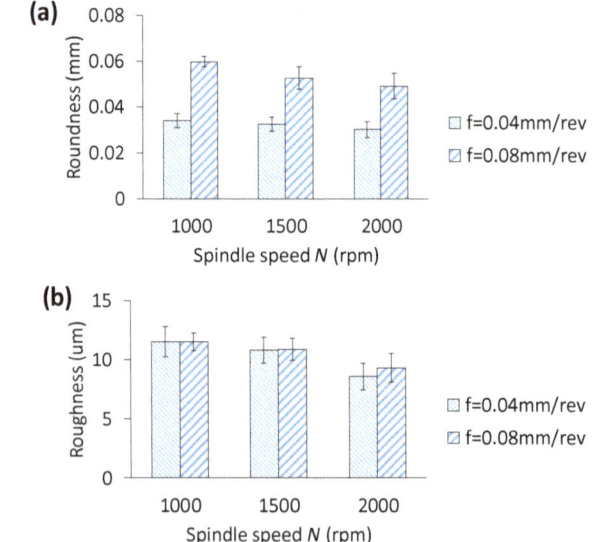

Figure 10. Effect of feedrate and spindle speed feedrate on (**a**) hole roundness and (**b**) surface roughness.

3.6. Chips Formation

Figure 11 shows cutting chips after 10th hole at different combination of spindle speed and feedrate. In most cases, the chips are continuous, entangled curly shape. It appears that the spindle speed has less influence on the chip generation, while the feedrate affects the most. As the feedrate increases from 0.04 to 0.08 mm/rev, the chips are not always continuous, but fractured and broken. Reduced edge sharpness due to wear at high feedrate is responsible for the broken and segmented chips. In other words, because of wear, the interaction between the rake face and the workpiece changes, which may result in the segmented chip generation. Also, when feedrate increases, shearing section becomes larger and chips become wider. Therefore, chips struggle to wind continuously due to large stiffness, and hence start to break into small segmented and/or spiral pieces. Furthermore, high spindle speed means high kinetic energy into chips, which may cause chip breakage at higher feedrate (Figure 11). In other words, chips flow through the flutes experience tremendous resistance due to the contact friction and break away. The similar finding is observed and reported via experimental and computational studies on machining of aerospace aluminum alloys [10,27]. It is to be noted that while segmented chips are favorable for easy evacuation and management of chips, the underlying process often deteriorate the generated hole quality. Therefore, the selection of drilling process parameters must be considered according to the desired hole-quality requirements, e.g., hole dimension, roundness, and finish.

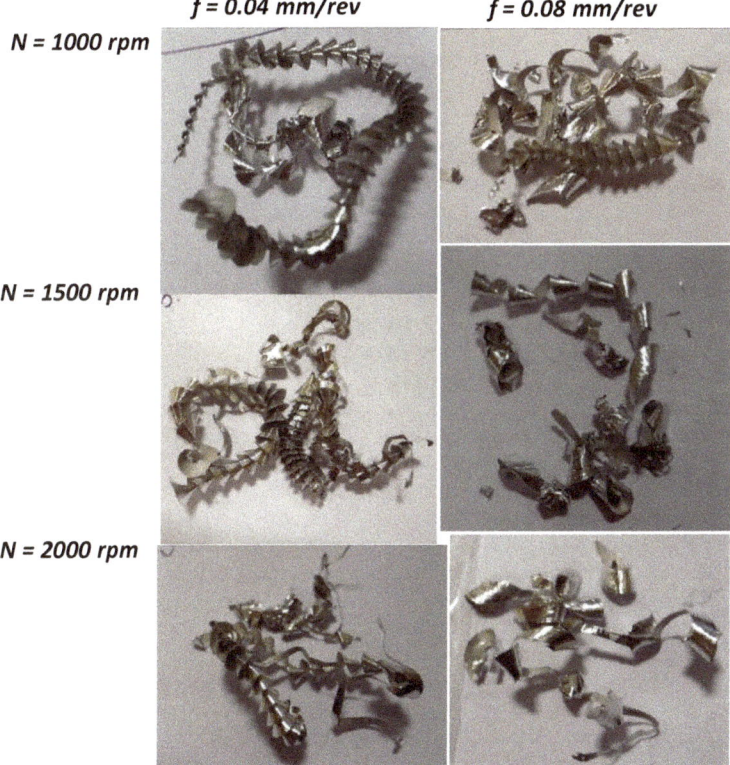

Figure 11. Cutting chips after drilling of 10th hole at different spindle speed and feedrate.

4. Conclusions

This paper presents an experimental investigation on the evaluation of hole quality in drilling of aluminum alloys. Compared to spindle speed, feedrate is the most dominant parameter, significantly affecting drilling behavior. For given spindle speed range, maximum increase of axial force and torque is 44.94% and 47.65%, respectively when feedrate increases from 0.04 mm/rev to 0.08 mm/rev. Stable, jerk-free cutting at feedrate of as low as 0.04 mm/rev is shown to result in hole dimensional error of less than 2%. Results of burr geometry and roundness follow the same trend, while roughness is minimally influenced by both spindle speed and feedrate. Built-up edge followed by abrasion and micro-chipping at the cutting edge produce noticeable wear mechanism, and their consequence may accelerate as the number of drilled holes further increases. This result is supported by chip morphology observation, i.e., more broken and segmented chips are noticed at a higher feedrate, as opposed to the continuous entangled chips at a lower feedrate.

It should be noted that coolant [27] and change of tool geometry [28], which may affect the drilling performance, is not taken into account in this study. While both may quantitatively change force and hole-quality metrics presented, it is expected that the qualitative trend will remain the same, and, as such, so do the conclusions of the paper.

Author Contributions: M.U. conceptualized idea, developed experimental, analyzed results, and wrote the manuscript. A.B. performed characterization of drill wear. A.P., S.S., C.P. and G.M.K. reviewed and provided feedback to improve the manuscript.

Funding: This research received no external funding.

Conflicts of Interest: The authors declare no conflict of interest.

References

1. Abdelhafeez, A.M.; Soo, S.L.; Aspinwall, D.K.; Dowson, A.; Arnold, D. Burr Formation and Hole Quality when Drilling Titanium and Aluminium Alloys. *Procedia CIRP* **2015**, *37*, 230–235. [CrossRef]
2. Pilný, L.; De Chiffre, L.; Píška, M.; Villumsen, M.F. Hole quality and burr reduction in drilling aluminium sheets. *CIRP J. Manuf. Sci. Technol.* **2012**, *5*, 102–107. [CrossRef]
3. Liu, D.; Tang, Y.; Cong, W.L. A review of mechanical drilling for composite laminates. *Compos. Struct.* **2012**, *94*, 1265–1279. [CrossRef]
4. Vilches, F.J.T.; Hurtado, L.S.; Fernández, F.M.; Gamboa, C.B. Analysis of the Chip Geometry in Dry Machining of Aeronautical Aluminum Alloys. *Appl. Sci.* **2017**, *7*, 132. [CrossRef]
5. Gautam, G.D.; Pandey, A.K. Pulsed Nd:YAG laser beam drilling: A review. *Opt. Laser Technol.* **2018**, *100*, 183–215. [CrossRef]
6. Rahamathullah, I.; Shunmugam, M. Analyses of forces and hole quality in micro-drilling of carbon fabric laminate composites. *J. Compos. Mater.* **2013**, *47*, 1129–1140. [CrossRef]
7. Niknam, S.A.; Davoodi, B.; Davim, J.P.; Songmene, V. Mechanical deburring and edge-finishing processes for aluminum parts—a review. *Int. J. Adv. Manuf. Technol.* **2018**, *95*, 1101–1125. [CrossRef]
8. Sultan, A.Z.; Sharif, S.; Kurniawan, D. Effect of Machining Parameters on Tool Wear and Hole Quality of AISI 316L Stainless Steel in Conventional Drilling. *Procedia Manuf.* **2015**, *2*, 202–207. [CrossRef]
9. Uddin, M.S.; Songyi, D. On the design and analysis of an octagonal–ellipse ring based cutting force measuring transducer. *Measurement* **2016**, *90*, 168–177. [CrossRef]
10. Wei, Y.; An, Q.; Ming, W.; Chen, M. Effect of drilling parameters and tool geometry on drilling performance in drilling carbon fiber–reinforced plastic/titanium alloy stacks. *Adv. Mech. Eng.* **2016**, *8*. [CrossRef]
11. Glaa, N.; Mehdi, K.; Zitoune, R. Numerical modeling and experimental analysis of thrust cutting force and torque in drilling process of titanium alloy Ti6Al4V. *Int. J. Adv. Manuf. Technol.* **2018**, *96*, 2815–2824. [CrossRef]
12. Ko, S.-L.; Chang, J.-E.; Yang, G.-E. Burr minimizing scheme in drilling. *J. Mater. Process. Technol.* **2003**, *140*, 237–242. [CrossRef]
13. Ko, S.-L.; Lee, J.-K. Analysis of burr formation in drilling with a new-concept drill. *J. Mater. Process. Technol.* **2001**, *113*, 392–398. [CrossRef]

14. Lauderbaugh, L.K. Analysis of the effects of process parameters on exit burrs in drilling using a combined simulation and experimental approach. *J. Mater. Process. Technol.* **2009**, *209*, 1909–1919. [CrossRef]
15. Nouari, M.; List, G.; Girot, F.; Coupard, D. Experimental analysis and optimisation of tool wear in dry machining of aluminium alloys. *Wear* **2003**, *255*, 1359–1368. [CrossRef]
16. Kurt, M.; Kaynak, Y.; Bagci, E. Evaluation of drilled hole quality in Al 2024 alloy. *Int. J. Adv. Manuf. Technol.* **2008**, *37*, 1051–1060. [CrossRef]
17. Chaanthini, M.K.; Murugappan, S.; Arul, S. Study on Hole Quality in Drilling AA 6063 Plate under Cryogenic Pre-Cooling Environment. *Mater. Today Proc.* **2017**, *4*, 7476–7483. [CrossRef]
18. Islam, M.N.; Boswell, B. Effect of cooling methods on hole quality in drilling of aluminium 6061-6T. *IOP Conf. Ser. Mater. Sci. Eng.* **2016**, *114*, 012022. [CrossRef]
19. Ramulu, M.; Branson, T.; Kim, D. A study on the drilling of composite and titanium stacks. *Compos. Struct.* **2001**, *54*, 67–77. [CrossRef]
20. Chen, W.S.; Ehmann, K.F. Experimental investigation on the wear and performance of micro-drills. In Proceedings of the 1994 International Mechanical Engineering Congress and Exposition, Chicago, IL, USA, 6–11 November 1994; Tribology in Manufacturing Processes: Chicago, IL, USA.
21. Ghasemi, A.H.; Khorasani, A.M.; Gibson, I. Investigation on the Effect of a Pre-Center Drill Hole and Tool Material on Thrust Force, Surface Roughness, and Cylindricity in the Drilling of Al7075. *Materials* **2018**, *11*, 140. [CrossRef]
22. Luo, B.; Li, Y.; Zhang, K.; Cheng, H.; Liu, S. A novel prediction model for thrust force and torque in drilling interface region of CFRP/Ti stacks. *Int. J. Adv. Manuf. Technol.* **2015**, *81*, 1497–1508. [CrossRef]
23. Songmene, V.; Khettabi, R.; Zaghbani, I.; Kouam, J.; Djebara, A. Machining and Machinability of Aluminum Alloys. *Alum. Alloys Theory Appl.* **2011**. [CrossRef]
24. Braga, D.U.; Diniz, A.E.; Miranda, G.W.A.; Coppini, N.L. Using a minimum quantity of lubricant (MQL) and a diamond coated tool in the drilling of aluminum–silicon alloys. *J. Mater. Process. Technol.* **2002**, *122*, 127–138. [CrossRef]
25. Bu, Y.; Liao, W.H.; Tian, W.; Shen, J.X.; Hu, J. An analytical model for exit burrs in drilling of aluminum materials. *Int. J. Adv. Manuf. Technol.* **2016**, *85*, 2783–2796. [CrossRef]
26. Haddag, B.; Atlati, S.; Nouari, M.; Moufki, A. Dry Machining Aeronautical Aluminum Alloy AA2024-T351: Analysis of Cutting Forces, Chip Segmentation and Built-Up Edge Formation. *Metals* **2016**, *6*, 197. [CrossRef]
27. Giasin, K. The effect of drilling parameters, cooling technology, and fiber orientation on hole perpendicularity error in fiber metal laminates. *Int. J. Adv. Manuf. Technol.* **2018**, *97*, 4081–4099. [CrossRef]
28. Zitoune, R.; Krishnaraj, V.; Collombet, F.; Le Roux, S. Experimental and numerical analysis on drilling of carbon fibre reinforced plastic and aluminium stacks. *Compos. Struct.* **2016**, *146*, 148–158. [CrossRef]

 © 2018 by the authors. Licensee MDPI, Basel, Switzerland. This article is an open access article distributed under the terms and conditions of the Creative Commons Attribution (CC BY) license (http://creativecommons.org/licenses/by/4.0/).

Article

Multiscale 3D Curvature Analysis of Processed Surface Textures of Aluminum Alloy 6061 T6

Tomasz Bartkowiak [1,*] and Christopher A. Brown [2]

1 Institute of Mechanical Technology, Poznan University of Technology, ul. Piotrowo 3, 60-965 Poznan, Poland
2 Surface Metrology Laboratory, Worcester Polytechnic Institute, Worcester, MA 01609, USA; brown@wpi.edu
* Correspondence: tomasz.bartkowiak@put.poznan.pl; Tel.: +48-61-665-24-52

Received: 12 December 2018; Accepted: 10 January 2019; Published: 14 January 2019

Abstract: The objectives of this paper are to demonstrate the viability, and to validate, in part, a multiscale method for calculating curvature tensors on measured surface topographies with two different methods of specifying the scale. The curvature tensors are calculated as functions of scale, i.e., size, and position from a regular, orthogonal array of measured heights. Multiscale characterization of curvature is important because, like slope and area, it changes with the scale of observation, or calculation, on irregular surfaces. Curvatures can be indicative of the topographically dependent behavior of a surface and, in turn, curvatures are influenced by the processing and use of the surface. Curvatures of surface topographies have not been well- characterized yet. Curvature has been used for calculations in contact mechanics and for the evaluation of cutting edges. Manufactured surfaces are studied for further validation of the calculation method because they provide certain expectations for curvatures, which depend on scale and the degree of curvature. To study a range of curvatures on manufactured surfaces, square edges are machined and honed, then rounded progressively by mass finishing; additionally, a set of surfaces was made by turning with different feeds. Topographic measurements are made with a scanning laser confocal microscope. The calculations use vectors, normal to the measured surface, which are calculated first, then the eigenvalue problem is solved for the curvature tensor. Plots of principal curvatures as a function of position and scale are presented. Statistical analyses show expected interactions between curvature and these manufacturing processes.

Keywords: surface; texture; machining; multiscale; aluminum alloy 6061 T6

---

1. Introduction

The objectives of this paper are to demonstrate the viability, and to validate, in part, how surface topographies can be characterized by curvature tensors calculated from areal topographic measurements of manufactured surfaces. In addition, two methods for specifying the scale are studied. The machined, honed, and mass finished surfaces have regular and irregular topographic components. The second-order curvature tensors vary with scale, position, and orientation, i.e., direction. They are calculated from regular arrays of measured surface heights, producing multiscale characterizations that are both position- and orientation-specific. The validation is tested by comparing the results with expectations, based on the machining, honing, and finishing processes.

Appropriate characterization of topographies is essential for discriminating with confidence surfaces with topographies that were created differently or that behave differently, and for discovering strong correlations between processing and topographies, or between topographies and behavior. The value of surface metrology for product and process design, i.e., the measurement and analysis of surface topographies, is largely founded on these abilities to discriminate and correlate [1,2].

Topographies can have components that are regular, like form, and components that are irregular, i.e., roughness. Sometimes the term roughness is used simply to refer to fine-scale topographies, even if they are highly regular, e.g., certain engineered surfaces. Surfaces that are essentially regular might be sufficiently characterized by a few measurements and parameters. Irregular surfaces can require millions of height measurements, multiscale geometric analyses, and statistics for sufficient characterizations [2].

Curvature is particularly attractive as a characterization method. Curvature is approximately the spatial derivative of the slope. No datum is required for the characterization of curvature, unlike heights or slopes. This aspect of curvature characterization can be especially valuable when the datum is not obvious, such as in characterizing redundant surfaces or voids, which can be measured with tomography.

Curvature is an essential parameter for characterizing edges and surfaces. The curvature of cutting edges has been discussed in the literature [3–5]. Curvature of peaks as a geometric property of surfaces is important in contact mechanics (e.g., [6–8]). Characterizing the valleys of topographies by their curvature also could be important for understanding crack initiation, fluid retention, and adhesion. Vulliez et al. presented a strong functional correlation (R^2 = 0.96) at a specific scale (610 μm) between the curvature of machined surfaces and their fatigue limit [9]. Logically, some kind of multiscale characterizations of curvatures of topographies could be used for discrimination and for correlation with processing and behavior, as has been done with area-scale analysis [10]. This characterization could be useful for surface research and for product and process design.

It is important to understand how the many ways of implementing multiscale in the analyses of characterization parameters can influence the results. Two methods are used here, and the results are compared.

Curvature has previously been calculated from profile measurements as a function of scale and position, where the height, z, as a function of position, x, such that $z = z(x)$ [9]. One method uses Heron's formula to calculate the curvature, based on three points from the profile. The scale is represented by the spacing in x of the three height measurements selected for the calculation. The curvature can vary with position and with scale along the profile.

Here it is shown how curvature can be calculated from areal measurements, i.e., on regular orthogonal arrays of heights, z, in x and y such that $z = z(x,y)$. Curvature can be characterized as a second-order tensor that can vary with position and scale. This is more complicated than calculating curvature from a profile, and the results can be more valuable. The result of an eigenvalue calculation gives the values and orientations of the maximum and minimum curvatures as a function of position and scale. In addition to the curvatures themselves, these results can be used to characterize the anisotropy, or directionality, of a surface, based on curvature orientation [11].

A study of the commonly used techniques for curvature estimation was presented by Petitjean [12]. Recent studies tend to concentrate on the triangular meshes because they are commonly used in representations in many computer-assisted design (CAD) or graphics programs. Many [13–15] use piecewise surface approximations, e.g., Bezier, quadratic, or polynomial. Thiesel et al. introduced a robust method for calculating the curvature tensor, based on vectors normal to the surface, which does not involve local surface interpolation [16]. That method calculates a curvature tensor with components that are constant within a certain triangular region on the surface. Coeurjolly et al. described a novel class of estimators of digital shapes, which are based on integral invariants [17]. They used a local approximation to convert discrete height data into continuous functions. An interesting study was presented recently by Foorginejad and Khalili, in which they introduced a method named umbrella curvature [18], which involved normal vectors and vectors between the point of estimation and their neighbors in order to estimate local curvature. Lai et al. described a method that connects profile and surface curvature. They searched principal curvatures by calculating profile curvature in multiple directions and looking for maximum and minimum values [19].

The physical determination of the height at a point on a real surface is problematic. During surface measurements, discrete heights are determined over lateral, or spatial, sampling zones, rather than at points. The height at a mathematical point, which is infinitesimally small, cannot be measured on a surface. Measured heights are determined at a certain lateral spacing, or sampling intervals, which might or might not exceed the size of the sampling zone. Nonetheless, the measured heights over zones are treated here, for the calculation, as an array of mathematical points, $z = z(x,y)$.

The actual, measured, areal surface, although continuous by the general definition of a surface, could represent an actual surface that is nowhere differentiable. However, digital representations of measured surfaces are commonly approximated as smooth. This approximation facilitates characterization by a series of curvatures that can vary with position and scale.

In the following development of the curvature calculation, a surface will be considered to consist of a collection of heights at points on a regular, spatial grid. The surface will be considered differentiable at the scales and locations required for the calculations of normal vectors to patches, or defined regions, on the surface.

The approach here is to use a representation of areal surfaces based on heights used to calculate vectors normal to the surface. The curvature tensor is considered constant over the size of the region that represents the scale. This method for curvature estimation uses normal vectors at points that represent the center of patches, over which the normal vectors are calculated.

Three ways to compute the normal vectors are given in [20]:

1. Covariance matrix, which computes the unit normal vectors from the neighboring points about a central point.
2. Average areal gradient, which uses horizontal and vertical differences of neighboring points around the central point.
3. Average depth change, which calculates horizontal and vertical differences from averaged neighbors.

In this work, the first method, covariance matrix, is used, because of its small estimation error in multiscale applications and the simplicity of implementation [21].

The characteristics of aluminum alloy 6061-T6, used in this study, may lead to the formation of built-up edge (BUE) when cut. This phenomenon increases the mechanical load on the cutting edge, making efficient chip flow difficult and the chip-removal process inefficient. Alloying elements (in particular, silicon present in this aluminum alloy), and the treatment methods, influence the machining properties [22,23]. The influence of machining parameters on the resulting surface texture for hard-to-cut materials was analyzed by Krolczyk et al. [24] and Twardowski et al. [25].

2. Materials and Methods

2.1. Preparation and Measurement of the Surfaces

Two different sets of surfaces were manufactured from aluminum alloy (6061 T6), the surface topographies were measured, and the measurements were analyzed. The topographies were measured with an Olympus LEXT 4100 OLS laser scanning confocal microscope (Olympus Corporation Shinjuku, Tokyo, Japan) with a 50× objective (NA 0.93). The measurement regions were cropped to 0.11 mm × 0.11 mm for the turned and 0.075 mm × 0.075 mm for the edge. The sampling interval is about 250 nm.

One set of surfaces was manufactured by turning a rod, initially 25.45 mm in diameter, on a Haas SL10 CNC lathe (Haas Automation, Inc., Oxnard, CA, USA), first to a diameter of 24.29 mm, then to 22.30 mm in a final pass, making final cutting depth, a_p, equal to 0.995 mm. The feeds were 0.2, 0.1, 0.05, 0.01 mm/rev. The spindle speed was 1000 rpm. Kennametal carbide inserts (VNMG 160404ms, KC5525, Kennametal Inc., Pittsburgh, PA, USA), with a tool nose radius of 0.4 mm, were used. The following tool geometry was applied: lead angle of 93°, both inclination angle and orthogonal rake angle equal to 0°.

Another set of surfaces was prepared by mass finishing a part with an edge that was milled and honed. First, two sides were side-milled to create an edge at a corner, with an angle of approximately

90°. The two faces were then honed by hand, using emery paper to remove a burr left from milling. The first measurement of the edge was taken after the honing. The part was subsequently placed in a BelAir FMSL 8T series centrifugal disk mass-finisher (Bel Air Finishing Supply Corporation, North Kingstown, RI, USA) and then measured after finishing for 2.5 and again after 7.5 min. The abrasive media was R-1000, a polyester pyramid with a height of 6.4 mm, with zirconia particles embedded.

2.2. Analysis for Estimation of the 3D Curvature Based on Vectors Normal to Surface Patches

This analysis calculates curvature tensors at each scale in the data set and each location on the surface where there are a sufficient number of measured heights for the calculation. The calculated curvatures are considered to be constant over the triangular patches, which are the regions used for the curvature calculations. The range of scales available in a measurement goes from the sampling interval to the size of the measured region.

First, three unit normal vectors are calculated, one for each vertex on the triangular patch that is used for the curvature tensor calculation (Figure 1). A covariance matrix method for computing unit normal vectors is used. At each vertex, the closest 3 × 3 neighborhood of measured heights is used for computing the normal vector. The edges of the measured region are excluded, due to insufficient measured heights, 2 × 2 or 3 × 2 neighborhoods of heights, instead of 3 × 3.

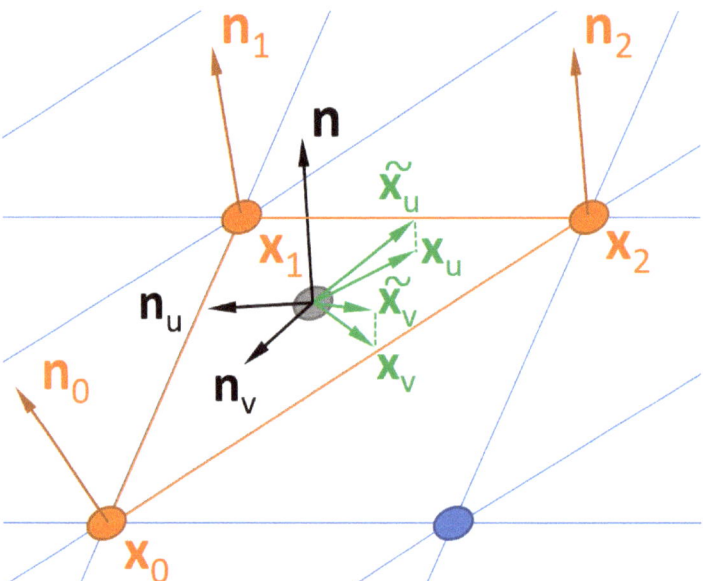

Figure 1. Visualization of x_u, x_v, n_u and n_v on a triangular patch.

Next, using the Weingarten curvature tensor, **T**, a symmetric 3 × 3 matrix is calculated, assuming that the surface is continuous and everywhere differentiable (within the patch) [26]. The resulting eigenvalues are, κ_1, κ_2, 0, where the first two represent the principal curvatures. The resulting eigenvectors include k_1, k_2, the corresponding principal directions for the principal curvatures, and **n**, the unit normal vector for the triangular patch.

Note that new local and global coordinate systems are introduced here for these calculations. Whereas measured, global heights were $z(x,y)$, as is usual in the literature, these are represented below as $x(u,v)$. Normal vectors are calculated from these global heights. Curvatures are calculated in a local coordinate system. This local coordinate system (u,v) is in the plane defined by the three points that constitute a single triangular patch. To calculate the curvature directions, transfers from the local systems (u,v) to the global (x,y) system are necessary.

At each scale, the surfaces that are considered here are completely defined by their partial directional derivatives and the partial directional derivatives of the unit normal vectors. Given the surface $x(u,v)$ and its partials $\mathbf{x}_u$ and $\mathbf{x}_v$, a unit normal vector $\mathbf{n}$ and its partials can be computed by the following formula:

$$\mathbf{n} = \frac{\mathbf{x}_u \times \mathbf{x}_v}{\|\mathbf{x}_u \times \mathbf{x}_v\|}, \quad \mathbf{n}_u = \frac{\partial \mathbf{n}}{\partial u}, \quad \mathbf{n}_v = \frac{\partial \mathbf{n}}{\partial v} \tag{1}$$

These four vectors have the following dependencies:

1. $\mathbf{x}_u, \mathbf{x}_v, \mathbf{n}_u, \mathbf{n}_v$ are coplanar.
2. $\mathbf{n}_u \mathbf{x}_v = \mathbf{n}_v \mathbf{x}_u$.

The computation of $\mathbf{T}$ from $\mathbf{x}_u, \mathbf{x}_v, \mathbf{n}_u, \mathbf{n}_v$ is a straightforward application of classical concepts of differential geometry [26]. The coefficients of the first and second fundamental form, or shape, tensor can be calculated as:

$$E = \mathbf{x}_u \cdot \mathbf{x}_u, \quad F = \mathbf{x}_u \cdot \mathbf{x}_v, \quad G = \mathbf{x}_v \cdot \mathbf{x}_v, \tag{2}$$

$$L = -\mathbf{n}_u \cdot \mathbf{x}_u, \quad M_1 = -\mathbf{n}_u \cdot \mathbf{x}_v, \tag{3}$$

$$M_2 = -\mathbf{n}_v \cdot \mathbf{x}_u, \quad N = -\mathbf{n}_v \cdot \mathbf{x}_v. \tag{4}$$

Then the Weingarten curvature matrix can be created,

$$\mathbf{W} = \begin{pmatrix} \frac{LG - M_1 F}{EG - F^2} & \frac{LG - M_1 F}{EG - F^2} \\ \frac{LG - M_1 F}{EG - F^2} & \frac{LG - M_1 F}{EG - F^2} \end{pmatrix} \tag{5}$$

with its eigenvalues κ_1, κ_2 and its corresponding eigenvectors:

$$\mathbf{w}_1 = \begin{pmatrix} w_{11} \\ w_{12} \end{pmatrix}, \quad \mathbf{w}_2 = \begin{pmatrix} w_{21} \\ w_{22} \end{pmatrix}. \tag{6}$$

The eigenvalues are used to calculate the Gaussian curvature, K, and the mean curvature, H; and the eigenvectors are used to calculate the principal directions $\mathbf{k}_1$ and $\mathbf{k}_2$ as it follows:

$$K = \kappa_1 \kappa_2, \quad H = \frac{1}{2}(\kappa_1 + \kappa_2), \tag{7}$$

$$\mathbf{k}_1 = w_{11} \mathbf{x}_u + w_{12} \mathbf{x}_v, \quad \mathbf{k}_2 = w_{21} \mathbf{x}_u + w_{22} \mathbf{x}_v. \tag{8}$$

Having all necessary components, curvature matrix $\mathbf{T}$ can be constructed:

$$\mathbf{T} = \mathbf{PDP}^{-1}, \tag{9}$$

where $\mathbf{P} = (\mathbf{k}_1, \mathbf{k}_2, \mathbf{n})$ and,

$$\mathbf{D} = \begin{pmatrix} \kappa_1 & 0 & 0 \\ 0 & \kappa_2 & 0 \\ 0 & 0 & 0 \end{pmatrix}. \tag{10}$$

Theisel et al. [16] presented a new technique for estimating curvature tensor $\mathbf{T}$ in a triangular mesh. That method shows better error behavior than a cubic fitting [13] and is independent of rotations of the mesh and does not involve any parameterization or fitting. The accuracy of Theisel's method depends primarily on the accuracy of the estimation of the unit normal vectors, which was the first part of the curvature computations above.

In that normal approach, only a single (non-degenerate) triangle, with the vertices x_0, x_1, x_2 and the corresponding normals n_0, n_1, n_2, are considered (see Figure 1). A point and normal vector on the triangle can be obtained by applying linear interpolation in local coordinates (u,v), with the origin x_0 and the base vectors $x_1 - x_0$ and $x_2 - x_0$:

$$\tilde{x} = \tilde{x}(u,v) = x_0 + u(x_1 - x_0) + v(x_2 - x_0), \tag{11}$$

$$\tilde{n} = \tilde{n}(u,v) = n_0 + u(n_1 - n_0) + v(n_2 - n_0), \tag{12}$$

The idea that stands behind the introduction of interpolated $\tilde{x}$ and $\tilde{n}$ is to use them for calculating vectors x_u, x_v, n_u, n_v and, subsequently, curvature matrix T. Unit normal vectors and their derivatives can be computed following Theisel et al. [16]:

$$n(u,v) = \frac{\tilde{n}}{\|\tilde{n}\|}, \quad n_u = \frac{\partial n}{\partial u}, \quad n_v = \frac{\partial n}{\partial v} \tag{13}$$

For the partials of the surface, we can obtain:

$$\tilde{x}_u(u,v) = \frac{\partial \tilde{x}}{\partial u} = x_1 - x_0, \quad \tilde{x}_v(u,v) = \frac{\partial \tilde{x}}{\partial v} = x_2 - x_0 \tag{14}$$

In order to assure that condition 1 (x_u, x_v, n_u, n_v are coplanar) is met, $\tilde{x}_u$ and $\tilde{x}_v$ are projected onto the plane defined by n_u and n_v:

$$x_u = \tilde{x}_u - (n\tilde{x}_u)n, \quad x_v = \tilde{x}_v - (n\tilde{x}_v)n. \tag{15}$$

Now, the curvature matrix T can be computed, by applying Equations (13)–(15) into Equations (2)–(10).

2.3. Multiscale Curvature Characterization Analysis

Multiscale characterizations can be achieved in several ways [2]. Two different methods of specifying the scale of the curvature analyses are described here. These two methods both apply to the selection of the measured heights that are used for the estimation of the normal vectors and to the selection of three points that form the triangular patches.

The first multiscale method here is down-sampling, shown in Figure 2. In this down-sampling, more measured heights are skipped with each iteration of the multiscale calculations, in order to achieve increasingly larger scales. At the finest, or nominal, scale, the spacing is the sampling interval. At two times the nominal scale, the spacing it is twice the sampling interval, for which every other measured height is used. At three times the sampling interval, every third measured height would be used. The scale here is the length of the horizontal interval in x and y, between the measured heights used in each analysis for determining the curvature.

The down-sampling is applied for both the selection of the measured heights for calculating the normal vectors and for selecting the points that define the triangular patches for the eigenvalue problem. To determine the position for calculating the curvature, even at the large scales, the iterations are performed at each location. That is, the calculation is indexed horizontally, one sampling interval for each locational calculation.

After heights are skipped in this down-sampling routine, just the nine, not-skipped, measured heights are used to calculate the normal unit vectors at each scale. These are the eight heights closest to the apexes of the triangular patches (Figure 2a) and the central point (apex) used for the eigenvalue problem. The spacing between the heights is scale-dependent and increases with the scale (Figure 2b). These triangles are always equilateral, right triangles in projection on a horizontal plane. The projected length of the short sides are equal to the scale. More details of the method and its application for multiscale analysis can be found in work by Bartkowiak and Brown [27]. The calculation of the curvature tensor is done for the next location distant from the previous, by using the original sampling interval (Figure 2b).

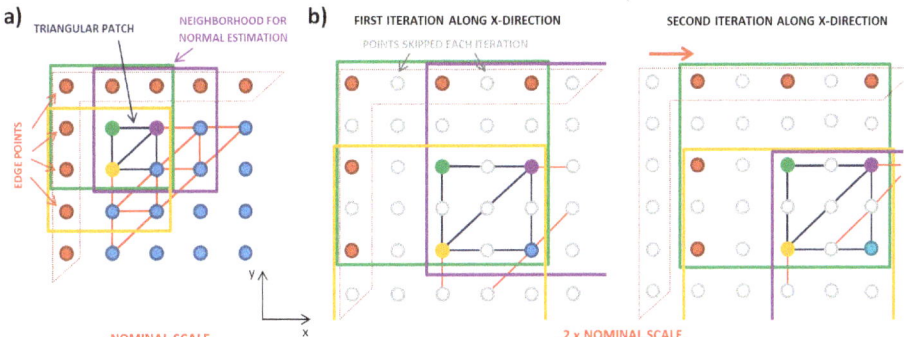

Figure 2. Down-sampling, one method of the multiscale analysis: (**a**) at the nominal scale, which is equal to the sampling interval, (**b**) at a two times the nominal scale. The colors of the squares, green, yellow, and purple, correspond to the color of the points that center the vertices of the triangles. Edge points are colored with red.

In the second, multiscale method considered here, no in-between heights are skipped in normal estimation, while the values for the scales are determined identically to the first method (Figure 3). In this way, each iteration by both methods includes the same measured regions, and they are signified by the same scale. The second method uses all the measured heights in the neighborhood, instead of just nine. The size of a neighborhood changes with scale. For the nominal scale, both methods use the same measured heights for the calculation of the unit normal vectors. For larger scales, the number of heights grows with the multiplication of the original sampling interval s, so that the neighborhood consists of $(1 + 2s) \times (1 + 2s)$ points. For instance, for a scale equal to three times the sampling interval, it is necessary to consider 7×7 heights, for calculating the unit normal vector for an apex that is centrally placed inside the neighborhood.

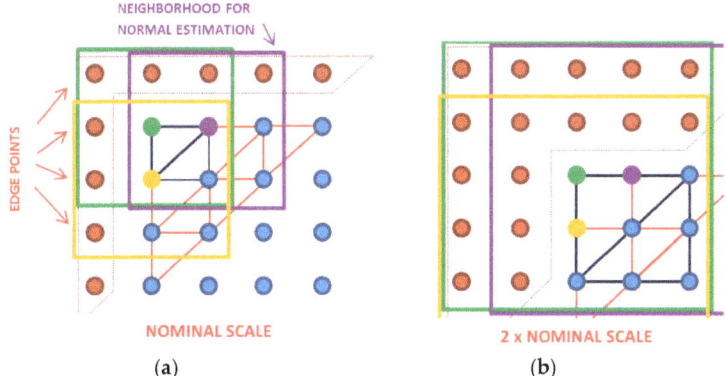

Figure 3. The second method of the multiscale analysis: (**a**) at the nominal scale, which is equal to the sampling interval, (**b**) at a two times the nominal scale.

In both methods, the normal unit vectors, and curvatures, are not estimated along the edges of the measured region, where entire neighborhoods cannot be formed. This second method can be time-consuming for the larger scales, because the covariance calculation includes more points.

3. Results

In the following sections, renderings of the measurements of the studied surfaces are shown at different scales. The calculated curvatures and Gaussian and mean curvatures are shown as a function of position and are plotted versus scale, along with the standard deviations.

3.1. Prepared Edge, Honed and Mass-Finished

Renderings of topographic measurements with three downsized scales of the machined and honed edge, as well as two finishing times, are presented in Figure 4. With the increasing scale, surfaces appear smoother, because the fine-scale details are skipped in the down-sampling.

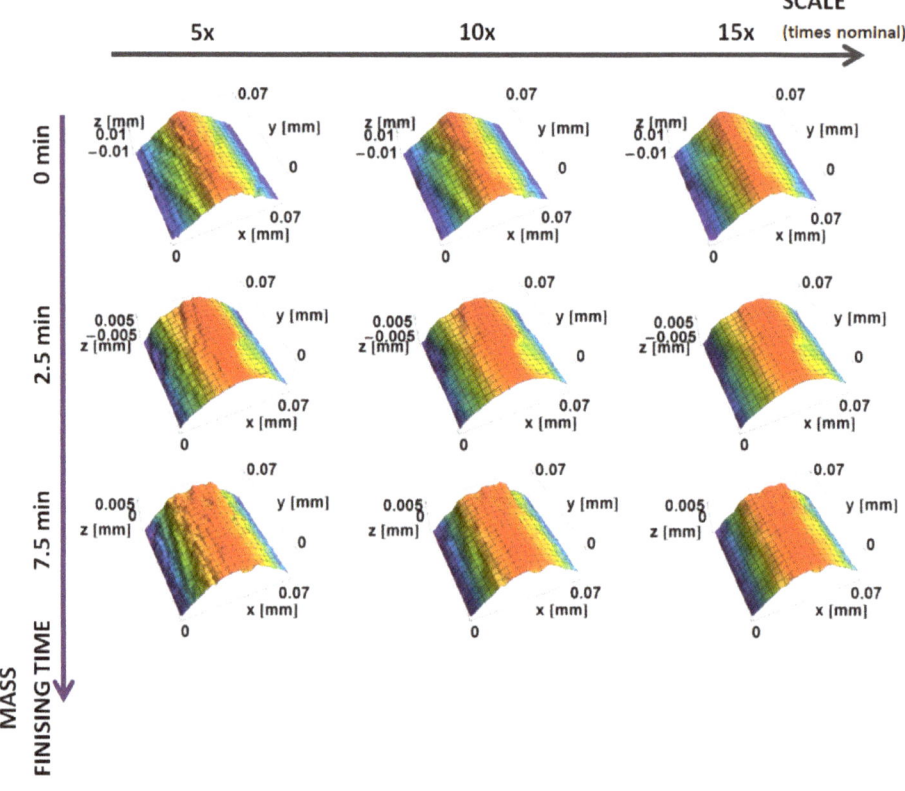

Figure 4. Renderings of topographic measurements of the edge machined and honed (0 min), and after 2.5 and 7.5 min of mass finishing, at three scales: 5× original sampling interval (1.25 µm), 10× (2.50 µm), and 15× (3.75 µm).

The principal, κ_1, curvatures, those with the largest magnitude, calculated by the down-sampling method, are shown as a function of position on the surface and the scale of calculation in Figure 5. Convex curvatures are negative and concave are positive, as usual.

At the largest scales, there is little variation in curvature. The magnitudes of the curvatures tend to increase with decreasing scale. Many small regions of convexity at the smallest scales are evident in Figure 5. The curvature of the prepared edges cannot be discriminated at the finest scales. At these fine scales, a multitude of fine features, with large principal κ_1 curvatures, masks the curvature of the edge. The curvatures on the honed part are clearly visible at 10× nominal scale. The curvature of the part that was mass-finished for 2.5 min is relatively uniform at the largest scales (10× and 15×). Concave features are clearly visible at 15×. The effect of mass-finishing is evident at all three scales.

The principal κ_2 curvatures, minimum in magnitude, calculated by the down-sampling method, are shown as a function of position on the surface and the scale of calculation in Figure 6. Similar to κ_1 at the large scales, there is little variation in curvature. The magnitudes of the curvatures tend to increase with decreasing scale. A positive curvature region is evident around the manufactured edge at 5× and 10× nominal scale. However, its value is significantly lower than the κ_1 curvature. The mass finishing process decreases the magnitude of κ_2 curvature for all three scales shown.

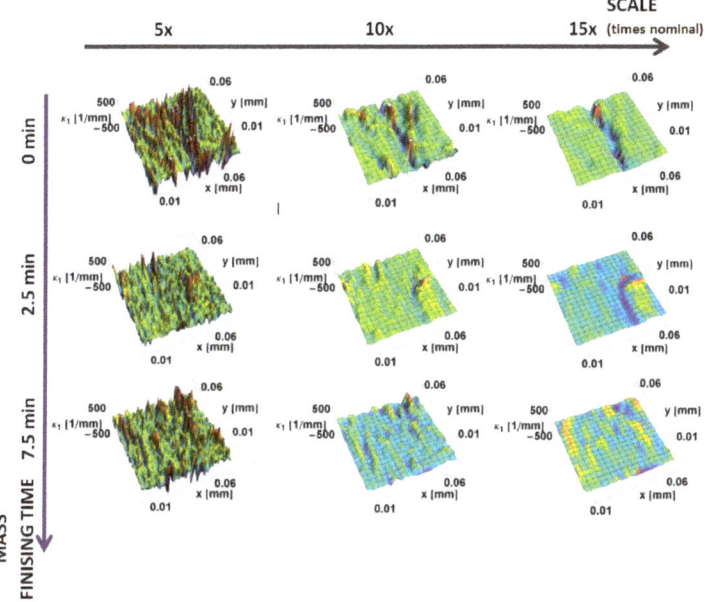

Figure 5. Maximum principal curvature, κ_1, on the prepared edge as a function of position, calculated using the down-sampling method for three scales.

The principal κ_1 and κ_2 curvatures, calculated by the increasing neighborhood method, are shown as a function of position on the surface and the scale of calculation in Figures 7 and 8, respectively. These show the same trends as the down-sampling method. The variability of curvatures for finer scales is higher for the increasing neighborhood method, when comparing the same scales. For the same subregions, both minimal and maximal curvature take greater values. For larger scales, more points are used in the calculation of the normal vectors, so artifacts and fine-scale features influence more points in their normal vectors estimation, which make these results more sensitive to local variations. The expected smoothing effect is less evident when compared to the down-sampling method.

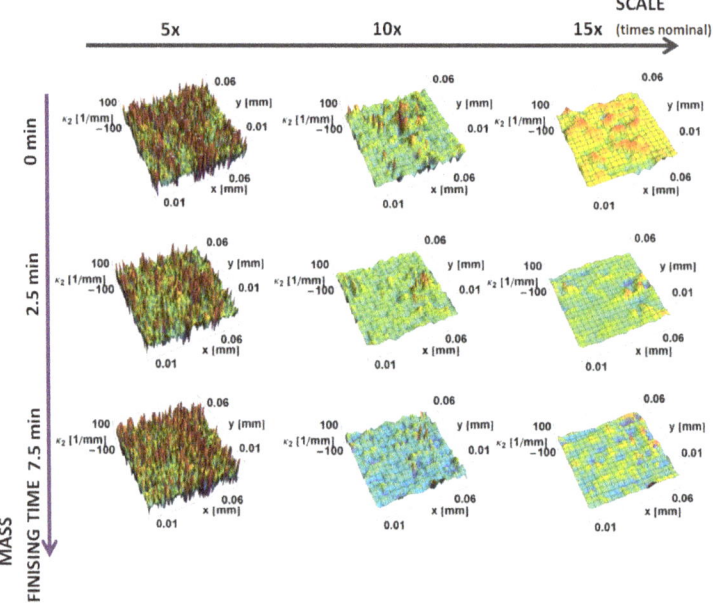

Figure 6. Minimum principal κ_2 curvature, as on the prepared edges, a function of position calculated, using the down-sampling method for three scales.

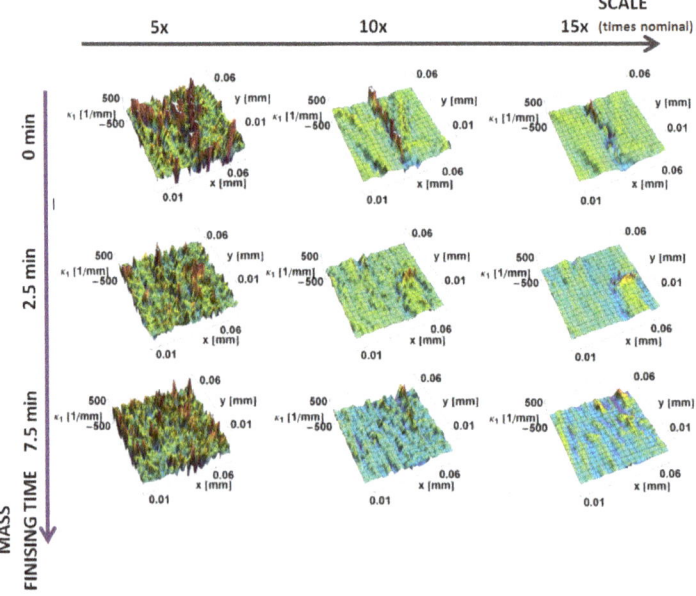

Figure 7. Maximum principal curvature, κ_1, on the prepared edge, as a function of position calculated, using the increasing neighborhood method, for three scales.

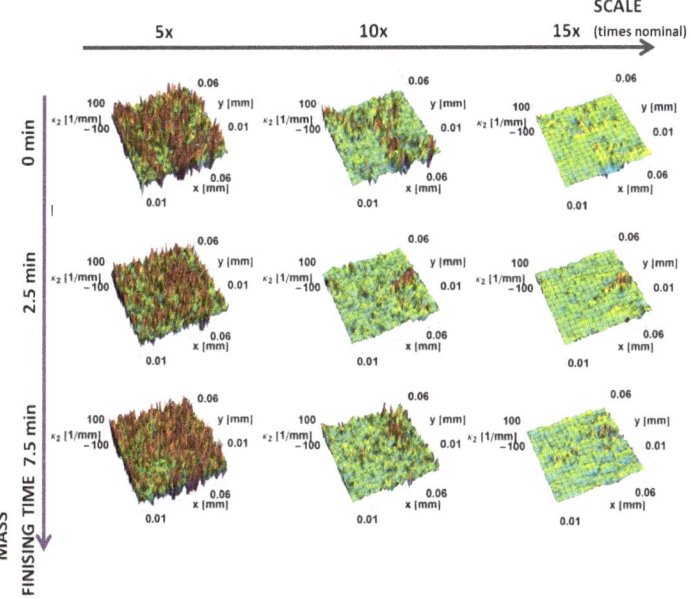

Figure 8. Minimum principal curvature, κ_2, on the prepared edge as a function of position calculated, using the increasing neighborhood method for three various scales.

Mean and standard deviations of the principal curvatures on the prepared edges, calculated by the down-sampling method, are shown, versus the log of the scale, in Figure 9a. The mean values of principal curvatures change with the scale for all the surfaces. The standard deviation of principal curvatures decreases regularly with increasing scale, i.e., the distribution of the curvatures is distinctly varied at the finer scales. For larger scales, the dispersion measure decreases, as fewer fine-scale surface features, characterized by high curvature, become evident, and mean values of principal curvature tend to indicate the general shape of the edge. The curvatures of microfeatures are generally greater than the overall form or waviness, which is quantified as larger values of standard deviations in comparison with the mean. It appears that the mean κ_1 discriminates the edges for scales greater than 4μm, with a sufficient sample size, because the variance is large. Mean values of κ_1 are negative for all calculated scales, which is consistent with the overall convexity of the surface perpendicular to the prepared edge. Logically, the means of the minimum principal curvatures, κ_2, are smaller. Their proximity to zero, particularly at large scales, is consistent with the straightness of the surfaces parallel to the prepared edge.

Statistics calculated by the method are presented in Figure 9b. It appears that the mean and standard deviations of κ_1 might be used to discriminate the prepared edges for some scales between 4 and 10 μm. The greatest differences between those two methods appears at the greatest scales. Similar trends appear for κ_2. Standard deviations of maximal curvatures take greater values when calculated by increasing the neighborhood method, which supports the effect of microfeatures that is evident with growing scales.

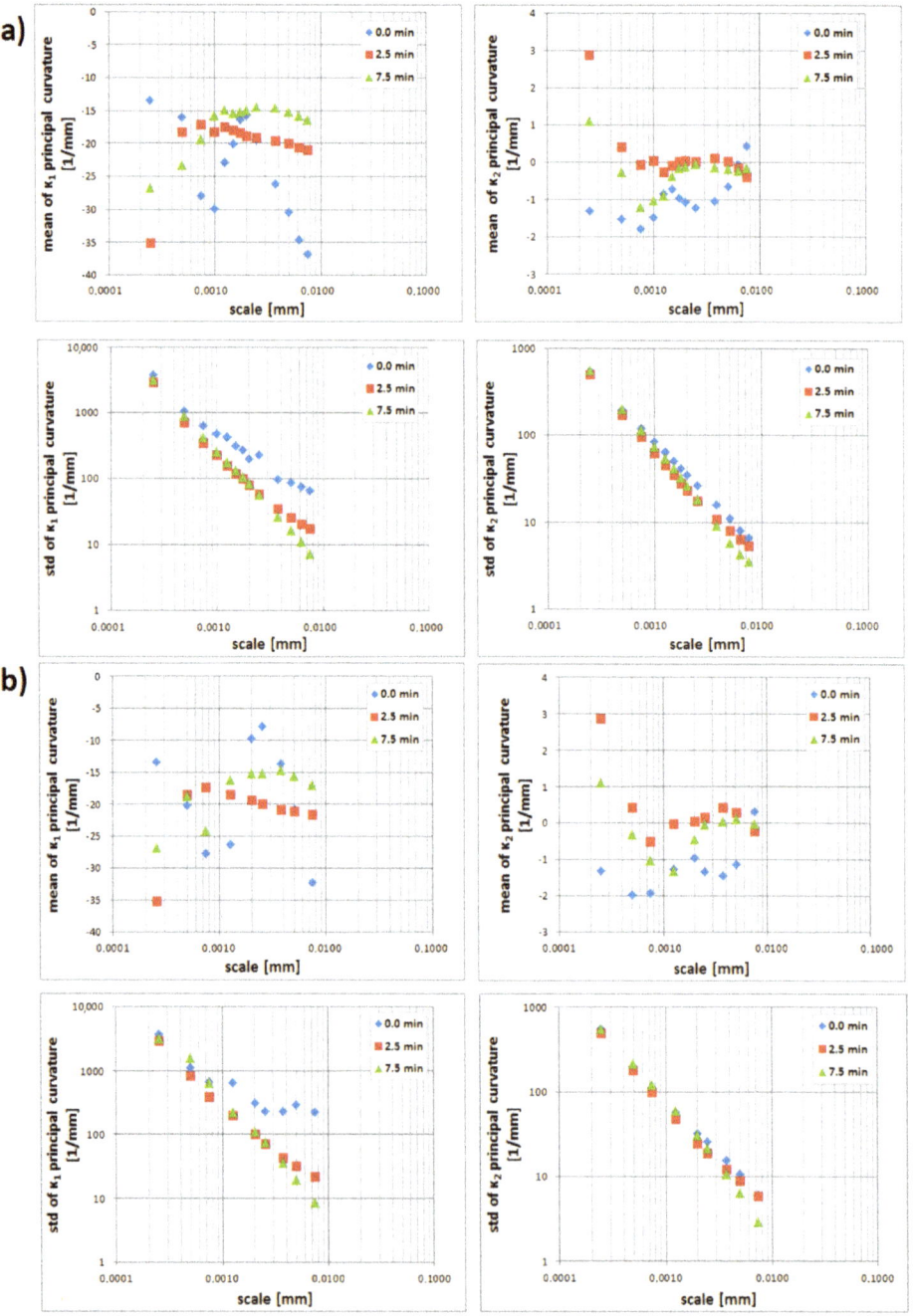

Figure 9. Statistics on the prepared edge as a function of scale: (**a**) mean and standard deviation of κ_1 and κ_2 principal curvatures, calculated for three scales by down-sampling, (**b**) mean and standard deviation (std, in short) of κ_1 and κ_2 principal curvatures, calculated for three scales by increasing the neighborhood.

3.2. Cutting Tool Edge and Turned Surfaces

Renderings of measurements representing three downsized scales and feed rates of the prepared surfaces and tool edge are presented in Figure 10. The figure presents surfaces for various scales. Both the turned surfaces and the tool edges are examples of clearly anisotropic textures. With increasing scales, fine-scale ridges and valleys tend to be smoothed. Both principal curvatures were calculated for the prepared region of 700 × 700 heights.

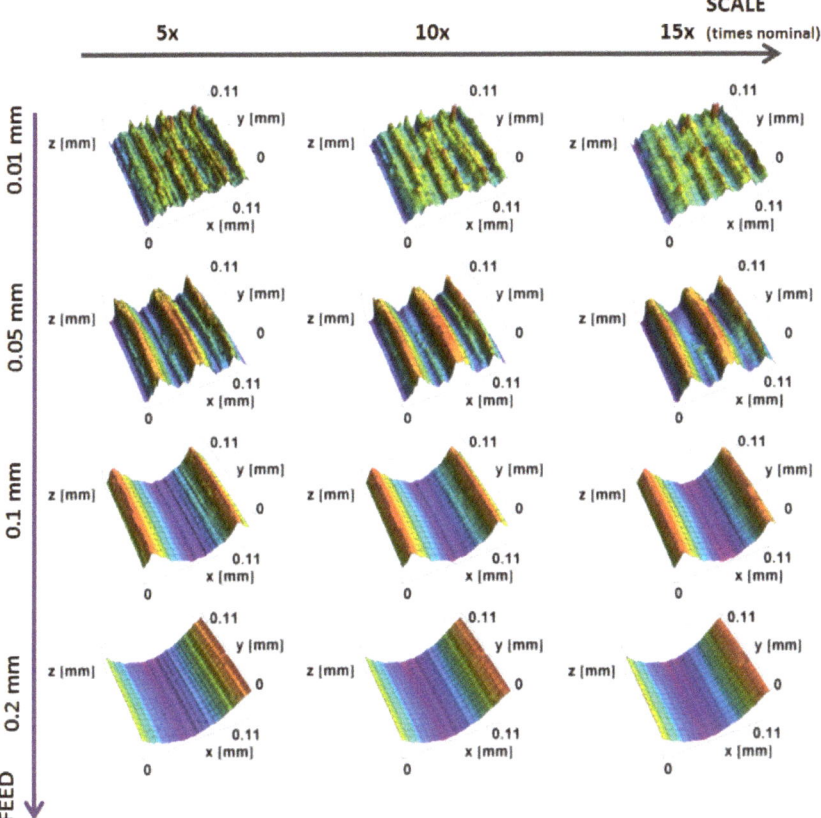

Figure 10. Renderings of surfaces turned at four feed rates, at three scales.

The principal, κ_1, curvatures, those with the largest magnitudes, calculated by the down-sampling method, are shown as a function of position on the surface and the scale of calculation in Figure 11. Similar to the mass finished surfaces, at the largest scales the variation in curvature decreases. The magnitudes of the curvatures have a tendency to increase with decreasing scale. The three top ridges on the surface, machined at 0.05 mm/rev, and the two ridges for the surface, machined at 0.1 mm/rev, are evident as blue stripes at larger scales. At finer scales, the anisotropic character of all of the machined surfaces becomes less visible. At finer scales, a multitude of fine features, with large principal κ_1 curvatures, masks the directional features. The curvatures of the main valleys are more evident at larger scales for surfaces machined at 0.05, 0.1 and 0.2 mm.

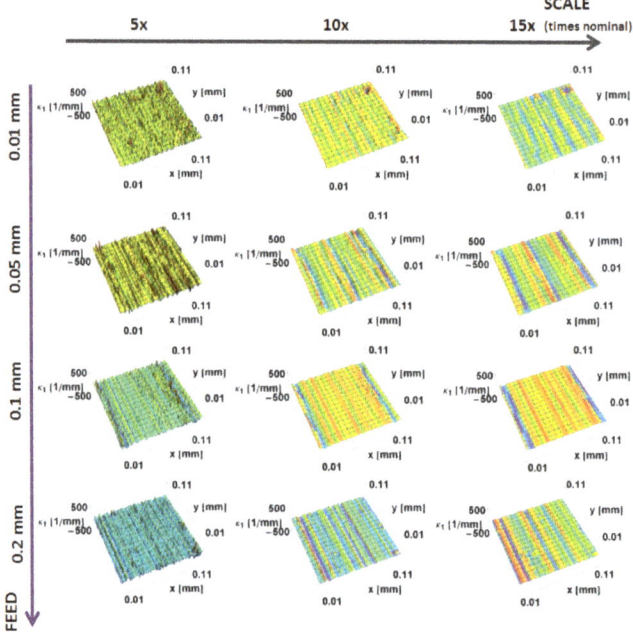

Figure 11. Maximum principal curvature, κ_1, as a function of position calculated for turned surfaces, using the down-sampling method for three scales.

The principal κ_2 curvatures, minimum in magnitude, calculated by the down-sampling method, are shown as a function of position on the surface and of the scale of calculation in Figure 12. Similar to κ_1 at the large scales, there is little variation in curvature. The magnitudes of the curvatures tend to increase with decreasing scale. At finer scales, the directional nature becomes more visible at the ridges, appearing as lines of high-magnitude curvature. At larger scales, for all the measured surfaces, the minimum curvature tends to zero, indicating that the surface is flat in one direction. The feed rate influences the magnitude of the κ_2 curvature at all three of the scales shown.

The principal κ_1 and κ_2 curvatures, calculated by the increasing neighborhood method, are shown as a function of position on the surface and the scale of calculation in Figures 13 and 14, respectively. These show the same trends as the down-sampling method. As in the previously studied surfaces, at larger scales, more heights are used in the calculation of the normal vectors, which makes these results more sensitive to artifacts and local variations. For the same subregions, principal curvatures take greater values for the increasing neighborhood method, and the smoothing effect is less evident.

Mean and standard deviations of the principal curvatures on the prepared edges, calculated by the down-sampling method, are shown versus the log of the scale in Figure 15a. The mean values of the principal curvatures change with the scale for all the surfaces. The standard deviations of the principal curvatures decrease regularly with increasing scale, i.e., the variance of the curvatures is larger at the finer scales. The mass-finished edge shows the same tendency.

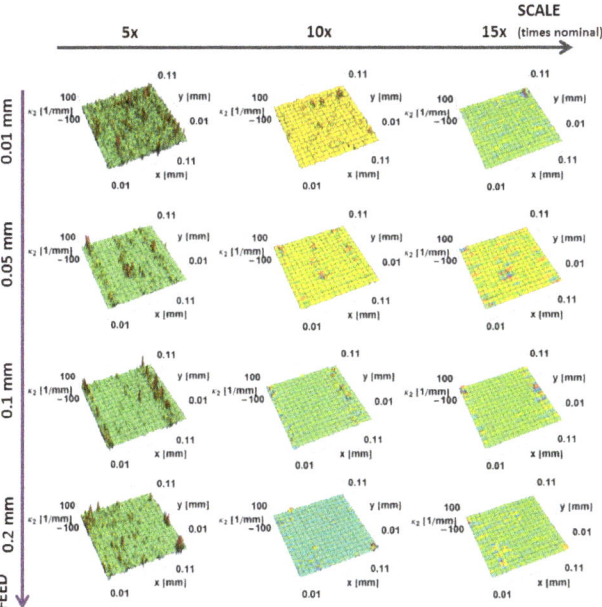

Figure 12. Minimum magnitude of the principal curvatures, κ_2, as a function of position for turned surfaces calculated, using the down-sampling method, for three scales.

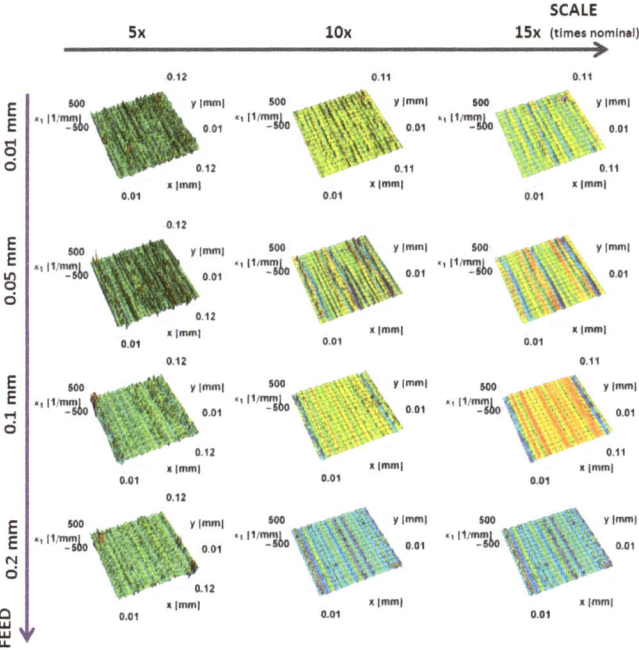

Figure 13. Maximum principal curvature, κ_1, as a function of position for turned surfaces, using the increasing neighborhood method for three scales.

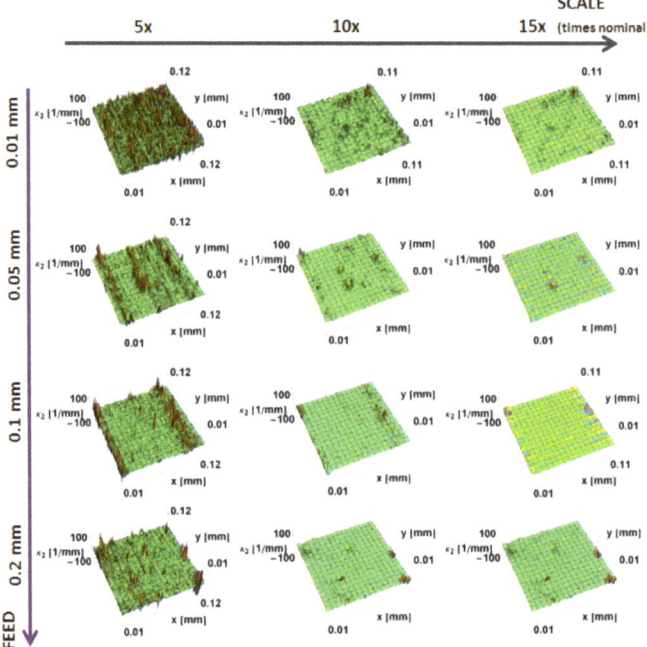

Figure 14. Minimum principal curvature, κ_2, as a function of position for turned surfaces, using the increasing neighborhood method for various scales.

The maximum curvatures decrease linearly ($R^2 > 0.93$) with feed rate, for scales between 0.75 and 1.25 µm. Other scales show the same trend, although the correlations are weaker ($0.42 > R^2 > 0.67$). In addition, the mean κ_1 discriminates the surfaces for scales between 0.75–1.25 µm, where the variance is large.

The mean and standard deviations of the minimum principal curvatures, κ_2, are smaller than for maximum curvature, κ_1. Their proximity to zero, particularly at large scales, is consistent with the straightness of the surfaces that are parallel to the prepared edge.

Standard deviations of both principal curvatures show mediocre to poor correlation with feed rate ($R^2 < 0.72$ for maximum and $R^2 < 0.17$ for minimum), suggesting that the variance of curvatures is not influenced strongly by feed. The strongest correlations were observed between the feed rates and the mean minimum curvatures ($R^2 > 0.8$ for scales between 0.75 and 3.25 µm, with a maximum of 0.982 at 2.75 µm). This suggests that minimum curvature is feed-dependent.

Statistics calculated by the method are presented in Figure 15b. It appears that the mean and standard deviations of κ_1 might be used to discriminate the prepared edges for some scales between 2.25 and 4.75 µm. The coefficients of determination R^2 for regression analysis for the same range take values greater than 0.81. These means of minimum curvatures correlate more weakly than when they are calculated using the down-sampling method (maximum $R^2 = 0.74$). The greatest differences in the statistical parameters between those two calculation methods appears at the largest scales. Similar trends appear for κ_2.

For both methods, the mean of principal curvatures of the tool edge take similar values for larger scales. The variation of the curvatures for the tool edges is greater than in the resulting turned surfaces. This suggests that large-scale features on the tool edge are transferred to the machined surface, whereas fine-scale details on the tool are not.

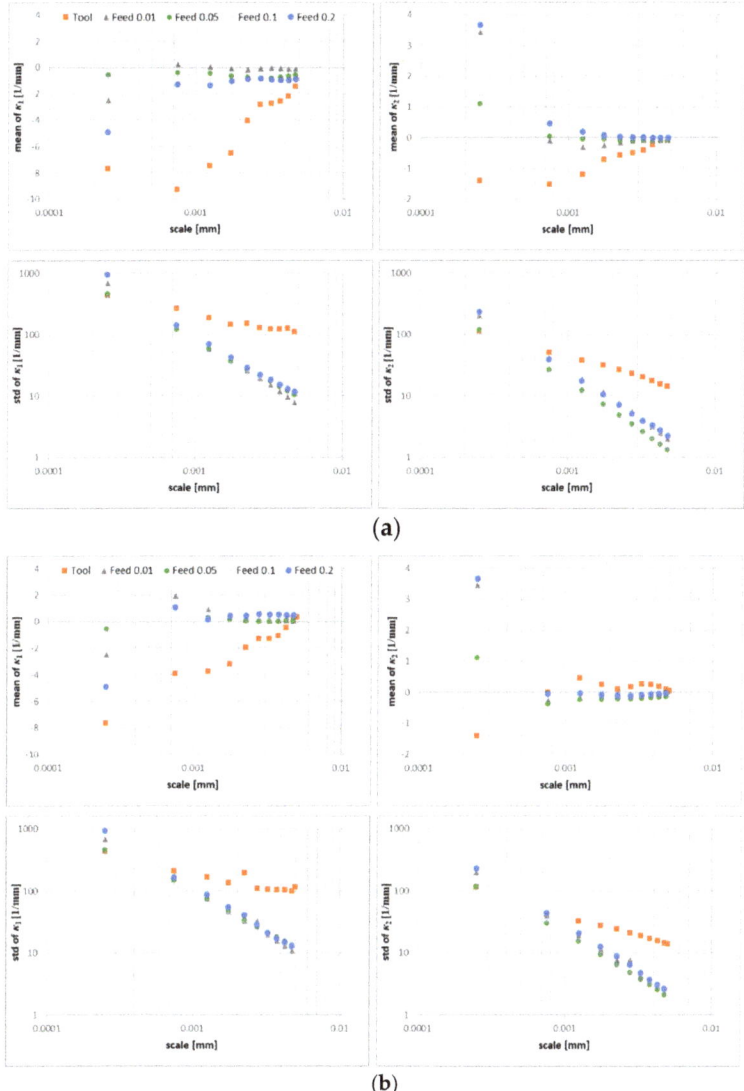

Figure 15. Statistics on the turned surfaces and tool as a function of scale: (**a**) mean and standard deviation of κ_1 and κ_2 principal curvatures, calculated for three scales by down-sampling, (**b**) mean and standard deviation of κ_1 and κ_2 principal curvatures calculated for three scales by increasing the neighborhood.

4. Discussion

The viability of this method of calculation of unit normal vectors and applying eigenvalue analyses to areal topographic measurements has been clearly demonstrated for all the variations studied. The results are consistent with the expected curvatures and tendencies with scale, feed, and mass-finishing times on the manufactured surfaces studied here. Curvatures in 3D can be calculated as a function of position and scale directly from a regular spatial array of heights. The presented method requires estimation of unit vectors normal to the surface prior to estimation of the curvature tensor. The method is sensitive to the quality of the estimation of the normal vectors. Two calculation methods present similar results for multiple scales; however, they vary in the computational time. The smoothing effect is less evident for the increasing neighborhood method, which can suggest its potential for discrimination. The down-sampling method takes, on average, 1/10 of the time in comparison with the increased-neighborhood method.

At the finer scales, the principle curvatures tend to increase with decreasing scale for all measured surfaces, because fine-scale details have larger magnitude curvatures. The principal curvatures calculated for particular scales tend to decrease with time in the mass finisher. The feed rate influences the curvature of the resulting topographies. In the cases studied, the mean of values for the principal curvatures are more appropriate for discrimination and correlation than standard deviations.

The important criterion for adding value with texture characterization methods is their ability to correlate with some phenomena of interest, such as processing or performance. The inclusion of multiscale methods in a characterization and analysis provides an important dimension for improvement, as has been demonstrated many times previously [2,9,28,29]. Some surface topographies cannot be well represented by smooth functions at all, or any, scales. Thus, a method of curvature calculation that does not require fitting smooth geometries, like quadratic forms, might be appropriate. The same can be said for the profiles.

The lengths of irregular profiles change with scale, as do the slopes [30]. Not surprisingly, then, the curvature also changes with scale and, naturally, with position, also. As with the area-scale analysis, correlations between curvature and topographically related behavior or processing might only be found over a narrow range of scales. Determination of appropriate scales for strong correlations and confident discriminations can follow the method previously reported for scale-based correlation, using multiscale regression and discrimination tests [9]. Multiscale characterizations of curvature have the potential to enable new, strong correlations and confident discriminations.

A recent study involving ENS-Cachan (Ecole normale supérieure de Cachan) and WPI (Worcester Polytechnic Institute) showed that multiscale analysis of curvatures of profiles could be successful for determining fatigue limits [9]. This is logical, because positive curvatures relate to stress concentrations that increase the likelihood of crack initiation. In addition, curvature has appeal as an appropriate geometric characterization for many kinds of contact mechanics. Because the curvature changes with position, appropriate statistical characterizations must be used as well [2].

Whitehouse [31] discusses sensitivity to scale and notes that it should be four to five times greater than the sampling interval. He also proposes that a better approximation can be found using a seven-point average to determine the slopes for the first step of the double-difference method. However, the sampling interval is often dictated by the measurement instrument, and it can be somewhat arbitrary with regard to the scale of the topographically related interactions of interest. A better approach could be to examine all the scales available in the measurement. Subsequently, these can be compared for regression and discrimination tests as a function of scale. This can lead to identification of the scales of interaction for the phenomena of interest.

The richness of the multiscale tensor curvature characterizations suggest that they have a strong potential for many kinds of applications in engineering, forensics, paleontology, physical anthropology, and archaeology.

5. Conclusions

The viability of these methods has been demonstrated for multiscale characterization of curvature tensors on measured topographies ($z = z(x,y)$). The analyses are based on calculating unit-normal vectors to the surface, at three proximal locations, and then using an eigenvalue approach to the problem of calculating the curvature tensors. The curvature tensors of measured topographies can be calculated and studied over a range of scales and positions. These methods are useful and feasible, and they have been demonstrated successfully for a variety of surfaces.

The validity of these methods has been furthered by the demonstrable consistency of the results with expectations on manufactured surfaces. These expectations include the nature of the curvature as shown by principal curvature values and their orientations relative to manufactured features on the surfaces. Mean curvature values and variance provide further validation of these methods by meeting expectations. The multiscale analyses and resulting multiscale characterization, using curvature tensors on areal measurements of topographies, also meets with the expectations based on the manufactured features.

The two methods studied here, for calculating the unit-normal vectors on the surface and systematically adjusting the scale of calculation, show the expected differences. The increasing neighborhood method, i.e., the second method, in contrast to the down-sampling method, has been shown to lead to a decrease in the variation of the curvatures. This suggests that the neighborhood method could be valuable when there is a concern that irregularity in the topographic data might be masking interesting tendencies. This comparison, which is consistent with expectations, demonstrates the viability and furthers the demonstration of the validity of both of these methods.

6. Patents

The multiscale curvature analysis in terms of outlier removal is the subject of patent application: Measurement equipment with outlier filter, US20180038687A1.

Author Contributions: Conceptualization, T.B. and C.A.B.; methodology, T.B.; software, T.B.; validation, T.B. and C.A.B.; formal analysis, T.B.; investigation, T.B.; resources, C.A.B.; data curation, T.B. and C.A.B.; writing—original draft preparation, T.B and C.A.B.; writing—review and editing, C.A.B. and T.B.; visualization, T.B.; supervision, T.B.; project administration, T.B.; funding acquisition, T.B and C.A.B.

Funding: Funding for this study was partially provided by the Kosciuszko Foundation and Polish Ministry of Science and Higher Education (subsidy number 02/22/DSPB/1431).

Acknowledgments: The authors are grateful to DigitalSurf for the use of MountainsMap software and Matthew A. Gleason, for measuring the samples.

Conflicts of Interest: The authors declare no conflict of interest. The funders had no role in the design of the study; in the collection, analyses, or interpretation of data; in the writing of the manuscript; or in the decision to publish the results.

References

1. Brown, C.A. Specification of surface roughness using axiomatic design and multiscale surface metrology. *Proced. CIRP* **2018**, *70*, 7–12. [CrossRef]
2. Brown, C.A.; Hansen, H.N.; Jiang, X.J.; Blateyron, F.; Berglund, J.; Senin, N.; Bartkowiak, T.; Dixon, B.; Le Goïc, G.; Quinsat, Y.; et al. Multiscale analyses and characterizations of surface topographies. *CIRP Ann. Manuf. Technol.* **2018**, *67*, 839–862. [CrossRef]
3. Childs, T.H.C.; Sekiya, K.; Tezuka, R.; Yamane, Y.; Dornfeld, D.; Lee, D.E.; Min, S.; Wright, P.K. Surface finishes from turning and facing with round nosed tools. *CIRP Ann. Manuf. Technol.* **2008**, *57*, 89–92. [CrossRef]
4. Denkena, B.; Lucas, A.; Bassett, E. Effects of the cutting edge microgeometry on tool wear and its thermomechanical load. *CIRP Ann. Manuf. Technol.* **2011**, *60*, 73–76. [CrossRef]
5. Moore, J.Z.; Malukhin, K.; Shih, A.J.; Ehmann, K.F. Hollow needle tissue insertion force model. *CIRP Ann. Manuf. Technol.* **2011**, *60*, 157–160. [CrossRef]
6. Adams, G.G.; Nosonovsky, M. Contact modeling forces. *Trib. Int.* **2000**, *33*, 431–442. [CrossRef]
7. Archard, J.F. Elastic deformation and the laws of friction. *Proc. Roy. Soc. London Ser. A* **1957**, *243*, 190–205.
8. Greenwood, J.A.; Williamson, J.B.P. The contact of nominally flat surfaces. *Proc. Roy. Soc. London Ser. A* **1966**, *295*, 300–319.
9. Vulliez, M.; Gleason, M.A.; Souto-Lebel, A.; Quinsat, Y.; Lartigue, C.; Kordell, S.P.; Lemoine, A.C.; Brown, C.A. Multi-scale curvature analysis and correlations with the fatigue limit on steel surfaces after milling. *Procedia CIRP* **2014**, *13*, 308–313. [CrossRef]

10. Berglund, J.; Brown, C.A.; Rosen, B.G.; Bay, N. Milled die steel surface roughness correlation with steel sheet friction. *CIRP Ann. Manuf. Technol.* **2010**, *59*, 577–580. [CrossRef]
11. Bartkowiak, T. Characterization of 3D surface texture directionality using multiscale curvature tensor analysis. In Proceedings of the ASME International Mechanical Engineering Congress and Exposition (IMECE), Tampa, FL, USA, 3–9 November 2017.
12. Petitjean, S. A survey of methods for recovering quadrics in triangle meshes. *ACM Comput. Surv.* **2001**, *34*, 211–262. [CrossRef]
13. Goldfeather, J.; Interrante, V. A novel cubic-order algorithm for approximation principal directions vectors. *ACM Trans. Gr.* **2004**, *23*, 45–63. [CrossRef]
14. Yang, X.; Zheng, J. Curvature tensor computation by piecewise interpolation. *Comput. Aided Des.* **2013**, *45*, 1639–1650. [CrossRef]
15. Zhihong, M.; Guo, C.; Yanzhao, M.; Lee, K. Curvature estimation for meshes based on vertex normal triangles. *Comput. Aided Des.* **2011**, *43*, 1561–1566. [CrossRef]
16. Theisel, H.; Rössl, C.; Zayer, R.; Seidel, H.P. Normal based estimation of the curvature tensor for triangular meshes. In Proceedings of the 12th Pacific Conference on Computer Graphics and Applications, Seoul, Korea, 6–8 October 2004.
17. Coeurjolly, D.; Lachaud, J.O.; Levallois, J. Multigrid convergent principal curvature estimators in digital geometry. *Comput. Vis. Image Underst.* **2014**, *129*, 27–41. [CrossRef]
18. Foorginejad, A.; Khalili, K. Umbrella curvature: A new curvature estimation method for clouds. *Proced. Technol.* **2014**, *12*, 347–352. [CrossRef]
19. Lai, P.; Samson, C.; Bose, P. Surface roughness of rock faces through the curvature of triangulated meshes. *Comput. Geosci.* **2014**, *70*, 229–237. [CrossRef]
20. Hulik, R.; Spanel, M.; Smrz, P.; Materna, Z. Continuous plane detection in point-cloud data based on 3D Hough Transform. *J. Vis. Commun. Image Represent.* **2014**, *25*, 86–97. [CrossRef]
21. Holz, D.; Holzer, S.; Rusu, R.B.; Behnke, S. Real-time plane segmentation using RGB-D cameras. In Proceedings of the 15th RoboCup International Symposium, Istanbul, Turkey, 5–11 July 2011.
22. Chandrasekaran, V.; Payton, L.N.; Hunko, W. Orthogonal turning of aluminum 6061 in liquid nitrogen cutting environment. In Proceedings of the ASME International Mechanical Engineering Congress and Exposition, Montreal, QC, Canada, 14–20 November 2014; Advanced Manufacturing. Volume 2B, p. V02BT02A037.
23. Reibenschuh, M.; Cus, F.; Zuperl, U. Turning of high quality aluminium alloys with minimum costs. [Tokarenje visoko kvalitetnih aluminijevih legura uz minimalne troškove]. *Tehnicki Vjesnik* **2011**, *18*, 363–368.
24. Krolczyk, G.M.; Maruda, R.W.; Krolczyk, J.B.; Nieslony, P.; Wojciechowski, S.; Legutko, S. Parametric and nonparametric description of the surface topography in the dry and MQCL cutting conditions. *Meas. J. Int. Meas. Confed.* **2018**, *121*, 225–239. [CrossRef]
25. Twardowski, P.; Wojciechowski, S.; Wieczorowski, M.; Mathia, T. Surface roughness analysis of hardened steel after high-speed milling. *Scanning* **2011**, *33*, 386–395. [CrossRef] [PubMed]
26. Weingarten, J. Ueber eine Klasse auf einander abwickelbarer Fläachen. *Journal für die Reine und Angewandte Mathematik* **1861**, *59*, 382–393. [CrossRef]
27. Bartkowiak, T.; Brown, C.A. A characterization of process-surface texture interactions in micro-electrical discharge machining using multiscale curvature tensor analysis. *J. Manuf. Sci. Eng. Trans. ASME* **2018**, *140*. [CrossRef]
28. Brown, C.A.; Siegmann, S. Fundamental scales of adhesion and area-scale fractal analysis. *Int. J. Mach. Tools Manuf.* **2001**, *41*, 1927–1933. [CrossRef]
29. Cantor, G.J.; Brown, C.A. Scale-based correlations of relative areas with fracture of chocolate. *Wear* **2009**, *266*, 609–612. [CrossRef]
30. Brown, C.A.; Johnsen, W.A.; Butland, R.M. Scale-sensitive fractal analysis of turned surfaces. *CIRP Ann. Manuf. Technol.* **1996**, *45*, 515–518. [CrossRef]
31. Whitehouse, D.J. *Handbook of Surface and Nano Metrology*, 2nd ed.; CRC Press: Boca Raton, FL, USA, 2010.

© 2019 by the authors. Licensee MDPI, Basel, Switzerland. This article is an open access article distributed under the terms and conditions of the Creative Commons Attribution (CC BY) license (http://creativecommons.org/licenses/by/4.0/).

MDPI
St. Alban-Anlage 66
4052 Basel
Switzerland
Tel. +41 61 683 77 34
Fax +41 61 302 89 18
www.mdpi.com

Materials Editorial Office
E-mail: materials@mdpi.com
www.mdpi.com/journal/materials

www.ingramcontent.com/pod-product-compliance
Lightning Source LLC
LaVergne TN
LVHW071944080526
838202LV00064B/6673